Remote Sensing for Crop Plants: Modeling Methods and Applications

Remote Sensing for Crop Plants: Modeling Methods and Applications

Editor: Callen Elliott

Callisto Reference
www.callistoreference.com

Callisto Reference,
118-35 Queens Blvd., Suite 400,
Forest Hills, NY 11375, USA

Visit us on the World Wide Web at:
www.callistoreference.com

ISBN: 978-1-64116-818-2 (Hardback)

Cataloging-in-Publication Data

Remote sensing for crop plants : modeling methods and applications / edited by Callen Elliott.
p. cm.
Includes bibliographical references and index.
ISBN 978-1-64116-818-2
1. Agriculture--Remote sensing. 2. Field crops--Remote sensing. 3. Plants, Cultivated--Remote sensing. I. Elliott, Callen.
S494.5.R4 R46 2023
577.82--dc23

Table of Contents

Preface

Remote sensing is the process of detecting and monitoring the physical characteristics of an area by measuring its reflected and emitted radiation at a distance from a satellite, drone or aircraft. It is used in several fields including photography, geology, and surveying. Remote sensing plays an instrumental role in the field of crop production and management. It is used extensively for forecasting crop production, assessing crop damage and crop progress, identifying pests and disease infestation, soil mapping, monitoring and managing irrigation, and collecting past and current weather data. Crop intensification is another important application of remote sensing that involves collecting important data related to crops including crop rotation needs, cropping pattern, and crop diversity over a given piece of land. The remote sensing data can also be used in crop models, which show the relationships among plant, water and soil. Such models help reveal water stress and its effects on crop growth and development. This book is a detailed explanation of the various applications of remote sensing in crop plants. It is a vital tool for all researching or studying the role of remote sensing in crop production as it gives incredible insights into emerging modeling methods and applications.

Various studies have approached the subject by analyzing it with a single perspective, but the present book provides diverse methodologies and techniques to address this field. This book contains theories and applications needed for understanding the subject from different perspectives. The aim is to keep the readers informed about the progresses in the field; therefore, the contributions were carefully examined to compile novel researches by specialists from across the globe.

Indeed, the job of the editor is the most crucial and challenging in compiling all chapters into a single book. In the end, I would extend my sincere thanks to the chapter authors for their profound work. I am also thankful for the support provided by my family and colleagues during the compilation of this book.

Editor

1

Sentinel 2-Based Nitrogen VRT Fertilization in Wheat: Comparison between Traditional and Simple Precision Practices

Marco Vizzari *, Francesco Santaga and Paolo Benincasa

Department of Agricultural, Food, and Environmental Sciences, University of Perugia, 06121 Perugia, Italy; francescosaverio.santaga@unipg.it (F.S.); paolo.benincasa@unipg.it (P.B.)

* Correspondence: marco.vizzari@unipg.it

Abstract: This study aimed to compare standard and precision nitrogen (N) fertilization with variable rate technology (VRT) in winter wheat (*Triticum aestivum* L.) by combining data of NDVI (Normalized Difference Vegetation Index) from the Sentinel 2 satellite, grain yield mapping, and protein content. Precision N rates were calculated using simple linear models that can be easily used by non-specialists of precision agriculture, starting from widely available Sentinel 2 NDVI data. To remove the effects of not measured or unknown factors, the study area of about 14 hectares, located in Central Italy, was divided into 168 experimental units laid down in a randomized design. The first fertilization rate was the same for all experimental units (30 kg N ha^{-1}). The second one was varied according to three different treatments: 1) a standard rate of 120 kg N ha^{-1} calculated by a common N balance; 2) a variable rate (60–120 kg N ha^{-1}) calculated from NDVI using a linear model where the maximum rate was equal to the standard rate (Var-N-low); 3) a variable rate (90–150 kg N ha^{-1}) calculated from NDVI using a linear model where the mean rate was equal to the standard rate (Var-N-high). Results indicate that differences between treatments in crop vegetation index, grain yield, and protein content were negligible and generally not significant. This evidence suggests that a low-N management approach, based on simple linear NDVI models and VRT, may considerably reduce the economic and environmental impact of N fertilization in winter wheat.

Keywords: NDVI; remote sensing; GIS; precision farming; variable rate technology; yield mapping; protein content

---

1. Introduction

In recent decades, population growth and the expanding demand of agricultural products for multiple purposes have constantly increased the environmental pressures on land and water resources. Worldwide concern of citizens and governments on environmental issues, together with the availability of improved and cost-effective methodologies and tools for spatial data acquisition, analysis, and modeling, have promoted the development of precision agriculture (PA) techniques. PA is based on innovative system approaches that use a combination of various technologies such as Geographic Information System (GIS), Global Navigation Satellite System (GNSS), computer modeling, Remote Sensing (RS), variable rate technology (VRT), yield mapping and advanced information processing for timely in-season and between season crop management [1].

Currently, a wide range of satellite data is available that varies in terms of acquisition cost, technique (active/passive), spatial resolution, spectral range, and viewing geometry [2]. A recent step forward was made thanks to the launches of Sentinel-2A (2015) and Sentinel-2B (2017) satellites by the European Space Agency (ESA), intended for Earth observation and monitoring of land surface variability. The Sentinel-2 satellites are equipped with a multispectral sensor (MSI) including 13 spectral bands, with a spatial resolution ranging from 10 m to 60 m, which provides relevant information for supporting precision agriculture [3]. Images provided by Sentinel-2 satellites are publicly available for free through the Copernicus Open Access Hub with 5 days temporal resolution averaging (2–3 days in mid-latitudes), in L1C or L2A processing levels, which make them attractive for time-series analysis and PA applications. The Level-2A product provides Bottom Of Atmosphere (BOA) reflectance images derived from the associated Level-1C product, which provides Top Of Atmosphere (TOA) reflectance images [4].

Satellite remote sensing of vegetation is mainly based on the green (495–570 nm) and red (620–750 nm) regions of the visible spectrum, the red-edge (680–730 nm), near and mid infrared bands (850–1700 nm). These bands are often combined with a range of algebraic formulas to obtain several Vegetation Indices (VIs) for assessing vegetation status and crop management (see e.g. references [5–7]). The Normalized Difference Vegetation Index (NDVI) [8], a normalized difference between reflectance of Red and Near-Infrared (NIR) spectral bands, is still the most used index worldwide because of its ease of calculation and interpretation. It varies between 0 and 1 in cropped areas, increasing with soil cover, LAI (Leaf Area Index), chlorophyll content, and plant N-nutritional status (see e.g. references [9–11]). The NDVI has been widely tested to assess wheat N-nutritional status and yield, with encouraging results [12–14].

N is the main nutrient supplied to most crops, including wheat, and may cause environmental impacts on near-surface and deep aquifers [15]. Increasing the N rate generally increases crop yield since it increases the grain number and size [16]. However, increasing the N rates reduces the N uptake efficiency and increases the amount of residual N in the soil which is exposed to leaching risks [9,17]. Countless studies are available on wheat N nutrition and several of the recent ones have dealt with precision N fertilization, but most of them propose quite complex approaches that are not widely adoptable by non-specialists of PA. For example, Bourdin et al. [18] propose a complex model starting from LAI, estimated by remote sensing, and yields from previous years; Basso et al. [19] use a crop simulated model (SALUS) based on weather and yields from previous years. Another approach consists in assessing spatial variability to delineate homogeneous sub-field areas overlapping various thematic spatial maps (such as yield, soil properties) or applying a multivariate geostatistical approach of the factors that affect yield and grain quality [20]. In this context Song et al. [21] delineated management zones on the basis of soil, yield data and remote sensing information derived from Quickbird imagery. Besides these methods, to improve the use of N prescription maps in PA, various simplified approaches based on NDVI have also been developed through user-friendly web interfaces (e.g. CropSAT [22], Agrosat [23], OneSoil [24]) but these approaches can be applied only in specific areas or lack either quantitative analysis or validation.

While traditional flat-rate spreaders typically tend to over- or under-apply fertilizers, with VRT spreaders it is possible to modify the distribution rate according to on-the-go proximal sensors measuring, in real-time, plant properties or prescription maps derived through more or less advanced models including various spatial factors (e.g.: vegetation indices, soil variables, yield maps, crop nutrition status) [25]. However, the majority of precision fertilization approaches are generally map-based, because on-the-go sensors are too expensive, not sufficiently accurate, or not available [26]. VRT spreaders are subject to errors depending on the technology and the intrinsic characteristics of the fertilizer [27].

Grain yield mapping aims at providing or increasing knowledge about the spatial heterogeneity of yields. This technology is based on the recording of georeferenced yield data thanks to the integration of different subsystems mounted on the combines [28]: (i) harvest quantity measurement sensors (mass or volume); (ii) GNSS location sensors; (iii) reference area measurement sensors (working width, speed, time); (iv) data recording and processing units. To generate yield maps, the harvested quantities are georeferenced and associated with the corresponding reference area which is calculated by multiplying the working width by the area length (working speed * time interval). Yield mapping accuracy is influenced by many factors including flow sensor calibration, combined speed changes, grain flow variations, grain moisture, and the level of smoothing applied for mapping [29].

While many studies combine VIs (from various remote or proximal sensors) and VRT crop fertilization, only a small quantity of research integrates yield mapping sensors, and, to our knowledge, probably no study has ever combined VIs, VRT fertilization, and yield mapping in a PA case study and, in particular, to assess the effects of different variable N-rate treatments on winter wheat.

In this framework, this study was aimed at comparing two VRT N fertilization treatments (based on Sentinel 2) versus a standard flat N rate in terms of crop NDVI trend, grain yield, and protein content. In order to promote the adoption of PA techniques among farmers and non-specialists in PA, a simplified approach, based on free and open-source software, the widely used NDVI, and an easily-applicable linear model, was applied to calculate the VRT N fertilization rates.

2. Materials and Methods

2.1. Crop Management and Experimental Treatments

The experiment was carried out in the cropping season 2017-2018, on a 14 ha plain field of the middle Tiber valley, near Deruta, Umbria, Italy (170 m a.s.l., 42°95′07″ N, 12°38′18″ E), provided by the Foundation for Agricultural Education of Perugia (Figure 1). The soil was loam with increasing sand content from the west to the east side. The climate is Mediterranean, characterized by a dry season between May and September and a cold and rainy season from October-November to March-April. The cropping season 2017–2018 was unusually rainy in December and March, while temperatures were generally higher than the poly-annual trend except for a very cold end of February (Figure 2).

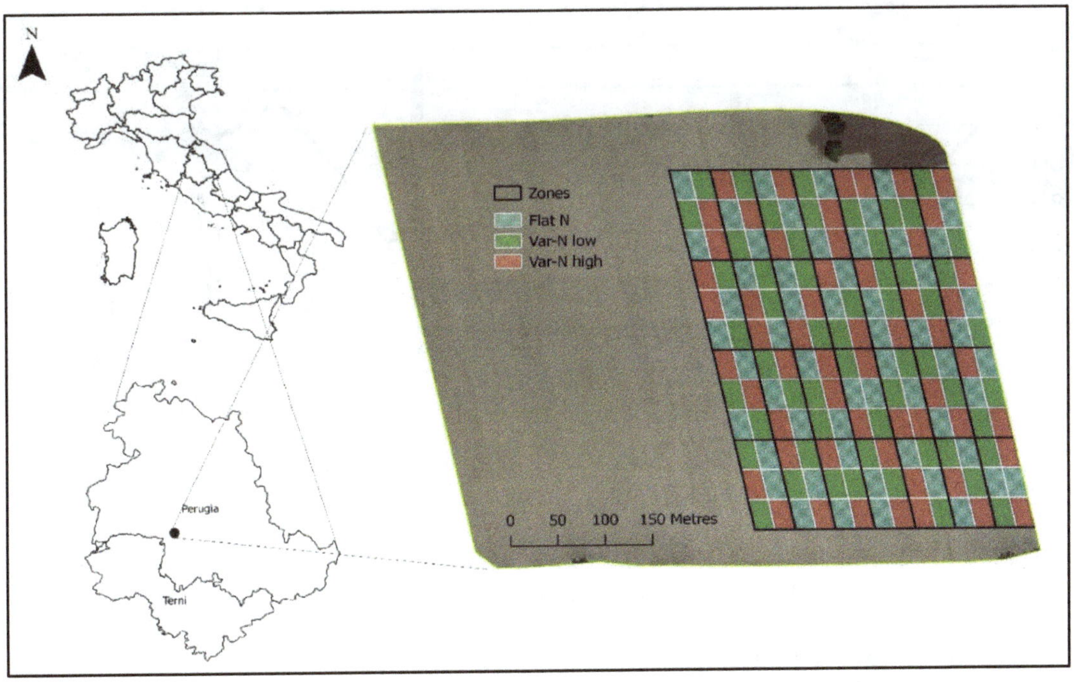

Figure 1. Geographical location of the study area and experimental layout. Flat-N (standard rate of 120 kg N ha^{-1}), Var-N-low (variable rate from 60 to 120 kg N ha^{-1}), Var-N-high (variable rate from 90 to 150 kg N ha^{-1}).

The experimental crop was a rainfed winter wheat (*Triticum aestivum* L., cv. PRR58) sown on 10 November 2017 at a nominal density of 450 viable seeds m^{-2}. The previous crop had been pea (*Pisum sativum* Asch et Gr). The crop was managed according to ordinary practices, while weeds and diseases were controlled chemically. The N rate was split in two application times in order to increase N uptake efficiency, support the formation of yield components and limit N leaching by fall-spring rainfall. The first N application (as urea) occurred on 18 January 2018 with 30 kg N ha^{-1}, while the second N fertilization (as urea), occurred on 26 March 2018 and was managed according to three experimental treatments: 1) a standard rate of 120 kg N ha^{-1} (Flat-N) derived by an N balance approach (the relatively high rate is justified by the very rainy winter); 2) a variable rate of 60 to 120 kg N ha^{-1}, based on NDVI, where the maximum rate was equal to the standard rate (Var-N-low); 3) a variable rate of 90 to 150 kg N ha^{-1}, based on NDVI, where the medium rate was equal to the standard rate (Var-N-high). An inverse linear relationship between NDVI and VRT N-rates was adopted in Var-N-low and Var-N-high on the assumption that NDVI and other correlated VIs (e.g. NIR/Red simple ratio), before the second N fertilization (7–8 Feekes' stage), are directly related to crop N nutritional status [10,13,30]. Thus, the Var-N-high was defined with the aim of keeping the average N-rate equal to the flat N-rate while optimizing the N fertilization according to the supposed N nutritional status. The range adopted in Var-N-high represents a variation of ±25% of the flat N rate, and a ±20% of the total N supplied (including the 30 kg N ha^{-1} of the first N application), which was considered as being appropriate to see appreciable effects on wheat yield. The Var-N-low was defined with the aim of testing a sensible N reduction, the results of which were very important in those areas where a more eco-compatible approach is required. So, the flat N-rate was considered as the maximum N level and the −50% of the flat N rate (−40% of the total rate, equal to 90 Kg/ha including the first N application) was adopted as a lowest reference level based on the evidence from a previous study in the same location [16].

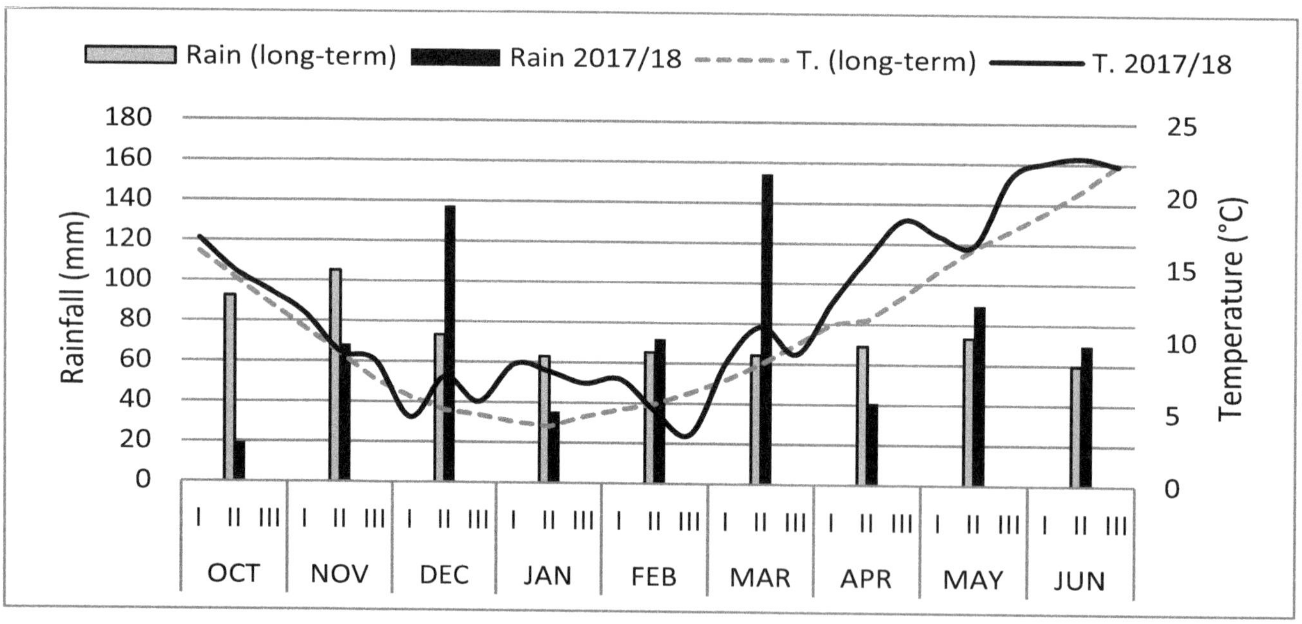

Figure 2. Monthly cumulated rainfall and average temperatures per decade during the wheat growing season 2017–2018 and in the long term (temperature 1951–2018, rainfall 1921–2018).

To account for the possible not measured or unknown factors, the 14 ha study area was divided into 168 plots of about 700 square meters each (35 m long, 21 m wide) grouped in 28 zones. The three treatments were laid down according to a randomized design with 2 replicates per treatment in each zone for a total of 56 replicates per treatment in the whole study area (Figure 1). All the precision on-field operations were performed using a tractor equipped with a Topcon GNSS automatic guide device connected to a Real Time Kinematic (RTK) network. The variable rate treatments were performed using a Sulky 40+ (ECONOV) VRT fertilizer spreader connected through ISOBUS (a widely used software protocol complaint to ISO 11783 standard) to the Topcon system console. The console, after each VRT fertilization, releases the output of each fertilization map showing the estimated distributed rates. This output was compared to the prescription map to investigate the accuracy of the N distribution.

2.2. Calculation of Rates for VRT Treatments

A Level 2A Sentinel-2 image collected on 22 March 2018 (i.e., four days before the second fertilization), georeferenced according to WGS84-UTM32, was downloaded from The Copernicus Open Access Hub. NDVI, calculated from bands 4 and 8 using QGIS Raster calculator (Figure 3a), was used for the N rates calculation of the two variable rate treatments. In both cases, a linear relationship was imposed between the average NDVI value calculated for all experimental units and fertilizer-N rates where the 5° percentile of NDVI value corresponded to the minimum fertilizer-N rates (60 or 90 kg N ha^{-1}) and the 95° percentile of NDVI value corresponded to the maximum fertilizer-N rate (120 or 150 kg N ha^{-1}) (Figure 3b). All the data processing and analysis were carried out using QGIS software, version 2.18.12 64bit [31] and MS Excel 2016. Average NDVI values were calculated using the SAGA grid statistics for the polygons algorithm included in the QGIS processing framework. To simplify this step, pixels overlapping the border of the experimental plots were not excluded from the calculation. To assess the effect of these pixels, average NDVI values were compared with those obtained while considering only non-overlapping pixels.

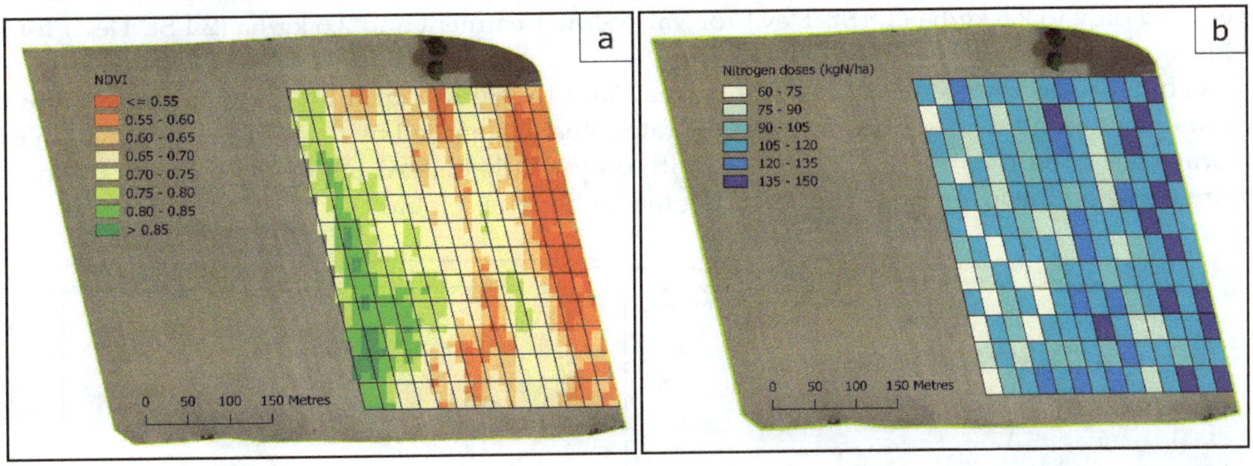

Figure 3. (**a**) NDVI calculated from Level 2A Sentinel-2 image collected on 22 March 2018; (**b**) Prescription N map developed by integrating the three different experimental treatments.

2.3. Determination of NDVI

To monitor and compare the NDVI of experimental treatments, all relevant level 2A Sentinel-2 images with no cloud cover on the study area, were collected from 22 March to the beginning of the senescence. Average NDVI values were calculated for each plot using the SAGA grid statistics for the polygons algorithm included in the QGIS processing framework. Six to nine S2 pixels fell within each experimental unit. To highlight possible differences between the effects of the experimental treatments while considering different crop vigor status before the second N fertilization, the NDVI time series analysis was performed by classifying plots according to three NDVI classes (NDVI $\leq$ 0.68; 0.68 < NDVI $\leq$ 0.79; NDVI > 0.79), as revealed by the S2 image from 22 March.

2.4. Determination of Grain Yield and Protein Content

Harvest was carried out on 26 June 2018 using a combined harvester Claas Lexion 780–740 equipped with a Topcon YieldTrakk system (processing data from the optical sensor measuring grain mass flow and moisture sensors), which produced a georeferenced yield map as an ESRI polygon shapefile. The combine harvester had a cutting width of 7.50 m and was equipped with a tilt sensor to correct the effect of slope on the sensor readings. An initial on-field calibration was performed on the combine to adjust for the actual working width and measure the unit weight of grain, which was used by the system to convert the measured mass flow (l/s) to Mg.

The output shapefile containing the georeferenced yield data was intersected with the experimental units to calculate the average yield expressed in Mg ha^{-1} which was used for the subsequent analysis. To produce an easy-to-read yield map, the shapefile was converted to a 5 m resolution raster. Subsequent gaussian kernel smoothing, based on a 50 m search radius, produced the final yield map.

Protein content was measured on 20 June 2018 (i.e., six days before final harvest). To reduce the number of samples, the sampling scheme was defined according to the three classes of NDVI previously defined for a total of 36 field samples, then merged two by two to obtain 18 lab samples, which were analyzed according to the official Kjeldahl method, a widely used chemical procedure for the quantitative determination of protein content in food, feed, feed ingredients and beverages [32].

3. Results

The NDVI values on 22 March show within-field differences mainly associated with the west-east textural gradient (Figure 3a). Average NDVI values including overlapping pixels with plot borders differ only slightly from those obtained while excluding these pixels (0.015 in average absolute value, 0.011 St. Dev.). According to the linear model used for N-rates calculation, these differences correspond only to 2.6 kg/ha (1.8 St. Dev.) for Var-N-low treatment and 2.6 kg/ha (2.1 St. Dev.) for the Var-N-high treatment.

A comparison between the prescription map and the Sulky output (Figure 4a) shows that the correlation between the prescribed rate and the rate actually supplied by the VRT machine within each plot was very good (R^2 = 0.91) (Figure 4b), with only a little bias which resulted in smoothing the extremes, raising the lower rates and lowering the higher ones (Table 1).

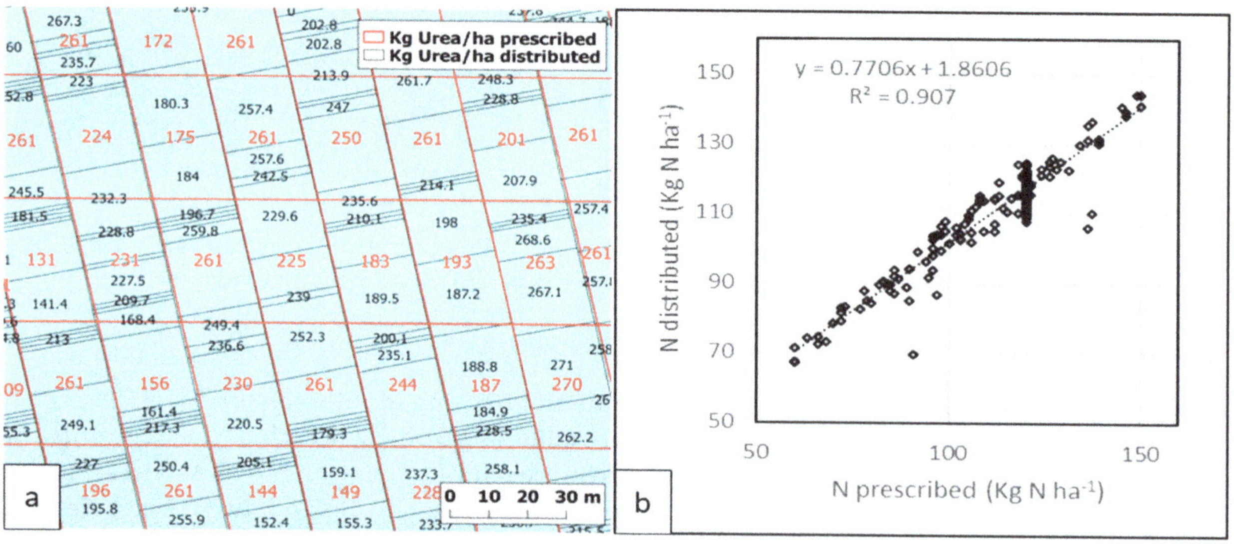

Figure 4. (**a**) Comparison between prescription map and the distributed Sulky output. (**b**) Scatter plot between the prescribed and distributed rate within each plot.

Table 1. Amounts of N prescribed and supplied with the second N application in the three treatments: Flat-N (standard rate of 120 kg N ha^{-1}), Var-N-low (variable rate from 60 to 120 kg N ha^{-1}), Var-N-high (variable rate from 90 to 150 kg N ha^{-1}).

Treatments	Average N prescribed (Kg N ha^{-1})	Average N supply (Kg N ha^{-1})	Average Δ (%)	Δ St. Dev. (Kg N ha^{-1})
Flat-N	120.0	116.5	−1.3%	0.03
Var-N-low	90.2	95.4	5.4%	0.06
Var-N-high	120.5	117.0	−3.0%	0.05

After the second fertilization, the trend of NDVI, considering the three vigor classes previously defined, was not affected by fertilization treatments (Figure 5). In the portion of the field where the NDVI was already high on 22 March (class III—about 0.85), all treatments showed a further moderate increase up to over 0.9, while in the portion of the field where NDVI was low (class I—just over 0.6), all treatments showed an increase up to 0.85 in about one month, and nearly 0.9 later on. As expected, an intermediate trend can be observed for treatments with NDVI falling within class II. Thus, whatever the crop vigor status was on March 22nd, all treatments reached a high NDVI one month after the second N application (Figure 5).

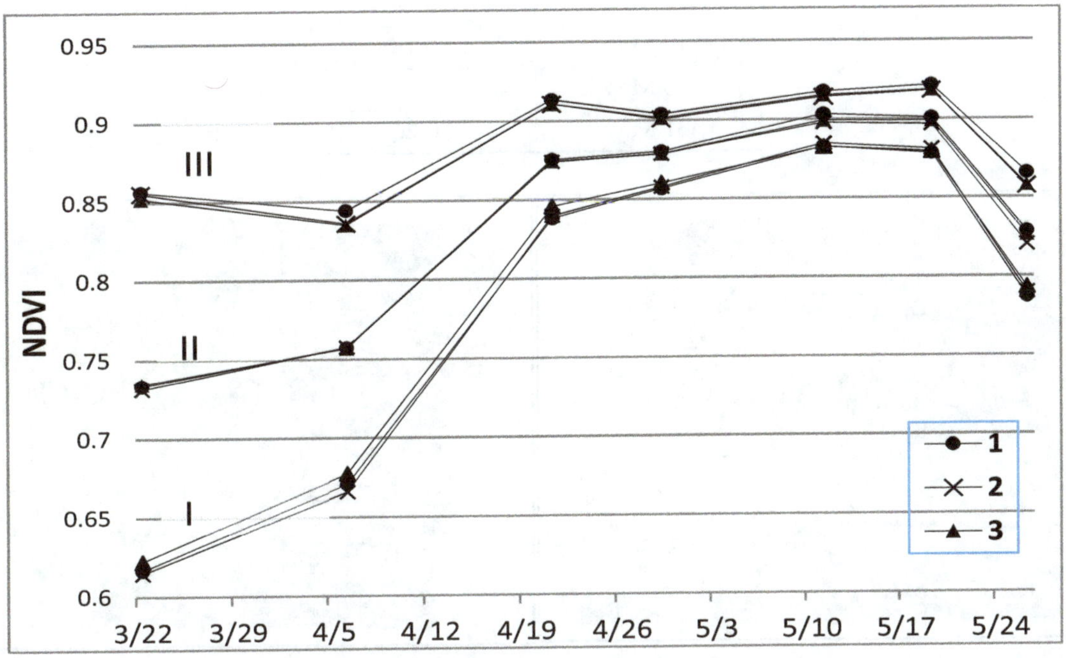

Figure 5. NDVI time series analysis, considering the three vigor classes as identified in the S2 image of March 22nd (I: NDVI ≤ 0.68; II: 0.68 < NDVI ≤ 0.79; III: NDVI > 0.79), for the three N fertilization treatments: 1) Flat-N (standard rate of 120 kg N ha^{-1}); 2) Var-N-low (variable rate from 60 to 120 kg N ha^{-1}); 3) Var-N-high (variable rate from 90 to 150 kg N ha^{-1}).

Grain Yield and Quality

The yield map used for the quantitative analysis is shown in Figure 6a, while the generalized version used only for visual analysis purposes is shown in Figure 6b. The three treatments did not differ significantly for total yield (Table 2) and no relationship was found between the N rate and yield (R^2 = 0.087). Similarly, no correlation was found between N treatments and protein content (R^2 = 0.001). However, the slight difference in protein content observed between Flat-N and Var-N-low (Table 2) was statistically significant (p = 0.02) even though it was not agronomically relevant. Finally, grain yield was weakly correlated to NDVI at any time of NDVI measurements (Table 3).

Table 2. Grain yield and protein content for the three N fertilization treatments: Flat-N (standard rate of 120 kg N ha^{-1}), Var-N-low (variable rate from 60 to 120 kg N ha^{-1}), Var-N-high (variable rate from 90 to 150 kg N ha^{-1}).

Treatment	Average Yield (Mg ha^{-1})	St. Dev. Yield (Mg ha^{-1})	Average Protein Content (%)	St. Dev. Protein Content (%)
Flat-N	6.74	0.36	9.4	0.31
Var-N-low	6.73	0.39	8.9	0.37
Var-N-high	6.76	0.38	9.2	0.48

Table 3. Correlation between yield and NDVI at any time of NDVI measurement plotted over all of the 168 plots.

Date	R^2
22 Mar	0.29
6 Apr	0.26
21 Apr	0.24
29 Apr	0.19
11 May	0.19
10 May	0.18
26 May	0.27
31 May	0.26

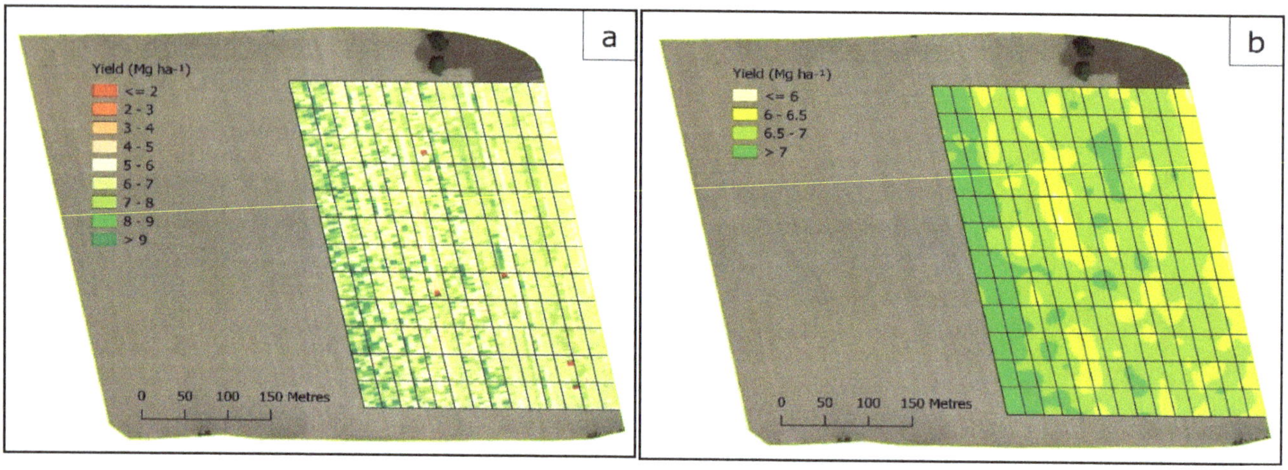

Figure 6. (**a**) yield map used for quantitative analysis; (**b**) generalized version of yield map.

4. Discussion

Our study investigated the differences between a flat N fertilization rate and two variable rates defined through a simplified method based on Sentinel-2 NDVI collected a few days before the second fertilization. One of the two variable rates was aimed at drastically reducing the overall N input (Var-N-low), the other one was aimed at maintaining the same overall input (Var-N-high) as in the flat rate (Flat-N), i.e., 120 kg N ha^{-1} while optimizing the N fertilization according to the supposed N nutritional status. Our results show that a VRT approach with a lower overall N rate may be more efficient, giving same grain yield and quality (Table 2) with a lower N input (Table 1). This result is consistent with the evidence of Raun et al. [33] which reports that the VRT method improves the NUE (Nitrogen Use Efficiency) by 15% compared to a flat rate. In vulnerable contexts, as in the study area, the N-rate reduction results environmentally and economically very relevant since it could reduce water pollution (still a critical issue in Umbria and all over the world) and decrease expenditures on N. However, differently from our case study, a Var-N-low treatment could determine a lower protein grain content. In this regard, the economic trade-off between lower N costs and reduced yield value due to lower grain protein content could be calculated by the farmers using their local costs and prices. Indeed, the treatment Var-N-low differed from the other two treatments by only about 20 kg N ha^{-1} on average, which was supposed to be not so important compared to the total amount available (i.e., about 160 vs 180 kg ha^{-1}), accounting for the residual N left by the previous pea crop (likely around 30 kg N ha^{-1}) and the amount of N supplied with the first mineral N application (30 kg N ha^{-1}). It is worth noting that about 180 kg ha^{-1} of available N was in line with the usual practice adopted by farmers in the Tiber valley of Umbria and adequate, if not limiting, in view of the high yielding potential of this area

(6–8 tons grains ha^{-1}). Nonetheless, the rainy fall-spring of 2017–2018 caused a relevant N loss from the soil, thus, in this case, a difference of 20 kg N ha^{-1} was expected to make a difference. Several reasons can be addressed to explain the lack of difference among treatments. The main one is that crop development was not associated to N availability but to the soil textural gradient, with NDVI values tendentially as lower as higher appeared the sand content in the soil. In such a condition, and considering the rainy season, forcing the N rate in plots with low NDVI (i.e., with higher sand content) resulted in losing N (mainly by leaching) without affecting yield. As reported by Asseng & Turner [34], NUE is mainly influenced by soil and rain. Thus, in our case, the N rate increase did not translate in a NUE increase as this was probably limited by the other factors. This would justify that, in some cases, the N rate should be directly related to the NDVI [35], i.e., increased where NDVI is higher and decreased where NDVI is lower, on the assumption that lower NDVI values could prove that other soil characteristics, besides N availability, are not suitable for allowing high yields. For example, a soil with low water retention is not suitable in case of dry grain filling periods (as it is in Mediterranean areas) due to water shortage and thus lower yields whatever the N availability [36–38]. The similar time course of NDVI observed in all the three treatments, distinguishing three different vigor status before the second fertilization (Figure 5), also proves that N was not the limiting factor, or at least, the 20 kg N ha^{-1} of difference among treatments was not relevant for crop NDVI.

The weak correlation between grain yield and NDVIs at any date of monitoring (R^2 always below 0.3) appears in contrast with the evidence of Sultana et al. [12] and Liu et al. [39] which show high correlation between NDVI and yield, especially for NDVI recorded during the milky-grain vegetative stage. However, the NDVI recorded in our experiment from 21 April onwards was quite high for all treatments, whereas it is known that the correlation with yield stands only when a wide range of NDVI is considered [16].

As discussed above, the textural gradient likely affected yield through N and water availability, more than the N fertilization treatments. This assumption is supported by the case study of Basso et al. [19] who observed a high correlation between yield and pedological conditions, in particular soil water capacity. Also, Zhao et al. [40] reported that fertilization, as well as the availability of water, significantly affected the content of wheat proteins, even though the latter was more influent, especially for those cultivars characterized by an intermediate protein content. Probably, further splitting the rate, with a third application at the end of shooting, would have prevented some N loss and increased the grain protein content [41], which was quite low compared to the standard of our cultivar. Again, the low protein content can be ascribed to the rainy season, as it was widespread for the harvest of 2018 in Umbria and all-over Central Italy.

Concerning the prescription map used in the study, including the overlapping pixels with plot borders in average NDVI calculation simplifies the GIS procedures without generating agronomically meaningful N-rates differences. Because of the high variability within field zones and among plots, the map could be considered as a kind of stress test for the VRT device. The very little differences between prescribed and distributed rates (Table 2) suggest that this technology is suitable for precision fertilization even with highly detailed prescription maps. The highlighted differences are clearly due to small plot size and random allotment of treatments. As a result, the VRT machine was subjected to abrupt rate variations due to the short time available to adapt the opening of the distribution valve.

The simple linear approach to calculate a N prescription map proposed in this research, even though requiring average GIS skills could be effective for VRT adoption by non-specialist farmers. Coherently with other similar experiences [22–24], to improve the overall usability of the method by non-specialists in PA and allow further validations and tuning, all the RS data management and GIS calculation could be implemented in an user-friendly web-GIS application where the user could upload (or digitize on aerial data) his own fields, choose the S2 NDVI reference date, and decide the most suitable approach for defining the N rate. NDVI S2 data could be conveniently accessed through the Sentinel Hub platform API [42].

Concerning the integration of NDVI from Sentinel 2, VRT fertilization, and yield mapping in PA, our experience confirms that these technologies, thanks also to both the web-based applications for calculating satellite-based prescription maps [22–24] and the user-friendly interfaces of system consoles, could be ordinarily used even in small or medium-sized farms. However, a preliminary cost-benefit analysis and a start-up training and support by a PA expert is often necessary.

5. Conclusions

Our results regarding crop vegetation index, grain yield, and protein content indicate that the adoption of a low-N management approach, based on simple linear models and VRT, may considerably reduce the economic and environmental impact of nitrogen fertilization in winter wheat. However, the rainfed nature of winter wheat in Mediterranean environments may cause unpredictable yield and quality variations depending on climatic trends and soil properties due to their effects on both soil nitrogen and water availability. In this view, VRT nitrogen fertilization can only partially mitigate the heterogeneity of production determined by such environmental factors. The alternative approach of providing a nitrogen supply proportional to the crop NDVI deserves to be considered when factors other than N fertilizer rate come into play, as it is with sandy soils where NDVI and yield may be limited by low N and water retention. Despite the known limitations of predictions and prescriptions based on remoted sensed vegetation indices, their use provides relevant information about within-field and between-fields variability. This information can support the implementation of crop fertilization management approaches based on GNSS/RTK and VRT technologies to replace the traditional flat-N rate, which provides results that are neither economically nor environmentally sustainable.

Author Contributions: All the Authors conceived the research and designed the methodology; M.V. and F.S. applied the methodology and analyzed the data; P.B. supervised the research; all the Authors wrote the paper.

Acknowledgments: The authors wish to thank Mauro Brunetti of the Foundation for Agricultural Education of Perugia for his very helpful and valuable support.

References

1. Liaghat, S.; Balasundram, S.K. A review: The role of remote sensing in precision agriculture. *Am. J. Agric. Biol. Sci.* **2010**, *5*, 50–55. [CrossRef]
2. Oza, S.R.; Panigrahy, S.; Parihar, J.S. Concurrent use of active and passive microwave remote sensing data for monitoring of rice crop. *Int. J. Appl. Earth Obs. Geoinf.* **2008**, *10*, 296–304. [CrossRef]
3. Zheng, Q.; Huang, W.; Cui, X.; Shi, Y.; Liu, L. New Spectral Index for Detecting Wheat Yellow Rust Using Sentinel-2 Multispectral Imagery. *Sensors* **2018**, *18*, 868. [CrossRef]
4. Copernicus Open Access Hub. Available online: https://scihub.copernicus.eu/ (accessed on 4 April 2019).
5. Silleos, N.G.; Alexandridis, T.K.; Gitas, I.Z.; Perakis, K. Vegetation indices: Advances made in biomass estimation and vegetation monitoring in the last 30 years. *Geocarto Int.* **2006**, *21*, 21–28. [CrossRef]
6. Xue, J.; Su, B. Significant remote sensing vegetation indices: A review of developments and applications. *J. Sens.* **2017**, *2017*. [CrossRef]
7. Mulla, D.J. Twenty five years of remote sensing in precision agriculture: Key advances and remaining knowledge gaps. *Biosyst. Eng.* **2013**, *4*, 358–371. [CrossRef]
8. Rouse, J.W.; Hass, R.H.; Schell, J.A.; Deering, D.W. Monitoring vegetation systems in the great plains with ERTS. *Third ERTS Symp. NASA* **1973**, *1*, 309–317.
9. Muñoz-Huerta, R.F.; Guevara-Gonzalez, R.G.; Contreras-Medina, L.M.; Torres-Pacheco, I.; Prado-Olivarez, J.; Ocampo-Velazquez, R.V. A review of methods for sensing the nitrogen status in plants: advantages, disadvantages and recent advances. *Sensors* **2013**, *13*, 10823–10843. [CrossRef]

10. Cabrera-Bosquet, L.; Molero, G.; Stellacci, A.; Bort, J.; Nogués, S.; Araus, J. NDVI as a potential tool for predicting biomass, plant nitrogen content and growth in wheat genotypes subjected to different water and nitrogen conditions. *Cereal Res. Commun.* **2011**, *39*, 147–159. [CrossRef]
11. Carlson, T.N.; Ripley, D.A. On the relation between NDVI, fractional vegetation cover, and leaf area index. *Remote Sens. Environ.* **1997**, *62*, 241–252. [CrossRef]
12. Sultana, S.R.; Ali, A.; Ahmad, A.; Mubeen, M.; Zia-Ul-Haq, M.; Ahmad, S.; Ercisli, S.; Jaafar, H.Z.E. Normalized Difference Vegetation Index as a Tool for Wheat Yield Estimation: A Case Study from Faisalabad, Pakistan. *Sci. World J.* **2014**, *2014*, 1–8. [CrossRef]
13. Zhu, Y.; Yao, X.; Tian, Y.C.; Liu, X.J.; Cao, W.X. Analysis of common canopy vegetation indices for indicating leaf nitrogen accumulations in wheat and rice. *Int. J. Appl. Earth Obs. Geoinf.* **2008**, *10*, 1–10. [CrossRef]
14. Cao, Q.; Miao, Y.; Feng, G.; Gao, X.; Li, F.; Liu, B.; Yue, S.; Cheng, S.; Ustin, S.L.; Khosla, R. Active canopy sensing of winter wheat nitrogen status: An evaluation of two sensor systems. *Comput. Electron. Agric.* **2015**, *112*, 54–67. [CrossRef]
15. Vizzari, M.; Modica, G. Environmental effectiveness of swine sewage management: a multicriteria AHP-based model for a reliable quick assessment. *Environ. Manage.* **2013**, *52*, 1023–1039. [CrossRef]
16. Benincasa, P.; Antognelli, S.; Brunetti, L.; Fabbri, C.A.; Natale, A.; Sartoretti, V.; Modeo, G.; Guiducci, M.; Tei, F.; Vizzari, M. Reliability of NDVI derived by high resolution satellite and UAV compared to in-field methods for the evaluation of early crop N status and grain yield in wheat. *Exp. Agric.* **2017**, *54*, 1–19. [CrossRef]
17. Spiertz, J.H.J. Nitrogen, sustainable agriculture and food security. A review. In *Sustainable Agriculture*; Springer: Dordrecht, The Netherlands, 2009.
18. Bourdin, F.; Morell, F.J.; Combemale, D.; Clastre, P.; Guérif, M.; Chanzy, A. A tool based on remotely sensed LAI, yield maps and a crop model to recommend variable rate nitrogen fertilization for wheat. *Adv. Anim. Biosci.* **2017**, *8*, 672–677. [CrossRef]
19. Basso, B.; Ritchie, J.T.; Cammarano, D.; Sartori, L. A strategic and tactical management approach to select optimal N fertilizer rates for wheat in a spatially variable field. *Eur. J. Agron.* **2011**, *35*, 215–222. [CrossRef]
20. Diacono, M.; Rubino, P.; Montemurro, F. Precision nitrogen management of wheat. A review. *Agron. Sustain. Dev.* **2013**, *33*, 219–241. [CrossRef]
21. Song, X.; Wang, J.; Huang, W.; Liu, L.; Yan, G.; Pu, R. The delineation of agricultural management zones with high resolution remotely sensed data. *Precis. Agricul.* **2009**, *10*, 471–487. [CrossRef]
22. CropSAT. Available online: https://cropsat.com (accessed on 4 April 2019).
23. Agrosat. Available online: https://www.agrosat.it (accessed on 4 April 2019).
24. OneSoil. Available online: https://onesoil.ai (accessed on 4 April 2019).
25. Nawar, S.; Corstanje, R.; Halcro, G.; Mulla, D.; Mouazen, A.M. Delineation of Soil Management Zones for Variable-Rate Fertilization: A Review. *Adv. Agron.* **2017**, *143*, 175–245.
26. Zhang, N.; Wang, M.; Wang, N. Precision agriculture—A worldwide overview. In Proceedings of the Computers and Electronics in Agriculture, Chicago, IL, USA, 1–3 May 2002; Volume 36, pp. 113–132.
27. Fulton, J.P.; Shearer, S.A.; Higgins, S.F.; Hancock, D.W.; Stombaugh, T.S. Distribution pattern variability of granular VRT applicators. *Trans. ASAE* **2013**, *48*, 2053–2064. [CrossRef]
28. Ross, K.W.; Morris, D.K.; Johannsen, C.J. A review of intra-field yield estimation from yield monitor data. *Appl. Eng. Agric.* **2008**, *24*, 309–317. [CrossRef]
29. Arslan, S.; Colvin, T.S. Grain yield mapping: Yield sensing, yield reconstruction, and errors. *Precis. Agric.* **2002**, *3*, 135–154. [CrossRef]
30. Vian, A.L.; Bredemeier, C.; Turra, M.A.; Giordano, C.P.d.S.; Fochesatto, E.; Silva, J.A.d.; Drum, M.A. Nitrogen management in wheat based on the normalized difference vegetation index (NDVI). *Ciência Rural* **2018**, *48*, 1–9. [CrossRef]
31. *QGIS Software, Version 2.18.12*; Quantum GIS Development Team: Berne, Switzerland, 2017.
32. *A guide to Kjeldahl Nitrogen Determination Methods and Apparatus*; ExpotechUSA: Houston, TX, USA, 1998; ISBN 9788578110796.
33. Raun, W.R.; Solie, J.B.; Johnson, G.V.; Stone, M.L.; Mullen, R.W.; Freeman, K.W.; Thomason, W.E.; Lukina, E.V. Improving nitrogen use efficiency in cereal grain production with optical sensing and variable rate application. *Agron. J.* **2002**, *94*, 815–820. [CrossRef]

34. Asseng, S.; Turner, N.C. Analysis of water- and nitrogen-use efficiency of wheat in a Mediterranean climate. *Plant Soil* **2001**, *233*, 127–143. [CrossRef]
35. Raun, W.R.; Solie, J.B.; Stone, M.L.; Martin, K.L.; Freeman, K.W.; Mullen, R.W.; Zhang, H.; Schepers, J.S.; Johnson, G.V. Optical sensor-based algorithm for crop nitrogen fertilization. *Commun. Soil Sci. Plant Anal.* **2005**, *36*, 2759–2781. [CrossRef]
36. Diacono, M.; Castrignanò, A.; Troccoli, A.; De Benedetto, D.; Basso, B.; Rubino, P. Spatial and temporal variability of wheat grain yield and quality in a Mediterranean environment: A multivariate geostatistical approach. *F. Crop. Res.* **2012**, *131*, 49–62. [CrossRef]
37. Bonciarelli, U.; Onofri, A.; Benincasa, P.; Farneselli, M.; Guiducci, M.; Pannacci, E.; Tosti, G.; Tei, F. Long-term evaluation of productivity, stability and sustainability for cropping systems in Mediterranean rainfed conditions. *Eur. J. Agron.* **2016**, *77*, 146–155. [CrossRef]
38. Miralles, D.J.; Slafer, G.A. Paper Presented at International Workshop on Increasing Wheat Yield Potential, CIMMYT, Obregon, Mexico, 20–24 March 2006. Sink limitations to yield in wheat: How could it be reduced? *J. Agric. Sci.* **2007**, *145*, 139–149. [CrossRef]
39. Liu, L.; Wang, J.; Bao, Y.; Huang, W.; Ma, Z.; Zhao, C. Predicting winter wheat condition, grain yield and protein content using multi-temporal EnviSat-ASAR and Landsat TM satellite images. *Int. J. Remote Sens.* **2006**, *27*, 737–753. [CrossRef]
40. Zhao, C.; Liu, L.; Wang, J.; Huang, W.; Song, X.; Li, C. Predicting grain protein content of winter wheat using remote sensing data based on nitrogen status and water stress. *Int. J. Appl. Earth Obs. Geoinf.* **2005**, *7*, 1–9. [CrossRef]
41. López-Bellido, L.; López-Bellido, R.J.; Redondo, R. Nitrogen efficiency in wheat under rainfed Mediterranean conditions as affected by split nitrogen application. *For. Crop. Res.* **2005**, *94*, 86–97. [CrossRef]
42. Sentinel Hub. Available online: https://www.sentinel-hub.com (accessed on 15 April 2019).

2

Development and Evaluation of a Leaf Disease Damage Extension in Cropsim-CERES Wheat

Georg Röll [1,*], William D. Batchelor [2], Ana Carolina Castro [3,4], María Rosa Simón [3,5] and Simone Graeff-Hönninger [1]

1 Department of Agronomy, Institute of Crop Science, University of Hohenheim, 70599 Stuttgart, Germany; Simone.graeff@uni-hohenheim.de

2 Biosystems Engineering Department, Auburn University, Auburn, AL 36849, USA; wdb0007@auburn.edu

3 Cerealicultura, Facultad de Ciencias Agrarias y Forestales, Universidad Nacional de La Plata, Av. 60 y 119, La Plata 1900, Argentina; ana.castro@agro.unlp.edu.ar (A.C.C.); mrsimon@agro.unlp.edu.ar (M.R.S.)

4 Consejo Nacional de Investigaciones Científicas y Técnicas (CONICET), Centro Científico Tecnológico (CCT), La Plata 1900, Argentina

5 Comisión de Investigaciones Científicas, Pcia. de Buenos Aires, 526, 10 y 11, La Plata 1900, Argentina

* Correspondence: georg.roell@uni-hohenheim.de

Abstract: Developing disease models to simulate and analyse yield losses for various pathogens is a challenge for the crop modelling community. In this study, we developed and tested a simple method to simulate septoria tritici blotch (STB) in the Cropsim-CERES Wheat model studying the impacts of damage on wheat (*Triticum aestivum* L.) yield. A model extension was developed by adding a pest damage module to the existing wheat model. The module simulates the impact of daily damage on photosynthesis and leaf area index. The approach was tested on a two-year dataset from Argentina with different wheat cultivars. The accuracy of the simulated yield and leaf area index (LAI) was improved to a great extent. The Root mean squared error (RMSE) values for yield (1144 kg ha^{-1}) and LAI (1.19 m^2 m^{-2}) were reduced by half (499 kg ha^{-1}) for yield and LAI (0.69 m^2 m^{-2}). In addition, a sensitivity analysis of different disease progress curves on leaf area index and yield was performed using a dataset from Germany. The sensitivity analysis demonstrated the ability of the model to reduce yield accurately in an exponential relationship with increasing infection levels (0–70%). The extended model is suitable for site specific simulations, coupled with for example, available remote sensing data on STB infection.

Keywords: wheat; disease; yield; septoria tritici blotch; leaf area index; crop modelling; decision support system for agrotechnology transfer (DSSAT); Cropsim-CERES Wheat

---

1. Introduction

Wheat (*Triticum aestivum* L.) is the second most important staple food crop for human nutrition. It is grown worldwide on approximately 220 million hectares under different climatic conditions. It is projected that wheat production must increase by 1.6% annually to meet the expected global demand by 2050 [1]. However, increasing temperatures and changing global rainfall patterns will likely influence breeding, management, fertilization and crop protection strategies for wheat [2] and also influence disease patterns [3,4]. Hence, crop protection measures will play an important role under future climate change, as rising temperatures and changes in rainfall pattern, will cause more favourable conditions for pests and diseases, especially in the warming north, where wheat production is predominant [2].

On a global scale, there are approximately 50 diseases and pests, which have the potential to damage wheat and reduce farmer's income [5–7]. On a global level, the most widely adapted wheat fungal diseases are leaf rust caused by "*Puccinia triticina* E.," stripe rust caused by "*Puccinia striiformis* W.," stem or black rust caused by "*Puccinia graminis* E.," powdery mildew caused by "*Blumeria graminis* P." and septoria tritici blotch (STB) caused by "*Zymoseptoria tritici* D." [1]. The infection by "*Zymoseptoria tritici* D." is the most economically damaging wheat disease worldwide [8]. It can cause yield reductions of 50% to 60% [9] by creating leaf lesions resulting in defoliation and reduced photosynthesis. It has been estimated that 70% of the annual usage of fungicides in Europe is related to the treatment of this disease [10].

During the past decade, there has been an increasing resistance of STB to azole and strobilurin fungicides in Europe [9–11]. Breeding for STB disease resistance is complicated, due to the variability of the pathogen reproduction cycle [12,13]. Researchers have studied different strategies including tillage, crop rotation, delayed sowing, fungicide application and a proper level of fertilizer application to reduce or control the infection of STB [14]. It appears that moderate fungicide application coupled with the right amount of fertilizer is a strategy that holds promise for environmentally friendly wheat production, while reducing at the same time STB infection.

Crop models are suitable for decision support and contribute to a better understanding in the development of new wheat production strategies. They can play a vital role in understanding plant growth processes, the impact of different weather scenarios as well as management strategies on disease outbreak, final yield and grain quality. Hence, crop models might help to spread the production of wheat in more economic and sustainable ways.

Crop models can also provide an insight into yield losses due to pests and diseases, including STB. Several mechanistic wheat crop growth models have been developed over the last several decades, including APSIM [15], WheatGrow [16], STICS [17], Sirius [18] and DSSAT [19]. These models were developed to study crop-environment interaction and to evaluate optimum management strategies.

The Cropsim-CERES-Wheat (CCW) model [20–22] included in the DSSAT version 4.6 [23] was developed to study the impact of genetics, management, weather and climate change on wheat growth and yield. The model simulates daily plant development based on daily maximum and minimum temperature, daylength and vernalisation requirements. Growth is computed on a daily basis using a radiation use efficiency approach. Carbon is allocated daily to different plant parts based on the development stage. The CCW model has been linked with remote sensing data [24] and was successfully tested with different cultivars, soil characteristics as well as in different climatic conditions including Canada [25], Argentina [26], Southern Italy [27] and the United Kingdom [28]. Currently the CCW model does not account for damage due to weeds, pests or diseases [29]. As a consequence, inaccurate simulations of crop growth and yield result when simulating datasets that include pests and diseases [30,31].

Developing and incorporating a disease damage extension would expand the use of the CCW model to simulate and study the impact of disease damage on crop growth and yield. Batchelor et al. [32] incorporated a pest damage into the CROPGRO [33] family of models distributed with the DSSAT [34]. In their approach, they defined pest coupling points as daily rate and state variable modifiers to simulate the impact of daily pest damage on leaf, stem, seed, shell and root state variables and daily photosynthesis rate based on daily inputs of pest damage. They tested this approach for different pest damage types for peanut and soybean crops. They evaluated this approach using a dataset to simulate the impact of velvetbean caterpillar on soybean.

Using the same approach, the purpose of this work was to: (i) develop a disease model extension for the simulation of STB in wheat, to (ii) evaluate the model performance using a dataset from Argentina; and (iii) to conduct a sensitivity analysis for the impact of different disease progress curves on leaf area index and yield using a dataset from Germany.

2. Materials and Methods

2.1. Model Development

Currently the CCW model does not account for competition with weeds, pests or diseases. To solve this problem, modifications of the current CCW version are necessary to include the impact of leaf diseases on final crop yield.

Plant dry matter accumulation and yield can be expressed as a function of leaf area index (LAI), radiation use efficiency and the loss of assimilates due to respiration. Pathogens can modify both leaf area index and daily photosynthesis [35].

The primary damage resulting from STB is defoliation, which reduces both leaf area and leaf photosynthetic rate [36,37].

To apply the damage theory, it was necessary to integrate the pest damage module [32] structure into the current CCW wheat model (Figure 1). These changes included the adding and linkage of the following subroutines to the original version: PEST.for, LINDM.for, PESTCP.for, VEGDM.for and OPPEST.for. A pest damage definition file was created to define the coupling point "leaf area" for the leaf disease STB (Figure 1), where daily damage could be applied to state and rate variables in the model. Percent cumulative leaf area destroyed (PCLA) was chosen as major coupling point in the model.

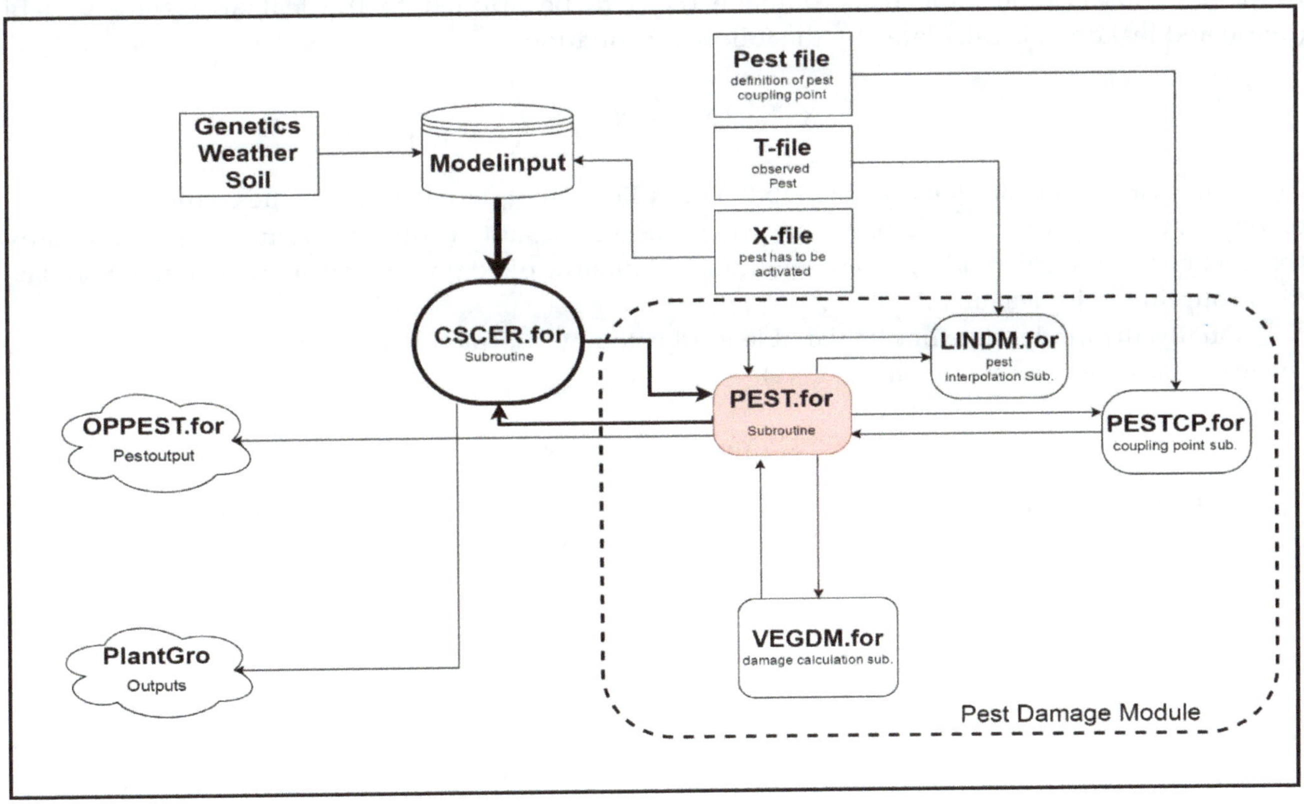

Figure 1. A simplified diagram of the Cropsim-CERES-Wheat (CCW) model with the pest damage module incorporated.

Field observed damage levels were entered in the time series file, referred to as File T in the DSSAT family of models. In this file, the year and day of year (DOY) are entered, along with STB infection in percent. Observed disease infection was linked to the percent cumulative leaf area destroyed (PCLA) coupling point. This damage type (i.e., PCLA) is defined in the pest damage definition model input file, which links field observed damage type and levels to the internal model pest damage coupling point. The model uses a linear interpolation to compute daily damage from periodic field observations. In this work, it was assumed that STB infection began ten days before the first infection symptoms were observed in the field. This assumption was made, based on Sánchez-Vallet et al. [38] who reported a latent period for STB between day 8 and 14 after infection depending on the environmental conditions. The daily percentage of damage (N_{pt}) was calculated between field observations using

$$N_{pt} = P_{rt*} + \frac{(P_{rt} - P_{rt*})}{(D_{pt} - D_{pt*})} \times (D_s - D_{pt*}) \quad (1)$$

N_{pt} = daily reported damage for damage type p at time t; D_{pt} = DOY of next field observation of damage (P_{rt}); D_{pt^*} = DOY of previous field observation of damage (P_{rt^*}); D_s = day of current simulation; P_{rt} = damage level reported in the next field observation; P_{rt^*} = damage level reported in the previous field observation.

The daily damage calculation, which was applied to the leaf area coupling point (P_{it}) is calculated in the PESTCP. for subroutine (Figure 1) by Equation (2):

$$P_{it} = (N_{pt})(C_{ip}) \quad (2)$$

The pest coefficient (C_{ip}) allows the model to convert units of damage into units used for the model state or rate variable that is being damaged.

After calculating the daily damage to be applied to the diseased leaf area based on interpolations from field observations, the daily damage (D_{ipt}) to be applied to the leaf area state variable (defoliated leaf area) is calculated by the following equation:

$$D_{ipt} = X_{it}{}^* - (Xt_{it} - Xs_{it}) \times \left(1 - \frac{P_{it}}{100}\right) \quad (3)$$

$X_{it}{}^*$ = state or model variable i on day t, before application of damage; X_{it} = state or model variable i on day t, after application of damage; Xt_{it} = cumulative amount of coupling point I; Xs_{it} = cumulative senescence of coupling point I; D_{ipt} = amount of damage applied on state or model variable i on day t; P_{it} = coupling point leaf area.

Finally, the model state or rate variable is adjusted by subtracting the computed defoliation from the leaf area state variable by Equation (4):

$$X_{it} = X_{it}{}^* - D_{ipt} \quad (4)$$

X_{it} = state or model variable i on day t, after application of damage; $X_{it}{}^*$ = state or model variable i on day t, before application of damage; D_{ipt} = amount of damage applied on state or model variable i on day t.

2.2. Field Trials

In this study, datasets from two different locations were used for model development. The first dataset was recorded on the Experimental Station Julio Hirschhorn in La Plata (34°56′ S, 57°57′ W, 15 m above sea level, 16.3 °C average temperature; 946 mm mean annual precipitation) National University of La Plata in Argentina. The second experiment was carried out at the Experimental Station Ihinger Hof (48°44′ N, 8°55′ E; 480 m above sea level, mean annual temperature 9.1 °C and 714 mm mean annual precipitation) University of Hohenheim in Germany.

The trial in Argentina was conducted in two consecutive years (2010 and 2011) and published by Castro and Simón [39]. The objective of this trial was to test the tolerance of ten different Argentinean wheat cultivars (*Triticum aestivum* L.) for STB and to evaluate the disease impact on grain yield and grain quality. The sowing dates were on 15th of July in 2010 and 16th of June in 2011. The soil type was a silty loam. Nitrogen was applied as urea at 100 kg N ha^{-1} at sowing and 80 kg N ha^{-1} at the end of tillering. Three different inoculation levels with *Zymoseptoria tritici* D. were performed. The first level was the control treatment, the second was considered to be a low inoculation level (with 5×10^5

spores mL^{-1} suspension), while the third treatment was considered as high inoculation treatment (5×10^6 spores ml^{-1} suspension). All inoculations were performed at growth stage 22 (beginning of tillering) [40] and at growth stage 39 (flag leaf emergence). For model development weather data (daily temperature, rainfall, solar radiation) from the weather station La Plata (34°56′ S, 57°57′ W), disease severity ratings (%) from three growth stages (GS 39, 60, 82), leaf area index (LAI) which was calculated from the green leaf area (GLAI) plus non-green leaf area (NGLAI), yield, soil properties and management information were collected.

The second trial in Germany was conducted in 2006 using the cultivar Monopol with three inoculation levels (control treatment; low inoculation 50%; high inoculation 100%) of *Zymoseptoria tritici*. Inoculation was imposed by spraying 50% or 100% of a spore suspension (1×10^6 spores per mL, strain CBS 292.69) onto the plots at growth stage 32 [41]. The sowing date was 22nd of October 2005 on a silty clay soil. Nitrogen in form of ammonium nitrate was applied at three growth stages: 100 kg N ha^{-1} at GS 30, 80 kg N ha^{-1} at GS 32 and 40 kg N ha^{-1} at GS 49. The objective of this field trial was to use different vegetation indices to determine the occurrence of plant diseases in winter wheat (*Triticum aestivum* L.). For model sensitivity analysis, data including temperature, rainfall and solar radiation from the weather station Ihinger Hof, as well as growth stages, yield monitoring data, disease severity ratings and the LAI at growth stages GS 31, 34 and 49 were collected. Further information on the trial layout can be found in Gröll [41].

2.3. Model Calibration and Evaluation

The modified CCW model extension was incorporated into the DSSAT 4.6 software. Model inputs were created for both datasets from Argentina and Germany. The dataset from La Plata of 2010, which included phenological, yield, soil data (Table 1) and weather data, was used for calibration to test the ability of the model to simulate the impact of STB on wheat growth and yield.

Table 1. Soil properties for experiments in La Plata and Ihinger Hof used in the simulation.

Location La Plata	Clay Content %	Sand Content %	Silt Content %	LLL *	DUL **	SAT ***
0–30 cm	20.7	28.9	50.4	0.226	0.457	0.561
30–60 cm	20.7	28.9	50.4	0.226	0.457	0.561
60–90 cm	20.7	28.9	50.4	0.226	0.457	0.561
Location Ihinger Hof						
0–30 cm	43.3	9.9	46.8	0.247	0.412	0.467
30–60 cm	43.3	9.9	46.8	0.247	0.412	0.467
60–90 cm	25.0	18.8	56.2	0.142	0.313	0.503

* Lower limit ≙ permanent wilting point (pF 4.2). ** Drained upper limit ≙ field capacity (pF 1.8). *** Saturated ≙ saturated water content (pF 0).

Genetic coefficients for growth and development were calibrated using the 2010 dataset and the control treatment for each cultivar. Calibration was performed by sequentially adjusting the genetic coefficients (Table 2) to minimize the error between measured and simulated values [19]. The existing species file was set as default and the existing ecotype UKWH01 as well as the cultivar file were modified. Coefficients for phenological development (P1V, P1D, P1–P5 and PHINT) were calibrated in the first step, followed by crop growth coefficients (G1, G2 and G3). The RMSE, index of agreement (d-Index) and modelling efficiency (EF) statistics were used to assess the quality of the calibration (see section statistical evaluation). After calibration of individual cultivars, the percentage infection with STB of the low and high inoculation was applied to test the model response. The dataset from La Plata of 2011 was used for model validation.

Table 2. Cultivar coefficients in Cropsim-CERES-Wheat (CCW) model used to simulate crop development, crop growth and crop yield for different cultivars.

Parameter	Definition	Ihinger Hof	La Plata						
		Monopol	**ACA801**	**B75 Aniversario**	**Buck Brasil**	**Buck Guapo * Baguette 10 * Klein Zorro * Relmó Centinela ***	**Klein Escorpion**	**Klein Flecha**	**Klein Chaja**
	Cultivar file								
P1D	Percentage reduction in development rate in a photoperiod 10 h shorter than the threshold relative to that at the threshold	110	100	100	100	100	100	100	100
P1V	Days at optimum vernalizing temperature required to complete vernalisation	90	20	20	20	20	20	20	20
G1	Kernel number per unit canopy weight at anthesis (kernels g^{-1})	15	25	25	20	22	22	22	25
G2	Standard kernel size under optimum conditions (mg)	48	25	30	30	30	35	30	39
G3	Standard, non-stressed dry weight of a single tiller at maturity (g)	2.0	4.0	4.0	4.0	4.0	4.0	4.0	4.0
P5	Duration of the grain filling phase (°C d)	600	420	430	420	420	420	420	420
PHINT	Interval between successive leaf tip appearances (°C d)	90	92	92	92	92	80	93	92
	Ecotype file								
P1	Duration of phase end juvenile to terminal spikelet	350	350	350	350	350	350	350	350
P2	Duration of phase from double ridges to the end of leaf growth (°C d)	200	250	250	250	250	250	250	250
P3	Duration of phase from the end of leaf growth to the end of spike growth (°C d)	300	220	220	220	220	220	220	220
P4	Duration of phase from the end of spike growth to the end of the grain fill lag (°C d)	380	300	300	300	300	300	300	300
PARUE	Conversion rate from photosynthetically active radiation to dry matter before the last leaf stage ($g\ MJ^{-1}$)	2.3	2.7	2.7	2.7	2.7	2.7	2.7	2.7
SLAS	Specific leaf area, standard first leaf (cm^2/g)	400	450	450	450	450	450	450	450

* various wheat cultivars with the same cultivar coefficients.

The second dataset from Ihinger Hof was used for sensitivity analysis to test the model on a different location to proof the concept and to test the responsiveness of the model of different STB infection levels. The calibration was performed in the same way as in La Plata with the control treatment by modifying the necessary cultivar coefficients (Table 2). We applied 0%, 10%, 20%, 30%, 50% and 70% damage rates at maximum LAI (GS 39) and started the damage application at growth stage GS 31 to estimate the corresponding yield loss. The different damage rates were used to test the model responsiveness on a broad disease range, which typically starts to impact yield after growth stage 31 [42].

Statistical Evaluation

The statistical model evaluation was conducted by comparing the simulated and observed LAI and yield of the different inoculation treatments (dataset from La Plata).

For statistical analysis, the root mean squared error (RMSE. Equation (5)), the index of agreement (d-Index, Equation (6)) [43] and the modelling efficiency (EF, Equation (7)) were used. The RMSE was used to quantify the amount of variation between simulated and measured values on a metric scale. The d-Index shows if the model is under -or over-estimating the measurements. The EF parameter compares simulated values with the average of the measurements. For a perfect fit between simulated and observed data, the RMSE should be at 0 and the d-Index and EF parameter should have a value of 1.0.

The statistical evaluation was done for simulation runs of the original CCW version and the modified CCW for LAI and yield from both years (2010; 2011) on the location La Plata over all cultivars and inoculation treatments.

Root mean square error (RMSE):

$$RMSE = \left[\frac{1}{n}\sum_{i=1}^{n}(S_i - O_i)^2\right]^{0.5} \tag{5}$$

Index of agreement (d):

$$d = 1 - \left[\frac{\sum_{i=1}^{n}(S_i - O_i)^2}{\sum_{i=1}^{n}\left(\left|S_i - \overline{O}\right| + \left|O_i - \overline{O}\right|\right)^2}\right] \tag{6}$$

Modelling efficiency (EF):

$$EF = 1 - \left[\frac{\sum_{i=1}^{n}(S_i - O_i)^2}{\sum_{i=1}^{n}(O_i - \overline{O})^2}\right] \tag{7}$$

where: O_i = observed values; S_i = simulated values; n = numbers of samples; $\overline{O}$ = mean of observed data

3. Results and Discussion

3.1. Model Calibration for La Plata

3.1.1. Leaf Area Index

The calibration was performed on the 2010 dataset by fitting the relevant genetic coefficients (Table 2) for phenology and growth. One essential prerequisite for model development is an accurate simulation of growth stages. In this study, growth stages (GS 39; GS 60; GS 82) were predicted by the model conclusively: For all ten cultivars the flowering date (GS 65) was documented approximately 110 days after sowing (DAS), the model simulated this growth stage 115 DAS. A similar result was obtained by comparing observed and simulated DAS of the early dough stage (GS 82) (observed approximately at 131 DAS, simulated 132 DAS).

The main focus of this model calibration was on the adjustment of leaf area as a major coupling point for disease damage. Figures 2 and 3 illustrate the simulated and the observed values for leaf area index and grain yield across different inoculation treatments along with the statistics (Table 3).

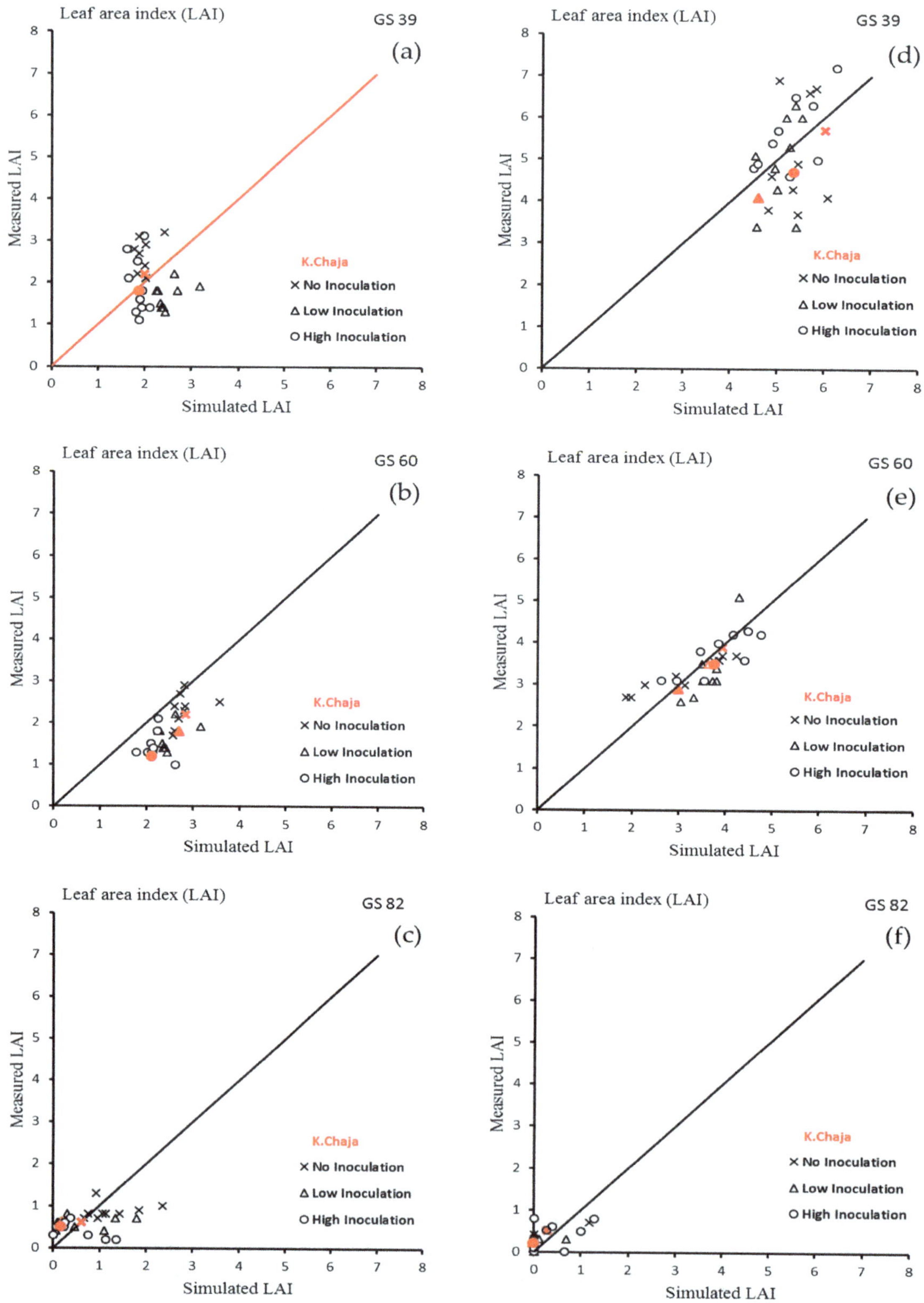

Figure 2. Simulated vs. measured leaf area index (LAI) for calibration (year 2010 **a**–**c**) and validation (year 2011 **d**–**f**) for all ten cultivars on the location La Plata. Different symbols represent the different inoculation treatments where × = No Inoculation; Δ = Low Inoculation and O = High Inoculation.

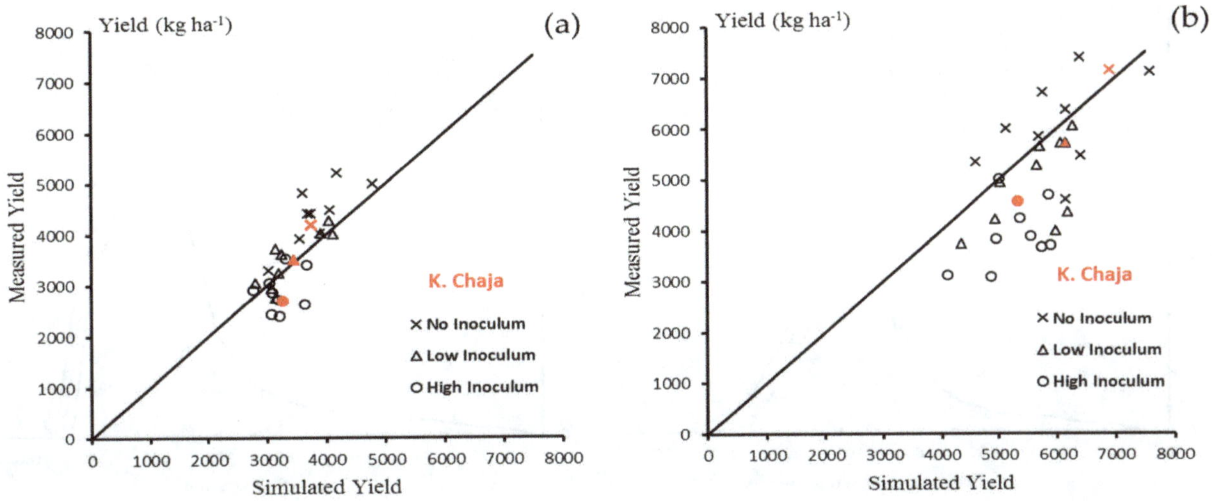

Figure 3. Simulated vs. measured yield (kg DM ha^{-1}) for calibration and validation for all ten cultivars on the location La Plata 2010 (**a**) and La Plata 2011 (**b**).

Table 3. Statistical evaluation of the simulation of leaf area index and grain yield of the original CCW model and the developed CCW model extension for diseases using root mean square error (RMSE), Willmott's d statistic (d-Index) and modelling efficiency (EF).

		Original CCW			Modified CCW		
Variable	**Experiment**	**RMSE**	**d-Index**	**EF**	**RMSE**	**d-Index**	**EF**
Leaf area index	La Plata 2010	1.19	0.33	−2.69	0.69	0.51	−1.07
	La Plata 2011	2.88	0.24	−0.98	1.11	0.70	0.68
Yield	La Plata 2010	1144	0.47	−1.19	499	0.81	0.58
	La Plata 2011	1755	0.50	−1.19	1285	0.66	−0.18

For demonstration of the overall model behaviour in regard to LAI changes induced by three different STB inoculation treatments over time, the wheat cultivar K. Chaja was selected. This cultivar was considered to be highly susceptible to STB infection [39]. Figure 4a shows the impact of disease infestation on LAI according to different inoculation treatments 90 days after sowing. All three simulation runs reached the maximum LAI at day 100. For the control treatment a maximum LAI of 3.5 was simulated. A difference of 0.5 LAI was found between the control and the high-inoculation treatment. Comparing simulated and observed LAI values, the model predicted the LAI over the vegetation period in an accurate manner (RMSE 0.47, d-Index 0.9).

Similar results are displayed in Figure 2a–c, which illustrates the simulated versus observed LAI across three different inoculation treatments for different cultivars. Regardless of susceptibility, in GS 39 all ten cultivars showed a homogenous distribution of all data points around the 1:1 line with no strong outliers. A slight tendency for an overestimation of LAI was given at GS 39 in the low and high inoculation treatments, whereas for the control treatment a slight underestimation was shown over all cultivars. In GS 60 and GS 82, a slight overestimation of LAI was found for all inoculation treatments.

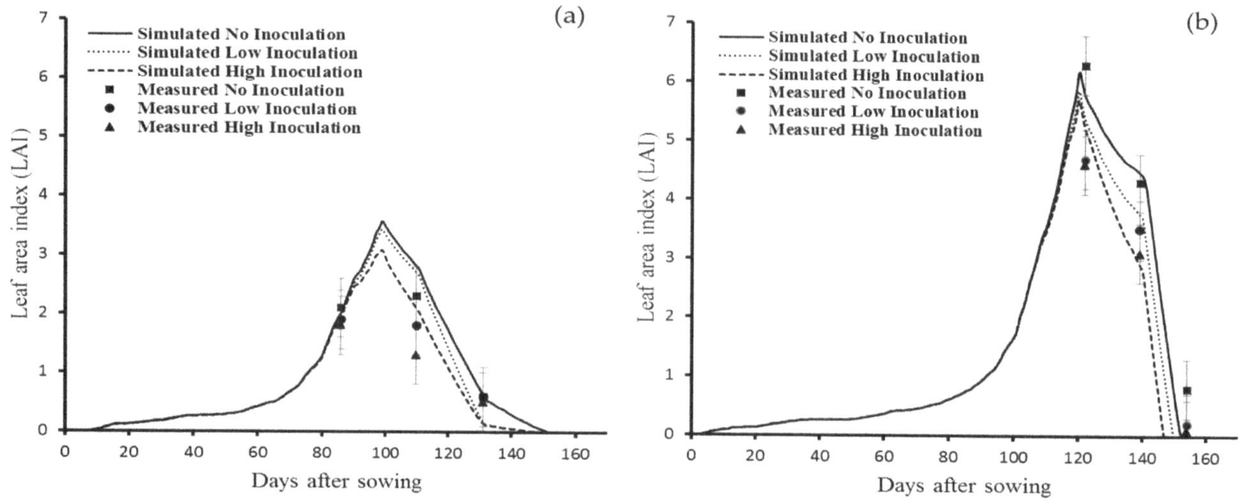

Figure 4. Simulated and measured leaf area index values for cultivar K. Chaja, year 2010 (**a**) and 2011 (**b**) including different inoculation treatments with septoria tritici blotch (STB). The error bars demonstrate the LSD of the leaf area index.

Finally, the modified CCW model (RSME 0.69; d-Index 0.51) performed better compared to the original CCW model (RSME 1.19; d-Index 0.33), as indicated by the corresponding statistics. Outliers in Figure 2a–c can be explained by the LSD ranging from 0.4 to 1.0 depending on the sampling date, reported from Castro and Simón [39].

Nevertheless, the model was able to account for all ten cultivars representing different tolerance levels to STB at different growth stages and disease severities accurately.

3.1.2. Yield

A reduction in LAI after infection with STB also leads to a reduction in yield (Figure 5a). Yield formation started for the cultivar K. Chaja (Figure 5a) on day 122 and was negatively correlated with the inoculation treatment. Yield of the control treatment (3800 kg ha^{-1}) was slightly underestimated and the high inoculation treatment (3400 kg ha^{-1}) showed a slight overestimation.

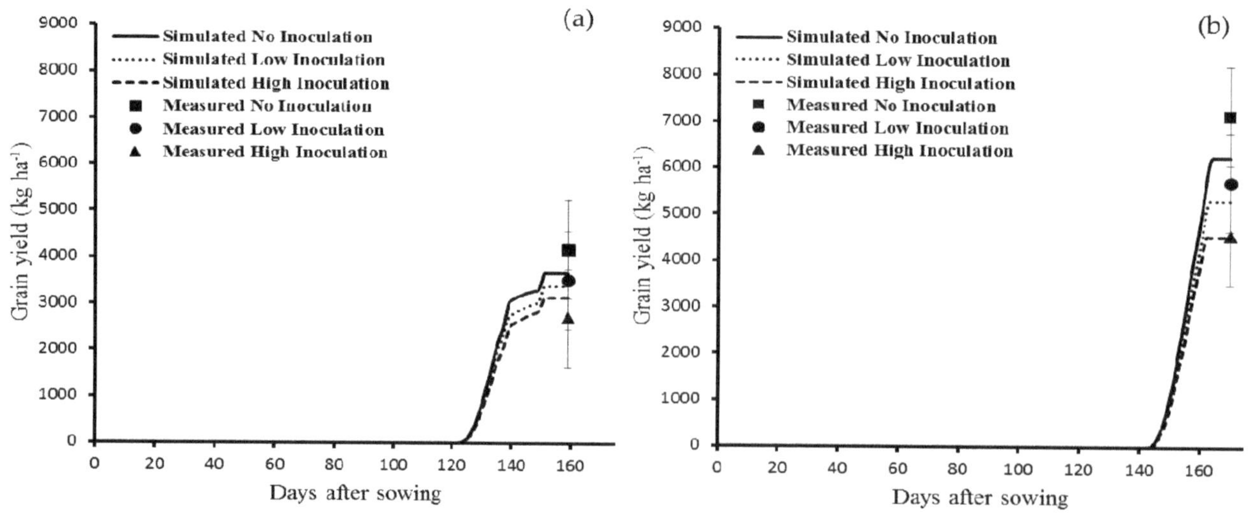

Figure 5. Simulated and measured grain yield values for cultivar K. Chaja year 2010 (**a**) and 2011 (**b**) including different inoculation treatments with STB. The error bars demonstrate the LSD of the yield.

Figure 3a represents the observed versus the simulated yield for all ten cultivars and showed a dense clustering of the different inoculation treatments around the 1:1 line. Overall, it indicated the highest simulated yield for the control and the lowest yield for the high inoculation treatment. The results illustrated the capability of the modified CCW model to account for disease damage. This is expressed in the statistical evaluation (Table 3), where a reduction of the RMSE from 1144 (original version) to 499 (modified version) was observed. The d-Index also underlined these findings, which increased from 0.47 (original version) to 0.81 (modified version).

The modification of the existing CCW showed very good results in LAI and yield simulation (Table 3). It indicated a clear improvement for all statistical parameters compared with the existing CCW included in the current DSSAT version. The calibration successfully minimized the error between measured and simulated data for both, LAI and yield.

3.2. *Model Validation for La Plata*

3.2.1. LAI

Illustrating LAI (Figure 2d–f) and yield (Figure 3b) for the cultivar K. Chaja and all cultivars in the validation year 2011.

In 2011, a maximum LAI of 6.3 was observed in the control treatment for K. Chaja (Figure 4b) at day 120. The model simulated a maximum LAI of 6.2 for the control treatment on the same day. For the low-inoculation treatment, a maximum LAI of 4.7 was observed, whereas the model simulated a maximum LAI of 5.8. Regarding the highest inoculation treatment, a LAI of 4.6 was observed in the field experiment. The model simulated for the same treatment a maximum LAI of 5.7. The model was capable to simulate the maximum LAI for the control treatment exactly but it slightly overestimated the maximum LAI both for the lowest and highest inoculation treatment.

In general LAI was higher in 2011 than in 2010 independent of cultivars, growth stages and inoculation treatments (Figure 2). For 2011 and GS 39, the 1:1 plot showed no strong outliers and a slight overestimation for the control treatment and a slight underestimation for the highest inoculation treatment. This can be caused by an earlier onset of disease in the inoculated treatments which was not reported and cause a slightly underestimation in the model. For GS 60 and GS 82 the model predicted the observed LAI values accurately.

3.2.2. Yield

Yield formation started 140 days after sowing for K. Chaja (Figure 5b), while full maturity was reached on day 165. A maximum yield of 6000 kg ha^{-1} was reached in the control treatment compared with the lowest yield of 5400 kg ha^{-1} in the high inoculation treatment. The corresponding error bars of the measured values were met by the simulated curves, which indicated a high accuracy of the simulation. Under consideration of all cultivars and inoculation treatments (Figure 3b) data points scattered around the 1:1 line on a broader range compared to the calibration (Figure 3a). An inverse relationship between inoculation level and yield was shown (Figure 5).

Overall, the developed model extension was able to account for STB disease damage. This is also shown by the statistics (Table 3), where a 30% improvement of the RMSE in the modified CCW version was achieved compared with the original model. This improvement was also shown by the d-Index and EF values. Further, the calibration showed a higher model accuracy when compared with the validation. Jing et al. [44] and Attia et al. [45] also reported a slightly weaker simulation accuracy regarding the validation dataset.

For 2010, a 20 days shorter growing period due to a 30 days later sowing date and a 130 mm lower precipitation compared to 2011 [39] was reported. Both factors resulted in a reduction of LAI and yield in 2010. Despite these differences the model performed very well for each inoculation treatment and showed its robustness when growing conditions differ between years. Measured yields in the inoculation treatments were simulated quite accurately, while the measured mean value of the control showed a 5% off-set. An explanation for this offset might be given in the way the disease ratings were performed and represented in the model. The model used the mean values from the disease ratings of all repetitions and did not represent each individual plot. It is also possible that the trials had a slight infection of other diseases, which were not measured and caused a slight model offset. Another conceivable reason is the defined onset of disease ten days before disease rating was reported. This assumption was made because of the reported latent period for STB between day 8 and 14 after infection from Sánchez-Vallet et al. [38].

For 2010 and 2011, the d-Index values of the original CCW version are in a similar range for both LAI and yield. In the developed model extension, the d-Index, which represents the model accuracy, increased strongly even though in different intensities for each year. This may point to one possible shortcoming of the current model extension, as it does not account for spore disposal [46]. Spore disposal model use different leaf layers, rain intensity thresholds, droplets, sporulation and concentrations of starting spore pools and can therefore extend the simulation accuracy further.

Nevertheless, other STB models show a strong performance, if a minimum dataset is provided, in which more inputs like leaf wetness are included [47] or in which the initial state of infection of the first leaves is known [46]. Magarey et al. [47] also reported the necessity of hourly weather data for many disease models. In previous datasets this information is not given [48] and they cannot be used for disease modelling, which is literally a loss of information for agricultural decision making. This clearly shows the advantage of the developed CCW extension, in which only the percentage of disease rating, weather- and soil data is needed as a minimum dataset. This leads to a more accurate simulation as shown in Table 3 and makes the CCW applicable for a broader use.

3.3. Sensitivity Analysis

In order to test the general responsiveness of the developed model extension, a sensitivity analysis was carried out by comparing different inoculation treatments with the corresponding disease infections [41]. The model was calibrated by using an independent dataset with disease infections from Germany [41]. The disease infections varied between 4%, 13% and 15%. Disease infection started in GS 31 (DAS 200). Figure 6a depicts the simulated curves for the three different inoculation treatments. A maximum LAI of 7.3 was reached 40 days after GS 31. Simulated curves illustrate a clear separation between the 4%, 13% and the 15% disease infection. A maximum LAI of 7.2, 6.3, 6.1 was reached at day 240 at 4%, 13% and the 15% disease infection rating. Comparing simulated infection scenarios with measured values, the model simulated the LAIs of the three different disease infection levels accurately.

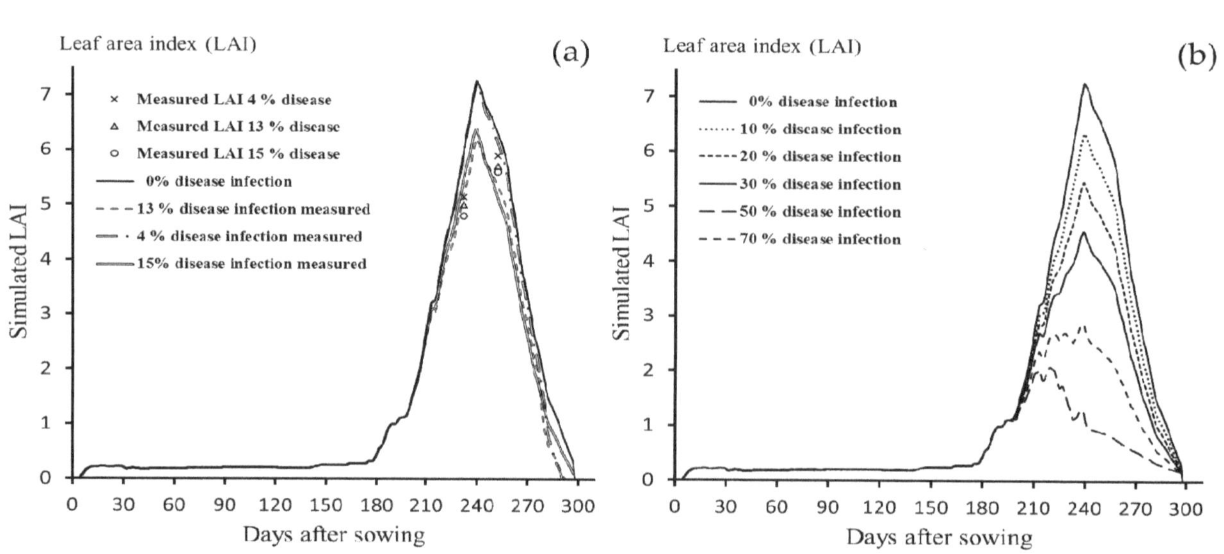

Figure 6. Sensitivity analysis of the CCW disease extension for measured disease infections (4%; 13%; 15%) (**a**) and for five infection scenarios (0%; 10%; 20%; 30%; 50%; 70% disease infection) (**b**) with STB at the location Ihinger Hof.

In the next step an artificial disease infection level of up to 70% was applied (Figure 6b) to test the general responsiveness and the boundaries of the developed model and to test the leaf damage theory on the leaf area coupling point (PCLA).

In Figure 6a a maximum LAI of 7.3 was reached 40 days after GS 31 in the control treatment. LAI increased almost linearly from day 200 to day 240 before the onset of senescence led to a constant decrease in LAI up to final harvest date. For the depicted disease infection scenarios of 10%, 20%, 30% and 50 % a maximum LAI of 7.3, 6.2, 5.4, 4.6, 2.9 and 1.3 was reached at day 240 (Figure 6b). The 70% disease infection scenario showed that a maximum LAI of 2.0 was reached earlier at 220 DAS (Figure 6b). Due to the massive destruction of leaf area, a shortage of assimilate production occurred, which affected in a next step the growth of new leaves. Simulated LAI reduction for maximum LAI in the different disease levels followed the magnitude of 12.5 % (10% diseased LAI), 24.8% (20% diseased LAI), 37.1% (30% diseased LAI), 60.3% (50% diseased LAI) and 82.4% (70% diseased LAI) (Table 4).

Table 4. Yield evaluation of the sensitivity analysis from the Ihinger Hof dataset, by comparing the percentage disease infection with STB and the corresponding simulated percentage yield reduction in kg DM ha^{-1} for the cultivar Monopol.

% Disease Infection	Simulated Yield kg ha^{-1}	Measured Yield kg ha^{-1}	% Yield Reduction	% LAI Reduction at Maximum LAI
0	4384		0	0
4 *	4332	4409	1.2	0.8
10	4190		4.4	12.5
13 *	4159	3934	5.1	14.5
15 *	4124	3965	5.9	12.0
20	3990		9.0	24.8
30	3737		14.8	37.1
50	3242		26.0	60.3
70	2380		45.7	82.4

* measured disease infection.

Table 4 shows the corresponding yields of the applied and measured disease infection levels. The maximum observed yield obtained with the disease infection level of 4% was 4409 kg ha^{-1}. Higher disease infection levels (13%; 15%) resulted in lower yield (3934 kg ha^{-1}; 3965 kg ha^{-1}). Simulated yield decreased gradually with higher infection levels. The control treatment resulted in a maximum yield of 4384 kg ha^{-1}, while the 70% disease infection level resulted in a total grain yield of 2380 kg ha^{-1} which corresponded to a 45.7% yield reduction. Over all tested disease infection levels, the simulated yield reduction followed an exponential shape, indicating that yield reductions became more severe and are more than doubled at higher disease infection levels. An exponential relationship between yield loss and disease infection was also shown by King et al. [49].

Comparing simulated and measured yield, the model showed a slightly underestimation for the 4% level and a slight overestimation for the 13% and 15% disease infection level. These variations are in an acceptable range.

Regarding the accuracy of the current simulation, similar results for yield reduction based on the occurrence of leaf diseases are reported by Ziv and Eyal [50]. Ziv and Eyal [50] tested different inoculation treatments in different spring wheat cultivars and reported yield losses of up to 53% at a disease infection of 73%. The developed CCW model extension gave comparable results to a previous study of Bhathal et al. [51] also at lower infection level scenarios. Bhathal et al. [51] tested different inoculation treatments in wheat to evaluate the relationship between disease infection and yield. Notably, they showed an onset of the disease as it was used in the sensitivity analysis of this study, at GS 31 and demonstrated a 10% yield loss due to a natural disease infection of 26%. King et al. [49], also confirmed this model theory on an independent dataset from the United Kingdom carried out at four different locations. Similar observations and an exponential yield loss curve due to STB disease were obtained. In addition, a yield loss of 30% by a disease infection of 55.1% as well as a yield reduction of 8% by a disease infection of 14.5% occurred. This confirmed the model theory and clearly showed the capability to simulate leaf disease infection with STB by using the coupling point leaf area (PCLA).

3.4. Future Model Applications

Despite the good simulation results, the developed model concept can currently only be used for STB. The concept was not tested on other wheat diseases like stripe rust caused by "*Puccinia striiformis* W.," stem or black rust caused by "*Puccinia graminis* E." or powdery mildew caused by "*Blumeria graminis* P." It can be assumed that this concept will also work for other diseases, by changing the pest coefficient in the pest file to account for different damage types. Bastiaans [52] showed a ß–value for STB, which represents the correlation between visible affected leaf area and the affected photosynthetic rate. A ß–value > 1 indicates a stronger effect on photosynthetic rate as it visually appears. For STB this value is close to 1 wherefore the pest coefficient in the CCW model extension was set to 1. Bastiaans [52] reported a ß–value of 8.7 for "*Erisyhe graminis*" or 1.3 for "*Puccinia recondita*" in winter wheat. It is assumed, that the pest coefficient has to be increased in a similar manner but it has to be proven by real data. However, the structure of the model extension is set up in a flexible way and has the possibility to be transferred to other leaf diseases.

Further, the disease extension routine can suite as a gateway between crop models and remote sensing data, like it was published by Thorp et al. [53]. Thorp et al. [53] showed an improvement of simulation results by updating the plant leaf area state variable with green LAI generated by remote sensing. This offers the opportunity to simulate a given field on a site-specific scale, which means the CCW model extension can be updated by the percentage diseased leaf area detected by for example, remote sensing. In this way, the model could serve as decision support tool to give farmers an economic advise on a field level as Ficke et al. [54] proposed.

4. Conclusions

In this study a disease extension for the CCW model was developed to simulate the damage effect of STB disease on LAI and yield in wheat. The model was tested successfully in a sensitivity analysis on a German dataset and on a dataset obtained from La Plata, Argentina. Results of the study clearly showed the effect of the implementation of the coupling point "PCLA" and on the corresponding LAI and yield for different locations. For the location La Plata, the obtained simulation results of the modified CCW model indicated a higher model accuracy which almost doubled and clearly showed an improved model behaviour. Especially for the cultivar K. Chaja, the CCW model extension showed a high modelling accuracy. The LAI and yield were simulated very accurate in both years. Furthermore the sensitivity analysis also displayed the flexibility of the CCW model extension to account for disease damage over a broad range between 0 and 70% of STB disease infection.

Nevertheless, further research is needed to test the developed model on other leaf diseases like leaf rust, powdery mildew or stripe rust in wheat. The model extension could be used in future studies as decision support system for example, coupled with remote sensing technologies to obtain the necessary disease ratings for the model input files.

Author Contributions: Conceptualization, G.R., W.D.B. and S.G.-H.; methodology, W.D.B., G.R. and S.G.-H.; software, G.R. and W.D.B.; validation, G.R.; formal analysis, G.R.; investigation, G.R., A.C.C. and M.R.S.; resources, G.R., A.C.C. and M.R.S.; data curation, G.R.; writing—original draft preparation, G.R.; writing—review and editing, G.R., W.D.B. and S.G.-H.; visualization, G.R.; supervision, W.D.B. and S.G.-H.; project administration, S.G.-H.; funding acquisition, S.G.-H.

Acknowledgments: The authors would like to thank the Universitätsbund Hohenheim e.V. for supporting the journey to Auburn University, Alabama.

References

1. Singh, R.P.; Singh, P.K.; Rutkoski, J.; Hodson, D.P.; He, X.; Jørgensen, L.N.; Hovmøller, M.S.; Huerta-Espino, J. Disease Impact on Wheat Yield Potential and Prospects of Genetic Control. *Annu. Rev. Phytopathol.* **2016**, *54*, 303–322. [CrossRef] [PubMed]
2. Olesen, J.E.; Trnka, M.; Kersebaum, K.C.; Skjelvåg, A.O.; Seguin, B.; Peltonen-Sainio, P.; Rossi, F.; Kozyra, J.; Micale, F. Impacts and adaptation of European crop production systems to climate change. *Eur. J. Agron.* **2011**, *34*, 96–112. [CrossRef]
3. Eigenbrode, S.D.; Binns, W.P.; Huggins, D.R. Confronting Climate Change Challenges to Dryland Cereal Production: A Call for Collaborative. *Prod. Engagem.* **2018**, *5*. [CrossRef]
4. Ahanger, R.A.; Bhat, H.A.; Bhat, T.A.; Ganie, S.A.; Lone, A.A.; Wani, I.A.; Ganai, S.A.; Haq, S.; Khan, O.A.; Junaid, M.J.; Bhat, T.A. Impact of Climate Change on Plant Diseases. *Int. J. Modern Plant & Anim. Sci. USA* **2013**, *3*, 105–115.
5. Weiss, M.V. *Compendium of Wheat Diseases*, 2nd ed.; APS Press: St. Paul, MN, USA, 1987.
6. Forrer, H.R.; Zadoks, J.C. Yield reduction in wheat in relation to leaf necrosis caused by Septoria tritici. *Neth. J. Plant Pathol.* **1983**, *89*, 87–98. [CrossRef]
7. Eyal, Z. The septoria tritici and stagonospora nodorum blotch diseases of wheat. *Eur. J. Plant Pathol.* **1999**, *105*, 629–641. [CrossRef]
8. Bearchell, S.J.; Fraaije, B.A.; Shaw, M.W.; Fitt, B.D.L.; Cowling, E.B. Wheat Archive Links Long-Term Fungal Pathogen Population Dynamics to Air Pollution. *Proc. Natl. Acad. Sci. USA* **2005**, *102*, 5438–5442. [CrossRef] [PubMed]
9. Eyal, Z.; Amiri, Z.; Wahl, I. Physiological Specialization of Septoria tritici. *Phytopathology* **1973**, *63*, 1087–1091. [CrossRef]
10. Fones, H.; Gurr, S. The impact of Septoria tritici Blotch disease on wheat: An EU perspective. *Fungal Genet. Biol.* **2015**, *79*, 3–7. [CrossRef] [PubMed]
11. Fraaije, B.; Cools, H.J.; Fountaine, J.; Lovell, D.J.; Motteram, J.; West, J.S.; Lucas, J. A Role of Ascospores in Further Spread of QoI-Resistant Cytochrome b Alleles (G143A) in Field Populations of Mycosphaerella graminicola. *Phytopathology* **2005**, *95*, 933–941. [CrossRef] [PubMed]
12. Estep, L.K.; Torriani, S.F.F.; Zala, M.; Anderson, N.P.; Flowers, M.D.; Mcdonald, B.A.; Mundt, C.C.; Brunner, P.C. Emergence and early evolution of fungicide resistance in North American populations of Zymoseptoria tritici. *Plant Pathol.* **2015**, *64*, 961–971. [CrossRef]
13. Simón, M.R.; Cordo, C.A.; Castillo, N.S.; Struik, P.C.; Börner, A. Population Structure of Mycosphaerella graminicola and Location of Genes for Resistance to the Pathogen: Recent Advances in Argentina. *Int. J. Agron.* **2012**, *2012*, 680275. [CrossRef]
14. Rodrigo, S.; Cuello-Hormigo, B.; Gomes, C.; Santamaria, O.; Costa, R.; Poblaciones, M.J. Influence of fungicide treatments on disease severity caused by Zymoseptoria tritici, and on grain yield and quality parameters of bread-making wheat under Mediterranean conditions. *Eur. J. Plant Pathol.* **2014**, *141*, 99–109. [CrossRef]
15. Keating, B.A.; Carberry, P.S.; Hammerb, G.L.; Probert, M.E.; Robertson, M.J.; Holzworth, D.; Huth, N.I.; Hargreaves, J.N.G.; Meinke, H.; Hochman, Z.; et al. An overview of the crop model APSIM. *Eur. J. Agron.* **2003**, *18*, 267–288. [CrossRef]
16. Lv, Z.; Liu, X.; Cao, W.; Zhu, Y. Agricultural and Forest Meteorology Climate change impacts on regional winter wheat production in main wheat production regions of China. *Agric. For. Meteorol.* **2013**, *171–172*, 234–248. [CrossRef]
17. Brisson, N.; Gary, C.; Justes, E.; Roche, R.; Mary, B.; Ripoche, D.; Zimmer, D.; Sierra, J.; Bertuzzi, P.; Burger, P.; et al. An overview of the crop model Stics. *Eur. J. Agron.* **2003**, *18*, 309–332. [CrossRef]
18. Jamieson, P.D.; Semenov, M.A.; Brooking, I.R.; Francis, G.S. Sirius a mechanistic model of wheat respinse to environmental variation. *Eur. J. Agron.* **1998**, *8*, 161–179. [CrossRef]
19. Jones, J.W.; Hoogenboom, G.; Porter, C.H.; Boote, K.J.; Batchelor, W.D.; Hunt, L.A.; Wilkens, P.W.; Singh, U.; Gijsman, A.J.; Ritchie, J.T. The Dssat Cropping System Model. *Eur. J. Agron.* **2003**, *18*, 235–265. [CrossRef]
20. Hunt, L.A.; Pararajasingham , S. CROPSIM—WHEAT: A model describing the growth and development of wheat. *Can. J. Plant Sci.* **1995**, 619–632.

21. Ritchie, J.T.; Singh, U.; Godwin, D.C.; Bowen, W.T. Cereal growth, development and yield. *Underst. Opt. Agric. Prod.* **1998**, 79–98. [CrossRef]
22. Ritchie, J.T.; Otter, S. Description and performance of CERES-Wheat: A user-orientes wheat yield model. *ARS Wheat Yield Proj.* **1985**, *38*, 159–175.
23. Hoogenboom, G.; Jones, J.W.; Wilkens, P.W.; Porter, C.H.; Boote, K.J.; Hunt, U.S.; Lizaso, J.I.; White, J.W.; Uryasev, O.; Ogoshi, R.; et al. *Decision Support System for Agrotechnology Transfer (DSSAT) [CD-ROM]*; University of Hawaii: Honululu, HI, USA, 2015.
24. Thorp, K.R.; Hunsaker, D.J.; French, A.N.; White, J.W.; Clarke, T.R. Evaluation of the CSM-CROPSIM-CERES-Wheat Model as a Tool for Crop Water Management. *Trans. ASABE* **2010**, *53*, 1–17. [CrossRef]
25. Chipanshi, A.C.; Ripley, E.A.; Lawford, R.G. Large-scale simulation of wheat yields in a semi-arid environment using a crop-growth model. *Agric. Syst.* **1999**, *59*, 57–66. [CrossRef]
26. Savin, R.; Satorre, E.H.; Hall, A.J.; Slafer, G.A. Assessing strategies for wheat cropping in the monsoonal climate of the Pampas using the CERES-Wheat simulation model. *Field Crops Res.* **1995**, *42*, 81–91. [CrossRef]
27. Sardinia, S.; Dettori, M.; Cesaraccio, C.; Motroni, A.; Spano, D.; Duce, P. Field Crops Research Using CERES-Wheat to simulate durum wheat production and phenology. *Field Crops Res.* **2011**, *120*, 179–188. [CrossRef]
28. Bannayan, M.; Crout, N.M.J.; Hoogenboom, G. Application of the CERES-Wheat model for within-season prediction of winter wheat yield in the United Kingdom. *Agron. J.* **2003**, *95*, 114–125. [CrossRef]
29. Gbegbelegbe, S.; Cammarano, D.; Asseng, S.; Robertson, R.; Chung, U.; Adam, M.; Abdalla, O.; Payne, T.; Reynolds, M.; Sonder, K.; et al. Baseline simulation for global wheat production with CIMMYT mega-environment specific cultivars. *Field Crops Res.* **2017**, *202*, 122–135. [CrossRef]
30. Batchelor, W.D.; Jones, J.W.; Boote, K.J.; Porter, C.H. *Pest and Disease Damage Module*; University of Florida: Gainsville, FL, USA, 2004.
31. Boote, K.J.; Bennet, J.M.; Jones, J.W.; Jowers, H.E. On-farming testing of peanut and soybean models in north Florida. *Paper Am. Soc. Agric. Eng. USA* **1989**. Available online: http://agris.fao.org/agris-search/search.do?recordID=US9165910 (accessed on 10 January 2019).
32. Batchelor, W.D.; Jones, J.W.; Boote, K.J.; Pinnschmidt, H.O. Extending the use of crop models to study pest damage. *Trans. Am. Soc. Agric. Eng. Gen. Ed.* **1993**, *36*, 551–558. [CrossRef]
33. Boote, K.J.; Jones, J.W.; Hoogenboom, G.; Pickering, N.B. The CROPGRO model for grain legumes. In *Understanding Options for Agricultural Production*; Tsuji, G.Y., Hoogenboom, G., Thornton, P.K., Eds.; Springer: Dordrecht, The Netherlands, 1998; pp. 99–128. ISBN 978-94-017-3624-4.
34. Andarzian, B.; Hoogenboom, G.; Bannayan, M.; Shirali, M.; Andarzian, B. Determining optimum sowing date of wheat using CSM-CERES-Wheat model. *J. Saudi Soc. Agric. Sci.* **2015**, *14*, 189–199. [CrossRef]
35. Waggoner, P.E.; Berger, R.D. Defoliation, Disease, and Growth. *Phytopathology* **1987**, *77*, 1495–1497.
36. Robert, C. Analysis and modelling of effects of leaf rust and Septoria tritici blotch on wheat growth. *J. Exp. Bot.* **2004**, *55*, 1079–1094. [CrossRef] [PubMed]
37. Robert, C.; Bancal, M.O.; Lannou, C.; Ney, B. Quantification of the effects of Septoria tritici blotch on wheat leaf gas exchange with respect to lesion age, leaf number, and leaf nitrogen status. *J. Exp. Bot.* **2006**, *57*, 225–234. [CrossRef] [PubMed]
38. Sánchez-Vallet, A.; McDonald, M.C.; Solomon, P.S.; McDonald, B.A. Is Zymoseptoria tritici a hemibiotroph? *Fungal Genet. Biol.* **2015**, *79*, 29–32. [CrossRef] [PubMed]
39. Castro, A.C.; Simón, M.R. Effect of tolerance to Septoria tritici blotch on grain yield, yield components and grain quality in Argentinean wheat cultivars. *Crop Prot.* **2016**, *90*, 66–76. [CrossRef]
40. Zadoks, J.C.; Chang, T.T.; Konzak, C.F. A decimal code for the growth stages of cereals. *Weed Res.* **1974**, *14*, 415–421. [CrossRef]
41. Gröll, K. *Use of Sensor Technologies to Estimate and Assess the Effects of Various Plant Diseases on Crop Growth and Development*; Universty of Hohenheim: Stuttgart, Germany, 2008; Available online: http://opus.uni-hohenheim.de/volltexte/2008/296/ (accessed on 10 January 2019).
42. Thomas, M.; Cook, R.; King, J. Factors affecting development of Septoria tritici in winter wheat and its effect on yield. *Plant Pathol.* **1989**, 246–257. [CrossRef]
43. Willmott, C.J. Some Comments on the Evaluation of Model Performance. *J. Appl. Phys.* **1982**, *36*, 1309–1313. [CrossRef]

44. Jing, Q.; Qian, B.; Shang, J.; Huffman, T.; Liu, J.; Pattey, E.; Dong, T.; Tremblay, N.; Drury, C.F.; Ma, B.L.; et al. Assessing the options to improve regional wheat yield in eastern canada using the csm–ceres–wheat model. *Agron. J.* **2017**, *109*, 510–523. [CrossRef]
45. Attia, A.; Rajan, N.; Xue, Q.; Nair, S.; Ibrahim, A.; Hays, D. Application of DSSAT-CERES-Wheat model to simulate winter wheat response to irrigation management in the Texas High Plains. *Agric. Water Manag.* **2016**, *165*, 50–60. [CrossRef]
46. Robert, C.; Fournier, C.; Andrieu, B.; Ney, B. coupling a 3D virtual wheat (Triticum aestivum) plant model with a Septoria tritici epidemic model (Septo3D): A new approach to investigate plant-pathogen interactions linked to caonpy architecture. *Funct. Plant Biol.* **2008**, *35*, 997–1013. [CrossRef]
47. Magarey, R.D.; Sutton, T.B.; Thayer, C.L. A Simple Generic Infection Model for Foliar Fungal Plant Pathogens. *Phytopathology* **2005**, *95*, 92–100. [CrossRef] [PubMed]
48. Donatelli, M.; Magarey, R.D.; Bregaglio, S.; Willocquet, L.; Whish, J.P.M.; Savary, S. Modelling the impacts of pests and diseases on agricultural systems. *Agric. Syst.* **2017**, *155*, 213–224. [CrossRef] [PubMed]
49. King, J.E.; Jenkins, J.E.E.; Morgan, W.A. The estimation of yield losses in wheat from severity of infection by Septoria species. *Plant Pathol.* **1983**, *32*, 239–249. [CrossRef]
50. Ziv, O.; Eyal, Z. Assessment of Yield Component Losses Caused in Plants of Spring Wheat Cultivars by Selected Isolates of Septoria tritici. *Phytopathology* **1977**, *68*, e796. [CrossRef]
51. Bhathal, J.S.; Loughman, R.; Speijers, J. Yield reduction in wheat in relation to leaf disease from yellow (tan) spot and septoria nodorum blotch. *Eur. J. Plant Pathol.* **2003**, *109*, 435–443. [CrossRef]
52. Bastiaans, L. Ecology and Epidemiology Ratio Between Virtual and Visual Lesion Size as a Measure to Describe Reduction in Leaf Photosynthesis of Rice Due to Leaf Blast. *Phytopathology* **1991**, *81*, 611–615. [CrossRef]
53. Thorp, K.R.; Hunsaker, D.J.; French, A.N. Assimilating leaf area index estimates from remote sensing into the simulations of a cropping systems model. *Trans. ASABE* **2010**, *53*, 251–262. [CrossRef]
54. Ficke, A.; Cowger, C.; Bergstrom, G.; Brodal, G. Understanding Yield Loss and Pathogen Biology to Improve Disease Management: Septoria Nodorum Blotch—A Case Study in Wheat. *Plant Dis.* **2018**, *102*, 696–707. [CrossRef] [PubMed]

3

Workflow to Establish Time-Specific Zones in Precision Agriculture by Spatiotemporal Integration of Plant and Soil Sensing Data

Elia Scudiero [1,2,*], Pietro Teatini [3], Gabriele Manoli [4], Federica Braga [5], Todd H. Skaggs [2] and Francesco Morari [6]

1. Department of Environmental Sciences, University of California Riverside, 900 University Ave., Riverside, CA 92521, USA
2. USDA-ARS, United States Salinity Laboratory, 450 West Big Springs Rd., Riverside, CA 92507, USA; todd.skaggs@ars.usda.gov
3. Department of Civil, Environmental, and Architectural Engineering (ICEA), University of Padua, Via Trieste 63, 35121 Padua, Italy; teatini@dmsa.unipd.it
4. ETH Zurich, Department of Civil, Environmental and Geomatic Engineering, Stefano-Franscini-Platz 5, 8093 Zürich, Switzerland; manoli@ifu.baug.ethz.ch
5. Institute of Marine Sciences (ISMAR), National Research Council, Arsenale Tesa 104, Castello 2737/F, 30122 Venezia, Italy; federica.braga@ve.ismar.cnr.it
6. Department of Agronomy, Food, Natural resources, Animals, and Environment (DAFNAE), University of Padua, Viale dell'Università 16, 35020 Legnaro, Italy; francesco.morari@unipd.it

* Correspondence: elia.scudiero@ars.usda.gov or elia.scudiero@ucr.edu

Abstract: Management zones (MZs) are used in precision agriculture to diversify agronomic management across a field. According to current common practices, MZs are often spatially static: they are developed once and used thereafter. However, the soil–plant relationship often varies over time and space, decreasing the efficiency of static MZ designs. Therefore, we propose a novel workflow for time-specific MZ delineation based on integration of plant and soil sensing data. The workflow includes four steps: (1) geospatial sensor measurements are used to describe soil spatial variability and in-season plant growth status; (2) moving-window regression modelling is used to characterize the sub-field changes of the soil–plant relationship; (3) soil information and sub-field indicator(s) of the soil–plant relationship (i.e., the local regression slope coefficient[s]) are used to delineate time-specific MZs using fuzzy cluster analysis; and (4) MZ delineation is evaluated and interpreted. We illustrate the workflow with an idealized, yet realistic, example using synthetic data and with an experimental example from a 21-ha maize field in Italy using two years of maize growth, soil apparent electrical conductivity and normalized difference vegetation index (NDVI) data. In both examples, the MZs were characterized by unique combinations of soil properties and soil–plant relationships. The proposed approach provides an opportunity to address the spatiotemporal nature of changes in crop genetics × environment × management interactions.

Keywords: remote sensing; proximal sensing; crop modeling; soil; plant; management zone; spatial variability; temporal variability; precision agriculture

1. Introduction

Crop yields and resource use efficiency (e.g., nutrients and water) have a strong spatial component, which can be observed over a wide range of scales, from regional to subfield [1]. At the field scale, yield variability in uniformly managed fields is often related to the spatial variability of soil properties and their impact on plant growth [2–6]. This variability can be addressed using precision agriculture practices [7], such as variable rate management (VRM) [8]. According to VRM principles, efficiency or crop production can be increased by varying agronomic inputs over a field according to varying soil and crop conditions [9]. Within a field, areas with similar soil properties (e.g., texture, topography, water holding capacity) that can be managed uniformly are commonly called management zones (MZs) [10]. To justify the subdivision of a field into MZs, there should be sizeable difference in soil properties between MZs [9,11]. The use of MZs for VRM has been shown to increase productivity, decrease costs, and/or reduce environmental impacts of agronomic practices [12,13]. Several authors have shown that within-MZ management should change over time [14–20], such an approach is often referred to as "dynamic VRM" [14].

Shanahan et al. [21], Long et al. [22], and Quebrajo et al. [23] indicated that coupling soil and plant data should increase the efficiency of MZ-based precision agriculture. Several sensor measurements can be used as indicators of soil spatial variability, including: apparent electrical conductivity, gamma-ray spectrometry, visible and near-infrared reflectance, and penetrometry [24–26]. Throughout the growing season, field-scale crop canopy measurements from near-ground (e.g., tractor mounted sensors) and remote sensing (e.g., unmanned aerial vehicles—UAV, satellites) can be used to provide information about plant health [27]. In particular, visible, near-infrared, and thermal sensing can be used to infer crop water status, nitrogen deficiency, and the effect of other biotic/abiotic stressors [28–31] and as an indicator of potential yield [16,32–35].

According to current practices, MZ-designs derived from soil maps, crop sensing, and/or historical yield maps are generally designed once and then implemented year after year. Nevertheless, several authors have reported inconsistent benefits for this precision management strategy [21,22]. Spatial patterns in yield tend to change from year-to-year, mostly because of changes in meteorological conditions [12,36–38]. In other words, the spatial patterns of most soil properties are fairly stable in time, but at different times plants may be limited in different ways at the same location because of the influence of transient factors affecting the soil–plant relationship, such as meteorological factors and agronomical management [12,38,39].

Sadler et al. [40] indicated that there is a need for accurate and inexpensive systems to delineate dynamic management zones, obtained by sensing within-field variability in real time, so that agricultural management can be controlled adaptively. Recent research strongly suggested that MZ designs should change over time, both intra- and inter-seasonally [19,20,38,41–44]. Myers [45] formally justified the need for time-specific and spatially dynamic VRM through the "fundamental theorem of precision agriculture production" where *Crop Yield* is a function of *Genetics* × *Environment* × *Management* × *Space* × *Time* interactions. According to this theorem, the spatiotemporal variability of crop performance should be addressed by adjusting the agronomic prescriptions over time and space. Several examples of protocols and data analysis workflow for static soil and/or plant-based MZ delineation are present in the scientific literature [46–49]. To our best knowledge, time-specific MZ delineation based on soil and in-season plant information has not been commonly discussed in the literature. Particularly, there is a lack of protocols and analytical workflow that farm managers, agricultural consultants, and scientists can use to take advantage of free/inexpensive in-season crop information (e.g., from UAV, Sentinel 2 satellite) and high-resolution soil maps.

We aim to present a novel workflow for the selection of time-specific MZs according to in-season spatial measurements of crop growth status and its relationship with high-resolution soil spatial information. The MZ should identify areas with homogeneous and unique (within a single field) soil–plant relationships. The MZ-delineation workflow will be described in detail. We will also provide two examples on how to implement the workflow. The first example is based on synthetic data. The second example uses data from a maize (*Zea mays* L.) field in northeastern Italy.

2. Materials and Methods

2.1. Time-Specific MZ-Delineation Workflow

The time-specific MZ delineation can be implemented as follows:

- STEP 1. Soil and time-specific plant spatial information acquisition, pre-processing, interpretation, and interpolation
- STEP 2. Time-specific sub-field soil–plant modeling
- STEP 3. Time-specific MZ delineation with cluster analysis
- STEP 4. Evaluation/interpretation of time-specific MZ design

2.1.1. Soil and Time-Specific Plant Spatial Information

In STEP 1, high-resolution spatial measurements for target soil properties and in-season plant-canopy information, such as canopy reflectance, are acquired, pre-processed, interpreted, and interpolated. In-season plant-canopy reflectance is acquired and used as an indication of crop status. Soil spatial information is used to interpret the crop canopy measurements.

Soil sensor data acquisition should be carried out according to established protocols [50–52] to increase the accuracy and consistency of the survey across large areas. Attention should be paid to selecting those sensors which represent the spatial variability of soil properties known or believed to influence a crop at the site of interest [53,54]. For information on sensor data pre-processing (e.g., conversion of spatial coordinates, removals of outliers), readers are referred to the first protocol step of Córdoba, Bruno, Costa, Peralta, and Balzarini [49]. Subsequent to the sensor surveys, soil sampling should be carried out across the root zone (e.g., 0–1.2 m) to calibrate/interpret the sensor readings [50]. Sensor-directed sampling schemes can be used to minimize the number of sampling sites [55,56]. Exploratory analyses, such as correlation analysis and principal component analysis (PCA) should be carried out to investigate the relationships between soil sensor and laboratory soil analyses. The strength of the relationships between collocated soil sensor values and laboratory soil analyses should be investigated. If these relationships are moderately to very strong, the sensor data can be considered as an indicator of spatial variability of the target soil properties. Then, sensor data can be used to generate maps of the selected soil properties [57,58]. Soil maps should be generated only if acceptable prediction errors [59] are obtained. Alternatively, for weak to moderately strong relationships, the soil sensor maps should be used as qualitative indicator of soil spatial variability. Córdoba, Bruno, Costa, Peralta, and Balzarini [49] describe how to practically process interpolated data to obtain a raster of desired block support (e.g., of the same resolution chosen for the MZ design) in the second step of their MZ-delineation protocol.

Free (e.g., Sentinel 2 satellite) or inexpensive (e.g., from UAV) crop measurements are available throughout the growing season with moderately-high temporal resolution. For example, the Sentinel 2 satellite from the European Space Agency provides multi-spectral canopy reflectance at the 10×10-m resolution, with a 5-day revisit time, free of charge. Remote sensing of crop canopy data can be used to calculate vegetation indices [34]. Point measurements, such as those from tractor-mounted active spectrometers [30], should be pre-processed and interpolated similarly to soil sensing measurements. High-resolution raster data, such as that from satellite imagery, may need to be re-gridded and scaled to the selected block support.

2.1.2. Time-Specific Sub-Field Soil–Plant Modeling

In STEP 2, soil information from STEP 1 is used to interpret in-season (i.e., time-specific) measurements of crop status. The interpretation of crop canopy sensing measurements is not straightforward—as they are influenced by species × growth-stage × stress levels × soil background interactions [60–62].

Moving window spatial regression modeling, such as geographically-weighted regression (GWR) [63,64], can be used to understand the local (i.e., sub-field scale) variation in the plant–soil relationship. With GWR, a regression is run for each grid location, rather than for the whole study area [65]. Soil map(s) are used as the independent (explanatory) variable(s) and the in-season crop status maps are the dependent variable. When multiple explanatory variables are available, one should consider standardizing them. The GWR allows non-stationarity of the regression equation parameters a (e.g., intercept, slope), estimating their values at each location i. For a dependent variable y the equation reads:

$$y_i = a_{0i} + a_{1i}x_{1i} + \ldots + a_{ki}x_{ki} + \varepsilon_i \quad (1)$$

where ε, is a random error term, a_0 is the regression intercept, and a_1 to a_k are the regression coefficient(s) for each explanatory variable(s) x_1 to x_k. The spatial variability of the soil–plant relationship can be described with the maps of local a_1 to a_k coefficients (i.e., the regression slope[s]). Maps of GWR slope coefficient(s) show the spatial variability of the impact (i.e., sensitivity) of the explanatory variable on the regression [65].

In the GWR framework, spatial weighting is determined by incorporating all the dependent and explanatory variables falling within a geographical kernel of each target feature [65]. The values of the regression parameters and goodness-of-fit of the GWR depend on how the kernel size is chosen [66]. Maps of the estimated dependent variable, local coefficient of determination (R^2), and local Pearson correlation coefficient r can be generated with the GWR [65]. The GWR is available in commercial GIS software platforms (e.g., ArcMap's Spatial Statistics package [65], version 10.5.1; ESRI, Redlands, CA, USA), and freeware (e.g., spgwr package in R, version 3.4.1; the R Foundation for Statistical Computing Platform, Vienna, Austria).

2.1.3. Time-Specific MZ Delineation with Cluster Analysis

In STEP 3, the MZ delineation is carried out using the soil map(s) and the time-specific GWR slope maps from STEP 2 as ancillary variables. As indicated by Córdoba, Bruno, Costa, Peralta, and Balzarini [49], fuzzy c-means (also known as "k-means") unsupervised clustering algorithms [67] can be employed to classify the data into MZs. ArcMap's Grouping Analysis tool (e.g., [68]), the Management Zone Analyst software [69] or the EZZone online tool [70] can be used to delineate MZs. Several MZ designs can be tested. The optimum number of MZs can be identified using cluster validity functions, including: the Calinski–Harabasz criterion [71], the fuzziness performance index [67], the normalized classification entropy index [67,72], and the Jenks optimization method [73]. The Calinski–Harabasz criterion (*CHC*), also known as pseudo F-statistic, describes the ratio between within-MZ similarity and between-MZ differences. It is defined as:

$$CHC = \frac{BMZSS/(MZn - 1)}{WMZSS/(N - MZn)} \quad (2)$$

where N is the number of data points, MZn is the number of considered MZs, $BMZSS$ is the between-MZ sum of squares, and $WMZSS$ is the within group sum of squares. The larger the value of the *CHC* the higher are the within-MZ homogeneity and between-MZ differences. Finally, one may consider smoothing the fuzzy c-means clustering results to reduce zone fragmentation [49,70].

2.1.4. Time-Specific MZ-Design Quality Control and Interpretation

In STEP 4, the quality of the MZ design from STEP 3 should be checked. Each MZ should identify a unique combination of soil and soil–plant relationship characteristics. To infer differences across MZs, parametric analysis of variance or the Kruskal–Wallis (KW) rank test [74] can be used. The KW test is a nonparametric analysis assessing if samples originate from the same distribution. The test can be used as alternative to the standard analysis of variance when assumptions for parametric testing are not met.

2.2. Synthetic Data Example

The proposed workflow can be demonstrated using synthetic data for soil clay content (percentage) and normalized difference vegetation index (NDVI) [75].

A spatial random field for clay content was generated using the RFsimulate function in the RandomFields package in R. The function estimated a Gaussian random field (RPgauss function) having an exponential covariance function (RMexp function with variance = 25 and scale = 300), a nugget effect (RMnugget function with variance = 5), and a pure trend model with covariance 0 (RMtrend function with mean = 15). The simulations were carried out over a square of size 250 × 250 cells. The random field was then converted into a 250 × 250-pixel raster with pixel size = 1 m using the function raster from the raster package in R. The clay content raster was then exported from R to a text file using the writeRaster function. We refer to this raster as the true clay content (*TCC*) map.

Next, we mimicked a field procedure for generating a soil clay content map, using *TCC* as the true, unknown, high resolution clay content. Accurate approximations of true soil properties can be obtained when high-resolution covariates, such as those from proximal-soil sensing [76], are available [77]. Soil sensor surveys can be calibrated to estimate the spatial variability of a target soil property by using laboratory measurements on collocated soil cores [58]. Heggemann et al. [78] reported that gamma-ray sensor readings can be calibrated to predict texture values with high accuracy (up to ~95 percent of observed variance in soil texture). We mimicked a typical sensor survey (e.g., Lesch [55]) in which 1870 data points (average nearest neighbor distance = 3.2 m) were spread across 24 nearly-parallel transects (average nearest neighbor distance = 9.9 m). Values from the *TCC* map were extracted at the sensor survey locations. A random error having mean = 0 and variance equal to 5 percent of the extracted *TCC* values was added to the sensor data. This simulated a realistic sensor calibration with goodness-of-fit with a R^2 close to 0.95. The spatial autocorrelation of the calibrated sensor measurements was described with a spherical semivariogram having range = 50.3 m, and nugget equal to 67% of the total sill. This spatial structure was similar to those reported by other authors in a 5-ha clay loam field in Italy [79] and in a cluster of fields with contrasting soil properties (from clay to gravelly) in Germany [80]. These calibrated sensor measurements were then interpolated using simple kriging in with ArcMap's Geostatistical Analyst package. The resulting sensor-derived clay content (*SCC*) map was retained for further analyses.

NDVI is a vegetation index calculated from surface reflectance in the red (RED) and near-infrared (NIR) regions of the electromagnetic spectrum. It is defined as:

$$\mathrm{NDVI} = \frac{(\mathrm{NIR} - \mathrm{RED})}{(\mathrm{NIR} + \mathrm{RED})} \tag{3}$$

NDVI ranges from −1 to +1. NDVI for agricultural crops usually ranges from ~0.1 to ~0.9, with lush vegetation generally having high NDVI [81,82]. The NDVI information was simulated with the understanding that, often, the spatial relationship between remote sensing canopy measurement and collocated soil properties has both deterministic and spatial random components [83–85]. The deterministic component of the relationship can be a linear model between the soil property and the available remote sensing plant information [84]. The spatial random component is often equal to the field of spatially correlated residuals from the deterministic linear model [57,83]. We simulated the NDVI as follow:

$$\mathrm{NDVI} = S \times (TCC + SpERR) + O \tag{4}$$

where *SpERR* was a spatial error raster, *S* was a scaling factor = 0.01, and *O* was an offset coefficient = 0.4. *SpERR* was generated in R, using the RandomFields package. We estimated a Gaussian random field based on a model with exponential covariance function (variance = 7.5 and scale = 50) and a nugget effect (variance = 5).

A GWR with 30-m bandwidth size was used to model the spatial variability of the synthetic NDVI image using the *SCC* map as explanatory variable. Then, MZs were delineated with a fuzzy c-mean unsupervised cluster analysis. Clustering was carried out in ArcMap using the Grouping Analysis tool from the Spatial Statistics package. The SCC and the time-specific GWR slope maps were used for the cluster analysis. No spatial constraints were set for the cluster analysis: the covariate data points could be grouped according to their value, disregarding the values of their geographical coordinates (i.e., points did not need to be neighbors to be part of the same MZ). The classifications were carried out for scenarios outputting two, three, and four MZs. Then, the optimal number of MZs was identified using the *CHC*. Differences in clay (from the *SCC* map), NDVI, and GWR slope across MZs were investigated with the KW test using STATISTICA (version 12, StatSoft Inc., Tulsa, OK, USA).

2.3. Field Data Example

2.3.1. Site Description

Another example is given using experimental data from researchers at the University of Padua, Italy [86–88]. Here, we describe two time-specific MZ designs for a maize field based on remote sensing imagery acquired during the late kernel blister stage (R-2), in two consecutive growing seasons.

The study site (Figure 1) is a 21-ha field located at Chioggia, Venice, Italy, along the southern margin of the Venice Lagoon (45°10′7.0″ N, 12°13′55.0″ E). Previous studies [86–89] discussed the relationship between soil and maize yield at this site. The site lies below average sea level, and the groundwater level is kept fairly shallow (approximately between −0.6 to −1.5 m below ground level) by a pumping station [87] to promote sub-irrigation [88]. The soil in the area is classified as Molli-Gleyic Cambisol [90] with two coarsely-textured paleochannels crossing it with SW-NE direction (Figure 1). The field is affected by soil salinity due to its proximity to the Venice Lagoon [86].

Here, we analyzed the two maize growing seasons (April to September), 2010 and 2011. The two growing seasons differed greatly in terms of meteorology. Compared to the average April to September rainfall from 1993 to 2012, 2010 was rainy (in the upper third quartile, 540 mm) and 2011 was a drought year (in the first quartile, 200 mm) [88]. The daily average reference evapotranspiration was 4.01 mm day^{-1} in 2010 and 4.42 mm day^{-1} in 2011. For the entire growing season, the reference evapotranspiration was 569 mm in 2010 and 672 mm in 2011 [88]. Agronomic management (base-dressing of 64 kg N ha^{-1}, and 94 kg P_2O_5 ha^{-1} and urea top-dressing of 184 kg N ha^{-1}) and maize hybrid (PR32P26, Pioneer Hi-Bred Italia, Gadesco-Pieve Delmona, Italy) were the same in the two years [86].

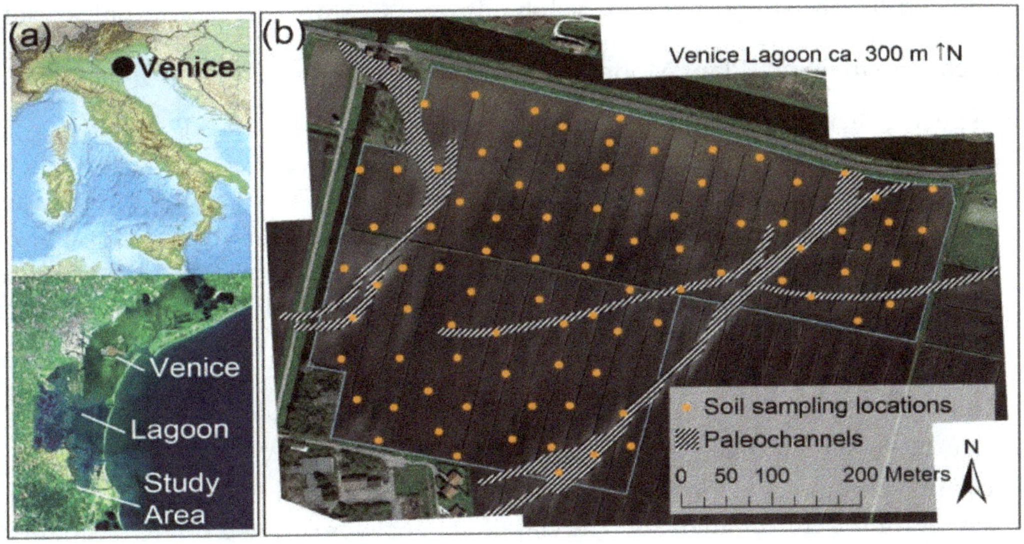

Figure 1. (**a**) Geographical location of the study area relatively to the Venice Lagoon, Italy, and (**b**) locations of soil sampling locations and coarsely textured paleochannels. Modified after Scudiero, Teatini, Corwin, Dal Ferro, Simonetti, and Morari [88].

2.3.2. Soil Maps and Plant Spatiotemporal Information

Geospatial measurements of soil apparent electrical conductivity (EC_a) were used as an indicator of soil spatial variability at the study site. EC_a over the 0–1.5 m soil profile (EC_a Deep) was measured in April of 2011 across the site with a frequency-domain electromagnetic induction sensor (CMD-1, GF Instruments, Brno, Czech Republic) at 20,470 geo-referenced locations. Soil samples were taken at 91 locations in May 2010 (Figure 1) down to 1.2 m with 0.3-m increments. Here we discuss salinity and texture for the 0–1.2-m soil profile. Texture was measured with a Mastersizer 2000 (Malvern Instruments Ltd., Great Malvern, UK) and salinity was measured as $EC_{1:2}$ (i.e., the electrical conductivity of a 1:2 soil–water extract) [91]. Principal component analysis (PCA) carried out with STATISTICA was used to visually assess similarities and differences between the EC_a Deep dataset and the selected soil properties [92]. The point measurements were preprocessed with a procedure comparable to that of Córdoba, Bruno, Costa, Peralta, and Balzarini [49] and spatially interpolated with ordinary kriging following Scudiero, Teatini, Corwin, Deiana, Berti, and Morari [86] on a 10 $\times$ 10-m block support—the desired support for MZ at the site.

Two remotely sensed measures of crop status were available. On 31 July 31 2010 (10:39:40 UTC) and 9 July 2011 (10:47:30 UTC) WorldView-2 (DigitalGlobe, Westminster, CO, USA) satellite scenes were acquired over the study area. For both years, acquisition dates corresponded to late R-2 [88]. Pre-processing procedures, including sensor calibration, atmospheric correction, and radiometric normalization, were applied according to Vicente-Serrano et al. [93]. Radiometric calibration was required to convert digital numbers to top-of-atmosphere radiances [W m^{-2} sr^{-1} μm^{-1}], using the absolute radiometric calibration factors and effective bandwidths for each band, according to the satellite data provider [94]. No topographic correction was applied to the images because the study area is relatively flat and the solar incident angles (28.11° in 2010; 23.58° in 2011) were quite similar at the two acquisition times. The top-of-atmosphere radiance was then transformed to surface reflectance through the *6S* code [95]. The values of aerosol optical thickness (AOT) at 550 nm collected from a nearby AERONET (Aerosol Robotic Network, https://aeronet.gsfc.nasa.gov/) [96] station (45°18′50.0″ N, 12°30′29.9″ E) were used as input for *6S*. The AOT measurements were obtained simultaneously to the satellite overpasses. WorldView-2 reflectance has 2 $\times$ 2-m spatial resolution over eight spectral bands, at wavelengths spanning from 400 to 1040 nm. The red band (RED, 630–690 nm) and the near-infrared band at 770–895 nm where used to calculate NDVI according to Equation (3). The NDVI maps were aggregated (i.e., the coarsened pixel is the average of the pixels within the aggregated cell) to the 10 $\times$ 10-m cell size, as suggested by Córdoba, Bruno, Costa, Peralta, and Balzarini [49].

2.3.3. Time-Specific Spatial Soil–Plant Modeling

The time-specific relationship between soil properties and in-season NDVI were described using geographically weighted regressions (GWRs). The GWRs were carried out using the NDVI maps as the dependent variable and spatial soil information (i.e., EC_a Deep) as the independent variable. GWRs were carried out with ArcMap using an adaptive kernel (i.e., moving window) of 70 neighbors. For locations with significant GWR, the observed-estimated NDVI relationship was used as indication of goodness-of-fit of the GWR models. The GWR slope map was selected as indicator of soil–plant relationship type.

2.3.4. Delineating MZs with Cluster Analysis

Clustering was carried out using the Grouping Analysis tool in ArcMap. Spatial soil information (i.e., EC_a Deep) and the time-specific GWR slope map were used for the cluster analysis. No spatial constraints were set for the cluster analysis. The classifications were carried out for scenarios outputting three, four, and five MZs. Then, the optimal number of MZs was identified using the *CHC*. To compare the proposed time-specific soil–plant based MZ delineation with a more traditional, static MZ design, EC_a Deep alone was used to select a static MZ design.

2.3.5. Evaluation and Interpretation of MZ Design

Differences across MZs in the two years were investigated using the KW test in STATISTICA. The following variables were tested for differences across MZs: EC_a Deep, GWR slope, GWR local *r*, $EC_{1:2}$, and sand and clay contents. Additionally, in-season potential yield estimations were used to evaluate the MZ designs. Appendix A reports on how yield maps (obtained from Scudiero, Teatini, Corwin, Dal Ferro, Simonetti, and Morari [88]) and NDVI measurements were used to calculate yield prediction maps.

3. Results and Discussion

3.1. Synthetic Data Example

Every time the crop status is mapped during the season (e.g., every week), a new set of MZ is delineated (i.e., time-specific MZ design). Clearly, this applies only for crop growth stages when crop canopy plays a relevant role in determining reflectance readings. For soil tillage, sowing, and early vegetation stages when surface reflectance is mostly determined by soil, a static MZ delineation approach based on the spatial variability of soil properties (e.g., [9]) may be more adequate.

Figure 2 outlines the proposed analytical protocol to delineate time-specific MZs from soil information and in-season crop information using the synthetic dataset. In Step 1 of the protocol, the soil and plant information were acquired and pre-processed. The clay content (percentage) map had mean = 18.6%, standard deviation = 4.5%, minimum = 9%, and maximum = 30.1%. The NDVI map had mean = 0.58, standard deviation = 0.04, minimum = 0.45, and maximum = 0.69. The NDVI values are typical of vegetative stages for crops such as maize [97] and soybean (*Glycine max* (L.) Merr.) [98].

A simple linear model between the two maps had $R^2 = 0.61$, with an intercept of 0.44 and a positive slope of 0.007. Similar goodness-of-fit values between soil properties and vegetation reflectance were reported by Gomez et al. [99] for clay and by Gomez et al. [100] for soil organic carbon. In Step 2, the clay map was used as the explanatory variable in a GWR with NDVI as dependent variable. Through the GWR analysis, clay explained 94.5% of the observed variance of NDVI. Such high goodness-of-fit can be found in real-world data: Scudiero, Corwin, Wienhold, Bosley, Shanahan, and Johnson [54] observed R^2 ranging between 0.83 and 0.94 for their GWR analyses between soil maps and remotely sensed winter wheat (*Triticum aestivum* L.) canopy reflectance. The GWR slope had mean = 0.009, standard deviation = 0.008, minimum = −0.018, and maximum = 0.038. In Step 3, the clay and the GWR slope maps were used to delineate MZs via unsupervised fuzzy c-means clustering. The *CHC* indicated that the best number of MZs was three.

In Step 4, the MZ-design was evaluated and interpreted. The three MZs identified unique combinations of clay content and GWR slope values. MZ 1 was characterized by low NDVI, low clay content, and large GWR slope values (indicating high sensitivity of NDVI to clay). MZ 2 was characterized by high NDVI, high clay, and moderately low GWR slope values. MZ 3 was characterized by low NDVI, low clay content, and the smallest GWR slope values.

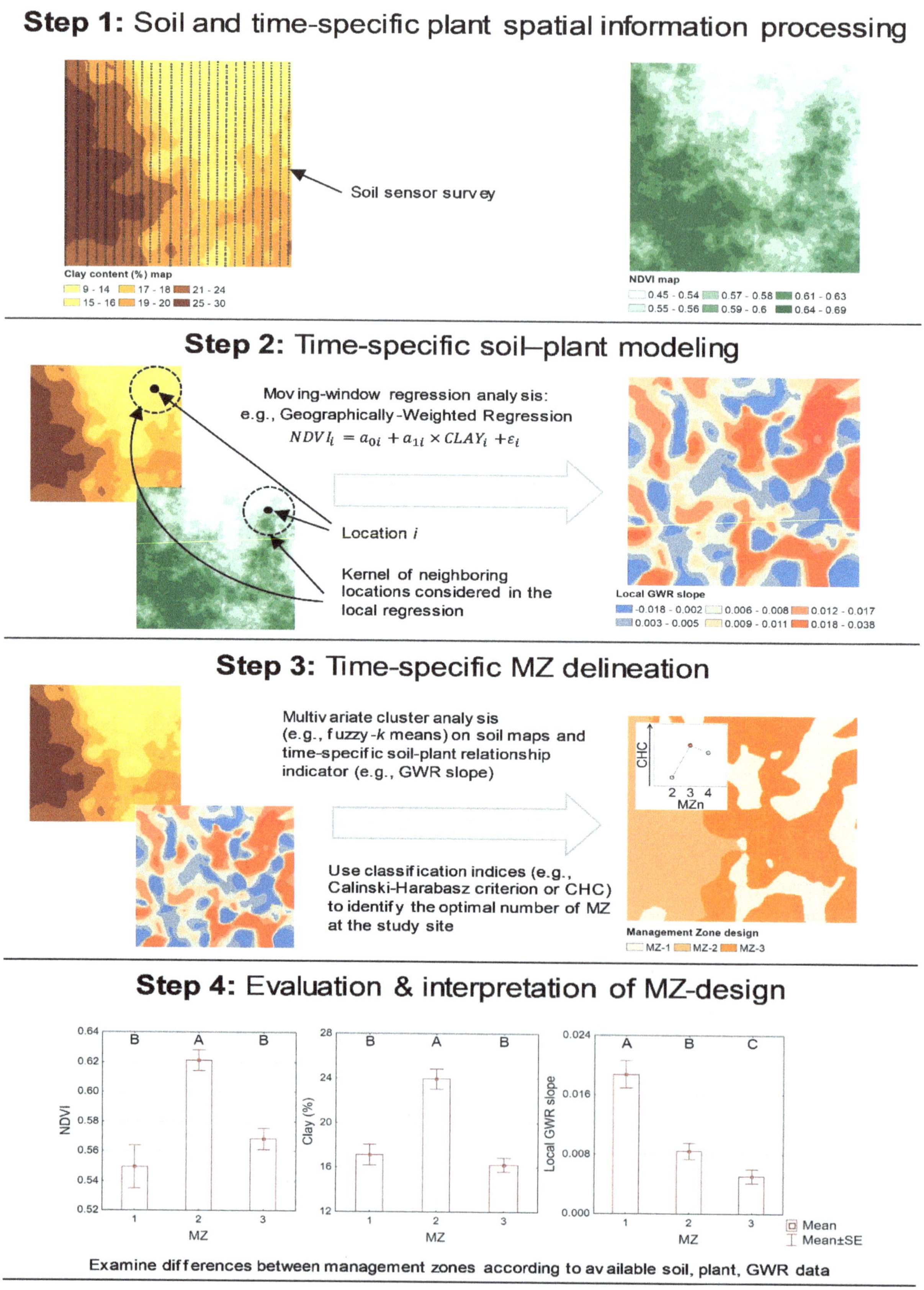

Figure 2. Diagram outlining the proposed analytical protocol to delineate time-specific management zones (MZs) from soil and (in-season) crop information. In Step 3, MZn refers to the number of considered management zones.

3.2. Field Data Example

3.2.1. Soil and Plant Information

Figure 3a shows an example of short-scale soil spatial variation at the site. Figure 3b depicts the frequency distribution of measured soil salinity. At this site, soils with $EC_{1:2}$ >0.79 dS m^{-1} should be considered salt affected, with $EC_{1:2}$ >1.4 dS m^{-1} indicating strongly saline soils [88]. Figure 3c reports the observed texture at the 91 soil sampling locations. Soil texture is predominantly in the loam and sandy loam classes. Few samples are in the clay loam and loamy sand class.

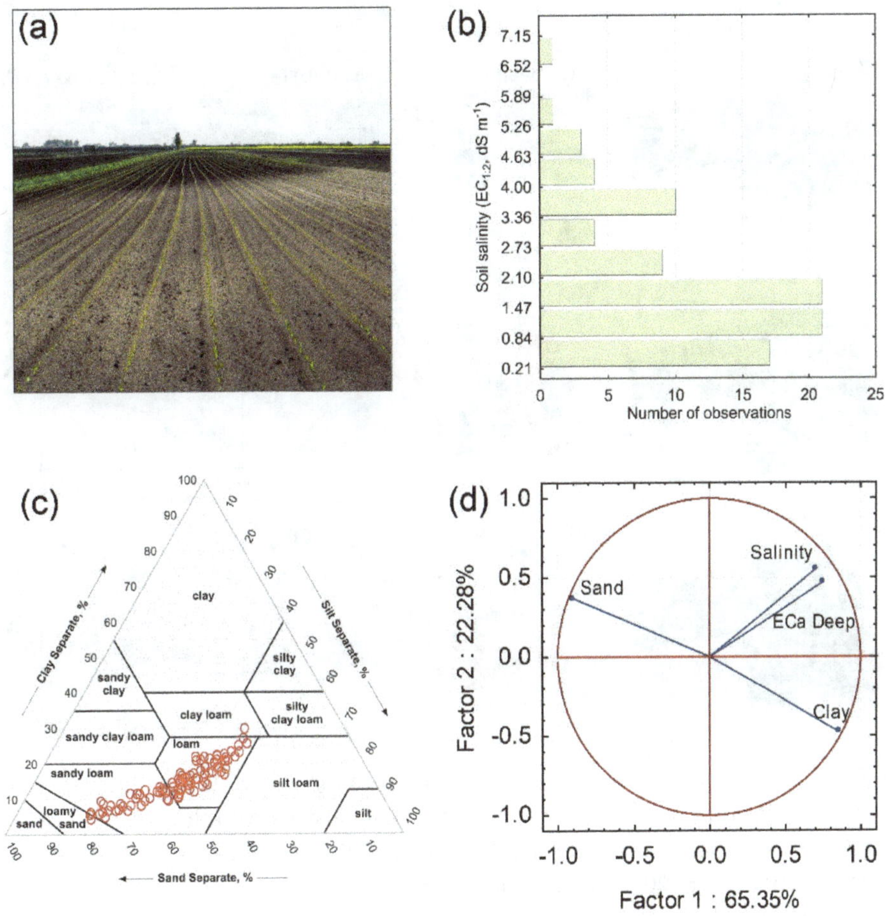

Figure 3. (**a**) Photograph showing the sharp textural change at the site (light vs. dark colors); (**b**) histogram for measured soil salinity ($EC_{1:2}$); (**c**) soil textural triangle; and (**d**) bi-plot of selected variables (soil apparent electrical conductivity [EC_a Deep], salinity, clay, and sand) on the two larger factors in the principal components analysis.

The EC_a Deep map is shown in Figure 4a. Descriptive statistics for the map are reported in Table 1. EC_a Deep showed a Pearson correlation coefficient of 0.89 with salinity, −0.51 with sand, and 0.39 with clay. The PCA extracted two factors (with eigenvalues of 2.61 and 0.89) explaining 65.3% (first component) and 22.3% (second component) of the total variance in the dataset. The bi-plot in Figure 3d—see Abdi and Williams [92] for interpretation guidelines—indicates that the first component contrasts the positive contribution of clay content and salinity on the EC_a Deep measurements with the negative correlation between sand content and EC_a Deep. In Figure 3d, clay content was not clustered with salinity and EC_a Deep, indicating that the three variables were not collinear, yet negatively correlated with sand content. According to the correlation and PCA analyses, high values of EC_a Deep were mainly interpreted as an indication of high salinity and fine soil texture. EC_a readings larger than 1–2 dS m^{-1} are most likely due to high soil salinity rather than other edaphic factors contributing to soil conductivity [101,102]. Conversely, lower EC_a Deep was interpreted as an indication of coarser soils with low salinity. Note that Scudiero, Teatini, Corwin, Deiana, Berti, and Morari [86] reported that the spatial patterns of EC_a Deep remained stable over time at the study area.

The two in-season NDVI maps for the late R-2 growth stage are reported in Figure 4b (2010) and Figure 4e (2011). Table 1 reports the descriptive statistics for the two NDVI maps. The 2010 NDVI was more heterogeneous (mean = 0.79, standard deviation = 0.046) than that of 2011 (mean = 0.83, standard deviation = 0.018). During the R-2 growth stage, environmental stressors (e.g., drought, heat, severe nutrient deficiency) may prevent kernels from developing properly (e.g., kernel abortion at the ear tip) [103,104].

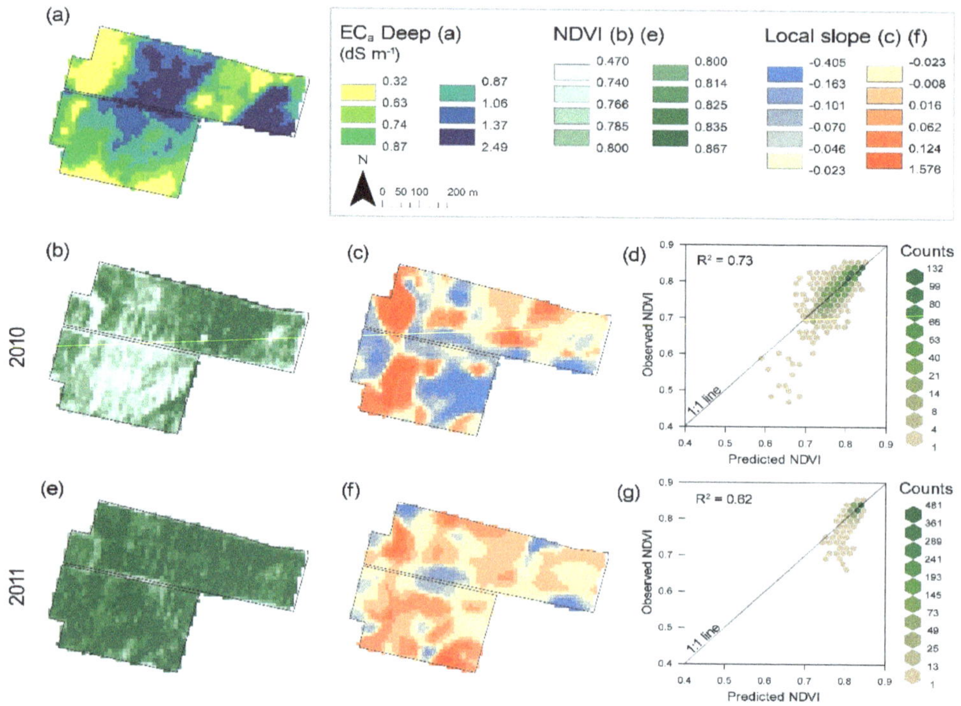

Figure 4. Maps of: (**a**) apparent electrical conductivity of the 0–1.5-m soil profile (EC_a Deep); Normalized Difference Vegetation Index (NDVI) for (**b**) 2010 and (**e**) 2011; local geographically-weighted regressions (GWRs) slope coefficient for (**c**) 2010 and (**f**) 2011; and observed versus estimated NDVI relationship for (**d**) 2010 and (**g**) 2011 on significant local GWRs.

Table 1. Descriptive statistics for the soil apparent electrical conductivity map (EC_a Deep), and the 2010 and 2011 Normalized Difference Vegetation Index (NDVI) maps.

Variable	Mean	Median	Minimum	Maximum	Standard Deviation
EC_a Deep (dS m^{-1})	0.96	0.86	0.32	2.49	0.36
2010 NDVI	0.790	0.80	0.470	0.867	0.046
2011 NDVI	0.829	0.83	0.664	0.858	0.018

3.2.2. Plant–Soil Relationship at Different Time Points

Figure 4 summarizes the GWR analysis between soil EC_a Deep and in-season NDVI for 2010 and 2011. For 2010, 1149 cells (62.5% of total) were characterized by a significant ($p < 0.05$) local Pearson correlation coefficient r. For 2011, 56.8% of the cells had a significant local r. At the locations with significant local r, the observed-estimated NDVI relationship was characterized by $R^2 = 0.73$ in 2010 (Figure 3d) and $R^2 = 0.62$ in 2011 (Figure 3g). There was a strong negative relationship in both years ($r = -0.36$ in 2010 and $r = -0.43$ in 2011) between the significant local r and EC_a Deep maps. The slope of this relationship was steeper in 2011 (−0.46 with standard error = 0.027) than in 2010

(−0.38 with standard error = 0.026). This may indicate that low EC_a Deep (e.g., indicating coarse texture) was a greater constrain in 2011 than 2010. Scudiero, Teatini, Corwin, Dal Ferro, Simonetti, and Morari [88] showed that water stress in areas with coarser soil texture is particularly limiting in dry years, such as 2011.

In areas where the significant local r values were consistently negative in both years, the EC_a Deep averaged 1.17 dS m^{-1} (standard deviation = 0.39 dS m^{-1}), whereas, in areas with negative local r only in 2010, it averaged 0.94 dS m^{-1} (standard deviation = 0.28 dS m^{-1}). This suggested that only very high EC_a Deep, which indicates high salinity, consistently limited crop growth in the two years. As discussed by other authors [105–107], the effects of high salinity on crops are generally stable throughout different growing seasons.

The GWR slope maps (Figure 4c for 2010, Figure 4f for 2011) changed remarkably between the two years, indicating a change in the sensitivity of plant NDVI to changes in soil properties. Changes of crop growth and yield spatial patterns are expected between seasons having widely different meteorology, as discussed by Maestrini and Basso [38] and McBratney, Whelan, Ancev, and Bouma [37].

3.2.3. Time-Specific MZ Delineation

The GWR slope and EC_a Deep maps were used to drive the delineation of in-season site-specific MZs. The *CHC* index indicated that a five-MZ configuration was optimal for 2010. The *CHC* was 1901.3 for three MZs, 1979.7 for four MZs, and 2002.7 for five-MZs. In 2011, the *CHC* index indicated that a four-MZ configuration was optimal. *CHC* was 1248.9 for three MZs, 1383.4 for four MZs, and 1374.0 for five MZs. The two MZ delineations are shown in Figure 5a (for 2010) and Figure 5b (for 2011).

Overall, 45.1% of the cells were classified in the same MZ at both times. The consistency rate was 94.3% for MZ I, 75.6% for MZ II, 20.1% for MZ III, and 48.5% for MZ IV. Conversely, changes in local GWR slope led to 54.9% of the cells being assigned to a different time-specific MZ. As reported by Maestrini and Basso [38], it is reasonable to expect portions of a field to have stable (e.g., consistently high yields) and unstable crop outputs. Management in the unstable areas should be addressed within each growing season to meet desired agronomic output goals [18,38]. The proposed methodology helps characterizing time-specific changes in the crop output according to soil spatial variability.

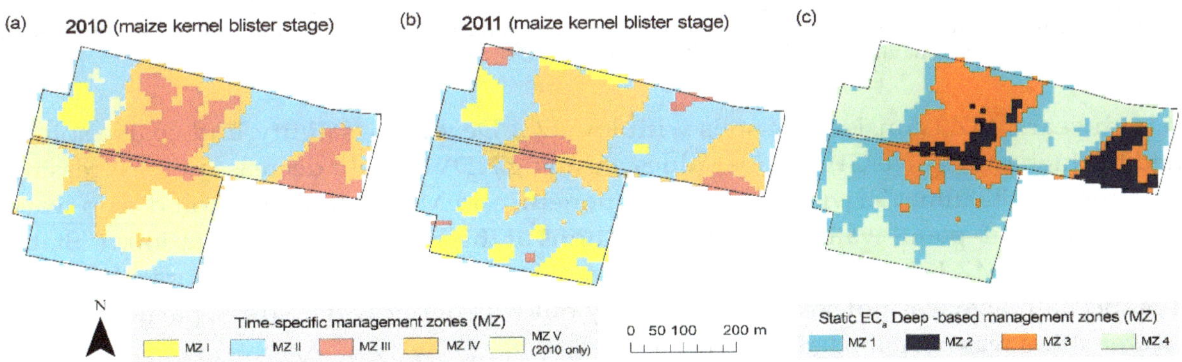

Figure 5. In-season delineation of time-specific management zones (MZs) for maize late kernel blister stage of (**a**) 2010 and (**b**) 2011. (**c**) Static MZ designs using soil apparent electrical conductivity for the 0–1.5 m soil profile (EC_a Deep) only.

3.2.4. Time-Specific MZ-Design Quality Control and Interpretation

In both years, each time-specific MZ was characterized by unique combinations of the variables used in MZ delineation. Moreover, the time-specific MZs differed greatly in terms of EC_a Deep, significant GWR r and slope, salinity, and texture (Table 2). Note that the local GWR r and slope, which can be used to interpret the EC_a Deep–NDVI relationship, remained fairly consistent over time at the different MZs.

Table 2. Count (n), mean, and standard deviation (Std. Dev.) of apparent electrical conductivity for the 0–1.5-m soil profile (EC_a Deep), local regression slope and Pearson correlation coefficient (r) obtained from significant geographically weighted regressions (GWR), soil salinity ($EC_{1:2}$), sand and clay contents, and potential yield estimations at each time-specific management zone (MZ) for 2010 and 2011 (n.a. = not available; NA = not assessed).

Variable (Unit)	MZ	2010				2011			
		n	Mean	Std. Dev.	KW [1]	n	Mean	Std. Dev.	KW [1]
EC_a Deep ($dS\ m^{-1}$)	I	53	0.44	0.10	e	263	0.65	0.21	d
	II	779	0.68	0.12	d	1164	0.78	0.15	c
	III	402	1.54	0.19	a	146	1.24	0.45	a
	IV	513	1.09	0.12	b	574	1.38	0.23	b
	V	400	0.80	0.12	c	n.a.	n.a.	n.a.	n.a.
GWR slope	I	53 (100.0% [2])	0.790	0.316	a	221 (84.0% [2])	0.073	0.040	a
	II	371 (47.6% [2])	0.065	0.113	b	488 (41.9% [2])	−0.024	0.028	b
	III	334 (83.1% [2])	−0.069	0.069	c	146 (100.0% [2])	−0.132	0.050	c
	IV	294 (57.3% [2])	−0.048	0.055	c	364 (63.4% [2])	−0.018	0.029	b
	V	397 (99.3% [2])	−0.176	0.092	d	n.a.	n.a.	n.a.	n.a.
GWR local r	I	53 (100.0% [2])	0.587	0.102	a	221 (84.0% [2])	0.433	0.126	a
	II	371 (47.6% [2])	0.028	0.411	b	488 (41.9% [2])	−0.279	0.285	b
	III	334 (83.1% [2])	−0.439	0.239	c	146 (100.0% [2])	−0.597	0.190	d
	IV	294 (57.3% [2])	−0.399	0.281	c	364 (63.4% [2])	−0.308	0.349	c
	V	397 (99.3% [2])	−0.537	0.150	d	n.a.	n.a.	n.a.	n.a.
$EC_{1:2}$ ($dS\ m^{-1}$)	I	1	0.21	NA	ab	11	1.73	1.33	b
	II	32	1.38	0.89	b	49	1.49	1.01	b
	III	17	3.67	1.46	a	3	3.16	1.32	ab
	IV	26	2.28	1.18	a	28	3.03	1.45	a
	V	15	1.35	0.85	b	n.a.	n.a.	n.a.	n.a.
Sand (%)	I	1	71.1	NA	ab	11	54.1	15.6	ab
	II	32	50.6	12.6	b	49	53.1	12.4	a
	III	17	42.8	10.9	b	3	37.5	13.6	ab
	IV	26	42.7	8.8	b	28	42.1	9.2	b
	V	15	63.8	8.5	a	n.a.	n.a.	n.a.	n.a.
Clay (%)	I	1	9.0	NA	ab	11	12.6	6.6	ab
	II	32	13.5	5.1	a	49	12.6	4.8	b
	III	17	17.4	5.7	a	3	20.8	9.3	ab
	IV	26	16.8	4.7	a	28	17.3	4.7	a
	V	15	8.7	3.2	b	n.a.	n.a.	n.a.	n.a.
Potential Yield ($Mg\ ha^{-1}$)	I	53	7.03	1.90	bc	263	10.0	1.7	b
	II	779	8.38	2.24	a	1164	10.5	1.2	a
	III	402	7.61	1.75	b	146	9.3	2.1	c
	IV	513	7.19	2.01	c	574	10.3	1.3	b
	IV	400	7.08	1.90	c	n.a.	n.a.	n.a.	n.a.

[1] Different letters are significantly different between MZs at the $p < 0.05$ level according to the Kruskal–Wallis (KW) test. [2] Percentage of cells having significant GWR within the MZs.

In both years, MZ I consisted of soils with low EC_a Deep, low salinity, and coarse soil texture. MZ I had the highest positive GWR slope values, meaning NDVI would decrease with increasing sand content. Scudiero, Teatini, Corwin, Dal Ferro, Simonetti, and Morari [88] monitored crop water and salt stress at five soil–plant–water monitoring stations at the site. One of their stations ("Station E") was located in our MZ I. Scudiero, Teatini, Corwin, Dal Ferro, Simonetti, and Morari [88] indicated that at Station E maize was not stressed by salinity but was under water stress, particularly in 2011, when water stress was described as "severe".

In contrast, MZ II in both years had GWR slope and r distributions overlapping with 0, indicating little to no influence of soil on crop NDVI variability. This is perhaps the reason why MZ II was consistently characterized by the highest potential yield estimations (Appendix A). The very saline MZ III (highest measured average EC_a Deep and $EC_{1:2}$) was characterized by lower potential yield estimations for both years in comparison with the other MZs (Table 2), as expected for (unmanaged) very saline portions of farmlands [106]. Similarly, the moderately saline MZ IV (second highest average EC_a Deep), was characterized by the second lowest yield predictions in 2010 and 2011, together with MZ I. MZ V was only selected in 2010. It grouped soils with moderately high EC_a Deep (third highest MZ average) together with the strongest negative GWR slope values and low potential yield estimations. In 2011, 89.5% of the 2010 MZ V locations were classified into MZ II. Previous research focusing on spatiotemporal variability of yield patterns [38], indicates that fields generally have areas of stable and unstable yield patterns. For unstable areas, in-season NDVI is suggested as the best predictor for yield spatial distribution [38]. Our proposed approach further refines the NDVI information by interpreting its spatial variability as a function of soil spatial variability.

A static MZ design based on EC_a Deep only (Figure 5c) would identify four areas with significantly different soil properties (Table 3), but very heterogeneous and inconsistent in terms of GWR local r and slope (Table 4).

Table 3. Count (n), mean, and standard deviation (Std. Dev.) of apparent electrical conductivity for the 0–1.5-m soil profile (EC_a Deep), soil salinity ($EC_{1:2}$), and sand and clay contents for the static management zones (MZs) delineated using EC_a Deep only.

Variable (Unit)	Static MZ	n	Mean	Std. Dev.	KW [1]
EC_a Deep (dS m^{-1})	1	718	0.92	0.10	c
	2	181	1.71	0.17	a
	3	444	1.31	0.12	b
	4	804	0.63	0.10	d
$EC_{1:2}$ (dS m^{-1})	1	28	1.98	1.04	b
	2	9	4.07	1.51	ab
	3	21	2.82	1.25	a
	4	33	1.07	0.68	c
Sand (%)	1	28	50.3	11.9	ab
	2	9	41.0	9.3	b
	3	21	41.1	9.8	b
	4	33	55.9	12.9	a
Clay (%)	1	28	13.3	5.0	bc
	2	9	18.3	5.7	ab
	3	21	17.9	5.0	a
	4	33	11.8	4.9	c

[1] Different letters are significantly different between MZs at the $p < 0.05$ level according to the Kruskal–Wallis (KW) test.

Table 4. Mean, standard deviation (Std. Dev.), and count (n) for the correlation coefficient (r) and slope obtained from significant geographically weighted regressions (GWRs) for the static management zones (MZs) delineated using soil apparent electrical conductivity for the 0–1.5-m soil profile (EC_a Deep) only.

GWR Variable	Static MZ	2010			2011		
		n	Mean	Std. Dev.	n	Mean	Std. Dev.
Local r	1	490 (68.2% [1])	−0.374	0.351	349 (48.6% [1])	−0.156	0.383
	2	163 (90.1% [1])	−0.474	0.158	148 (81.8% [1])	−0.542	0.158
	3	313 (70.5% [1])	−0.408	0.281	289 (65.1% [1])	−0.303	0.408
	4	483 (60.1% [1])	−0.098	0.472	433 (53.9% [1])	−0.041	0.420
Slope	1	490 (68.2% [1])	−0.086	0.148	349 (48.6% [1])	−0.009	0.039
	2	163 (90.1% [1])	−0.077	0.065	148 (81.8% [1])	−0.074	0.068
	3	313 (70.5% [1])	−0.058	0.065	289 (65.1% [1])	−0.023	0.045
	4	483 (60.1% [1])	0.066	0.304	433 (53.9% [1])	−0.002	0.079

[1] Percentage of cells having significant GWR within the MZ.

4. Conclusions

Characterizing the temporal variability of the soil–plant relationship within the growing season and over years is very relevant to developing best crop management practices with VRM. Soil influences plants according to complex interactions between factors that may change in time, such as meteorological conditions, nutrient availability, and water content. Static MZs are not ideal when spatial patterns of soil–plant relationship change in time—because of changing meteorology and/or other transient factors. We presented a time-specific MZ-delineation workflow to address such spatiotemporal variability.

The proposed approach for time-specific MZs should be further tested and evaluated in the field. Future work should focus on multi-field and multi-year comparisons of time-specific and static MZs and their relative impact on crop yields and resource use efficiency (e.g., water, nutrients).

Author Contributions: Conceptualization, E.S. and T.H.S.; methodology, E.S.; software, E.S. and F.B.; formal analysis, investigation, resources, data curation, writing—original draft preparation, writing—review and editing, All Authors; visualization, E.S.; funding acquisition, F.M. and P.T.

Acknowledgments: We acknowledge the outstanding work of Nicola Dal Ferro, Davide Piragnolo, Diego Mengardo, and Gianluca Simonetti who helped in the laboratory analyses and/or field surveys. Special thanks to John Shanahan and George Vellidis for providing valuable comments and advice on early versions of this manuscript.

Abbreviations

AOT, aerosol optical thickness; BMZSS, between-MZ sum of squares; *CHC*, Calinski–Harabasz criterion; $EC_{1:2}$, salinity measured as electrical conductivity of a 1:2 soil–water extract; EC_a, apparent electrical conductivity; EC_a Deep, apparent electrical conductivity measurements of the 0–1.5 m soil profile; GWR, geographically weighted regression; MZ, management zone; NDVI, normalized difference vegetation index; PCA, principal component analysis; *r*, Pearson's correlation coefficient; R^2, coefficient of determination; R-2, kernel blister stage of maize growth; *SCC*, (synthetic) sensor-derived clay content; *TCC*, (synthetic) true clay content; UAV, unmanned aerial vehicle; VRM, variable rate management; WMZSS, within-MZ sum of squares; WV2, WorldView-2 satellite.

Appendix A Interpreting NDVI as Potential Yield Indicator

In-season potential yield predictions can be very useful for the producers to decide on the feasibility of VRM. Throughout the season NDVI can be used as potential yield indicator [33,35]. Estimations can be made at different growth stages [32], provided the NDVI measurements are calibrated accordingly [33]. This is particularly relevant when historical yield data is available, so that yield-predicting NDVI functions can be developed and used in following years.

According to [16], the in-season potential yield can be estimated by fitting the NDVI values to the 68th percentile (i.e., sum of average and one standard deviation) of yield. The NDVI and yield data from 2010 and 2011 were clumped together. The yield data was obtained from Scudiero, Teatini, Corwin, Dal Ferro, Simonetti, and Morari [88]. The dataset was divided into fifty quantiles according to NDVI. The within-quantile average NDVI was then used as explanatory variable to estimate the within-quintile 68th percentile of yield. The estimation was done using a second-degree locally-weighted least squares regression. The regression modeling was carried out through the loess function (with span = 0.45) from the stats package in R. The span parameter was selected through a leave-one-out cross-validation using the loess.wrapper function from the bisoreg package in R. Figure A1a shows the scatterplot between NDVI (i.e., at late R-2) and maize yield, all points were clustered on a uniform grid of hexagons using the hexbin package in R.

The model describing the potential yield (Mean absolute error = 0.26 Mg ha^{-1}, mean error = −0.027 Mg ha^{-1}, and R^2 = 0.98) is reported with a solid red line in Figure A1. The locally-weighted least squares regression estimations were then applied to the two NDVI maps (Figure 4b,e) to obtain potential yield estimations across the whole study area for 2010 (Figure A1b) and 2011 (Figure A1c).

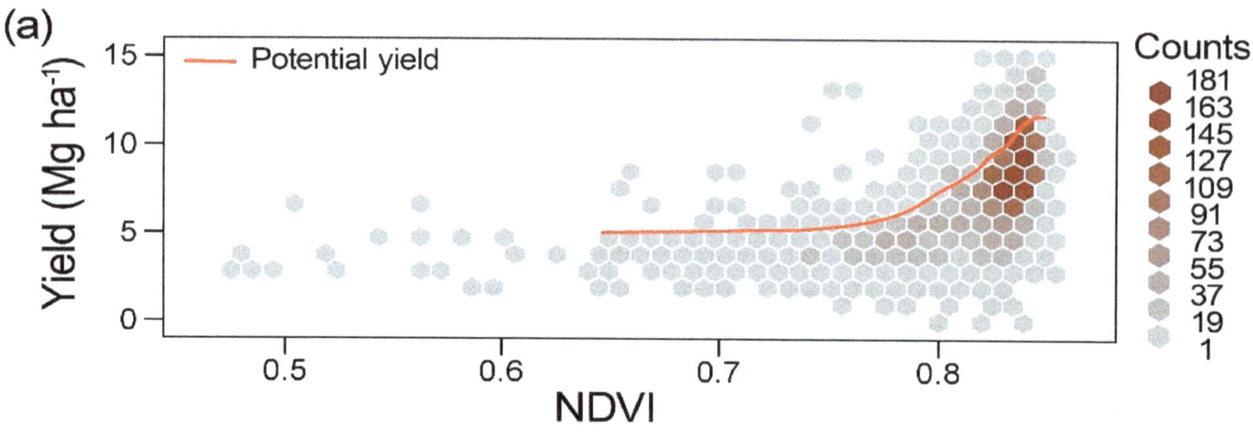

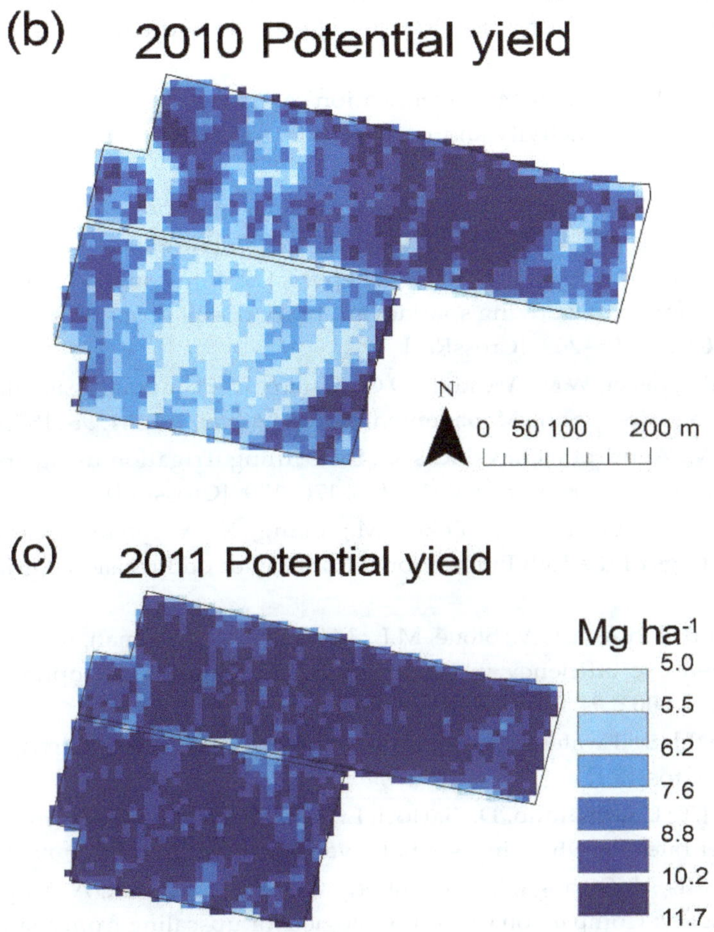

Figure A1. (**a**) Scatterplot between yield and WorldView-2 NDVI for both years. The modeled in-season potential yield is depicted with a solid red line. Potential yield maps calculated for (**b**) 2010 and (**c**) 2011.

References

1. Lobell, D.B.; Azzari, G. Satellite detection of rising maize yield heterogeneity in the US Midwest. *Environ. Res. Lett.* **2017**, *12*. [CrossRef]
2. Corwin, D.L.; Lesch, S.M.; Shouse, P.J.; Soppe, R.; Ayars, J.E. Identifying soil properties that influence cotton yield using soil sampling directed by apparent soil electrical conductivity. *Agron. J.* **2003**, *95*, 352–364. [CrossRef]
3. Kitchen, N.R.; Drummond, S.T.; Lund, E.D.; Sudduth, K.A.; Buchleiter, G.W. Soil Electrical Conductivity and Topography Related to Yield for Three Contrasting Soil–Crop Systems. *Agron. J.* **2003**, *95*, 483–495. [CrossRef]
4. Bobryk, C.W.; Myers, D.B.; Kitchen, N.R.; Shanahan, J.F.; Sudduth, K.A.; Drummond, S.T.; Gunzenhauser, B.; Gomez Raboteaux, N.N. Validating a Digital Soil Map with Corn Yield Data for Precision Agriculture Decision Support. *Agron. J.* **2016**, *108*, 957–965. [CrossRef]
5. Kaffka, S.R.; Lesch, S.M.; Bali, K.M.; Corwin, D.L. Site-specific management in salt-affected sugar beet fields using electromagnetic induction. *Comput. Electron. Agric.* **2005**, *46*, 329–350. [CrossRef]
6. Singh, G.; Williard, K.J.; Schoonover, J.E. Spatial Relation of Apparent Soil Electrical Conductivity with Crop Yields and Soil Properties at Different Topographic Positions in a Small Agricultural Watershed. *Agronomy* **2016**, *6*, 57. [CrossRef]
7. Zhang, N.Q.; Wang, M.H.; Wang, N. Precision agriculture—A worldwide overview. *Comput. Electron. Agric.* **2002**, *36*, 113–132. [CrossRef]
8. Fleming, K.L.; Westfall, D.G.; Wiens, D.W.; Brodahl, M.C. Evaluating Farmer Defined Management Zone Maps for Variable Rate Fertilizer Application. *Precis. Agric.* **2000**, *2*, 201–215. [CrossRef]

9. Fraisse, C.W.; Sudduth, K.A.; Kitchen, N.R. Delineation of site-specific management zones by unsupervised classification of topographic attributes and soil electrical conductivity. *Trans. ASAE* **2001**, *44*, 155–166. [CrossRef]
10. Moral, F.J.; Terrón, J.M.; Silva, J.R.M.D. Delineation of management zones using mobile measurements of soil apparent electrical conductivity and multivariate geostatistical techniques. *Soil Tillage Res.* **2010**, *106*, 335–343. [CrossRef]
11. Baveye, P.C.; Laba, M. Moving away from the geostatistical lamppost: Why, where, and how does the spatial heterogeneity of soils matter? *Ecol. Model.* **2015**, *298*, 24–38. [CrossRef]
12. Schepers, A.R.; Shanahan, J.F.; Liebig, M.A.; Schepers, J.S.; Johnson, S.H.; Luchiari, A. Appropriateness of management zones for characterizing spatial variability of soil properties and irrigated corn yields across years. *Agron. J.* **2004**, *96*, 195–203. [CrossRef]
13. Koch, B.; Khosla, R.; Frasier, W.M.; Westfall, D.G.; Inman, D. Economic Feasibility of Variable-Rate Nitrogen Application Utilizing Site-Specific Management Zones. *Agron. J.* **2004**, *96*, 1572–1580. [CrossRef]
14. Liang, X.; Liakos, V.; Wendroth, O.; Vellidis, G. Scheduling irrigation using an approach based on the van Genuchten model. *Agric. Water Manag.* **2016**, *176*, 170–179. [CrossRef]
15. Vellidis, G.; Liakos, V.; Porter, W.; Tucker, M.; Liang, X. A dynamic variable rate irrigation control system. In Proceedings of the 13th International Conference on Precision Agriculture, Monticello, IL, USA, 24–27 June 2018.
16. Raun, W.R.; Solie, J.B.; Johnson, G.V.; Stone, M.L.; Mullen, R.W.; Freeman, K.W.; Thomason, W.E.; Lukina, E.V. Improving nitrogen use efficiency in cereal grain production with optical sensing and variable rate application. *Agron. J.* **2002**, *94*, 815–820. [CrossRef]
17. Hedley, C.B.; Yule, I.J. Soil water status mapping and two variable-rate irrigation scenarios. *Precis. Agric.* **2009**, *10*, 342–355. [CrossRef]
18. Basso, B.; Ritchie, J.T.; Cammarano, D.; Sartori, L. A strategic and tactical management approach to select optimal N fertilizer rates for wheat in a spatially variable field. *Eur. J. Agron.* **2011**, *35*, 215–222. [CrossRef]
19. Cohen, Y.; Alchanatis, V.; Saranga, Y.; Rosenberg, O.; Sela, E.; Bosak, A. Mapping water status based on aerial thermal imagery: Comparison of methodologies for upscaling from a single leaf to commercial fields. *Precis. Agric.* **2017**, *18*, 801–822. [CrossRef]
20. Helman, D.; Bahat, I.; Netzer, Y.; Ben-Gal, A.; Alchanatis, V.; Peeters, A.; Cohen, Y. Using Time Series of High-Resolution Planet Satellite Images to Monitor Grapevine Stem Water Potential in Commercial Vineyards. *Remote Sens.* **2018**, *10*, 1615. [CrossRef]
21. Shanahan, J.F.; Kitchen, N.R.; Raun, W.R.; Schepers, J.S. Responsive in-season nitrogen management for cereals. *Comput. Electron. Agric.* **2008**, *61*, 51–62. [CrossRef]
22. Long, D.S.; Whitmus, J.D.; Engel, R.E.; Brester, G.W. Net Returns from Terrain-Based Variable-Rate Nitrogen Management on Dryland Spring Wheat in Northern Montana. *Agron. J.* **2015**, *107*, 1055–1067. [CrossRef]
23. Quebrajo, L.; Perez-Ruiz, M.; Pérez-Urrestarazu, L.; Martínez, G.; Egea, G. Linking thermal imaging and soil remote sensing to enhance irrigation management of sugar beet. *Biosyst. Eng.* **2018**, *165*, 77–87. [CrossRef]
24. Adamchuk, V.I.; Hummel, J.W.; Morgan, M.T.; Upadhyaya, S.K. On-the-go soil sensors for precision agriculture. *Comput. Electron. Agric.* **2004**, *44*, 71–91. [CrossRef]
25. De Lara, A.; Khosla, R.; Longchamps, L. Characterizing Spatial Variability in Soil Water Content for Precision Irrigation Management. *Agronomy* **2018**, *8*, 59. [CrossRef]
26. Badewa, E.; Unc, A.; Cheema, M.; Kavanagh, V.; Galagedara, L. Soil Moisture Mapping Using Multi-Frequency and Multi-Coil Electromagnetic Induction Sensors on Managed Podzols. *Agronomy* **2018**, *8*, 224. [CrossRef]
27. Gabriel, J.L.; Zarco-Tejada, P.J.; López-Herrera, P.J.; Pérez-Martín, E.; Alonso-Ayuso, M.; Quemada, M. Airborne and ground level sensors for monitoring nitrogen status in a maize crop. *Biosyst. Eng.* **2017**, *160*, 124–133. [CrossRef]
28. Irmak, S.; Haman, D.Z.; Bastug, R. Determination of Crop Water Stress Index for Irrigation Timing and Yield Estimation of Corn. *Agron. J.* **2000**, *92*, 1221–1227. [CrossRef]
29. Raun, W.R.; Solie, J.B.; Martin, K.L.; Freeman, K.W.; Stone, M.L.; Johnson, G.V.; Mullen, R.W. Growth stage, development, and spatial variability in corn evaluated using optical sensor readings. *J. Plant Nutr.* **2005**, *28*, 173–182. [CrossRef]

30. Solari, F.; Shanahan, J.; Ferguson, R.; Schepers, J.; Gitelson, A. Active sensor reflectance measurements of corn nitrogen status and yield potential. *Agron. J.* **2008**, *100*, 571–579. [CrossRef]
31. Franke, J.; Menz, G. Multi-temporal wheat disease detection by multi-spectral remote sensing. *Precis. Agric.* **2007**, *8*, 161–172. [CrossRef]
32. Shanahan, J.F.; Schepers, J.S.; Francis, D.D.; Varvel, G.E.; Wilhelm, W.W.; Tringe, J.M.; Schlemmer, M.R.; Major, D.J. Use of remote-sensing imagery to estimate corn grain yield. *Agron. J.* **2001**, *93*, 583–589. [CrossRef]
33. Teal, R.K.; Tubana, B.; Girma, K.; Freeman, K.W.; Arnall, D.B.; Walsh, O.; Raun, W.R. In-season prediction of corn grain yield potential using normalized difference vegetation index. *Agron. J.* **2006**, *98*, 1488–1494. [CrossRef]
34. Torino, M.S.; Ortiz, B.V.; Fulton, J.P.; Balkcom, K.S.; Wood, C.W. Evaluation of Vegetation Indices for Early Assessment of Corn Status and Yield Potential in the Southeastern United States. *Agron. J.* **2014**, *106*, 1389–1401. [CrossRef]
35. Tagarakis, A.C.; Ketterings, Q.M. In-Season Estimation of Corn Yield Potential Using Proximal Sensing. *Agron. J.* **2017**, *109*, 1323–1330. [CrossRef]
36. Blackmore, S.; Godwin, R.J.; Fountas, S. The analysis of spatial and temporal trends in yield map data over six years. *Biosyst. Eng.* **2003**, *84*, 455–466. [CrossRef]
37. McBratney, A.; Whelan, B.; Ancev, T.; Bouma, J. Future directions of precision agriculture. *Precis. Agric.* **2005**, *6*, 7–23. [CrossRef]
38. Maestrini, B.; Basso, B. Predicting spatial patterns of within-field crop yield variability. *Field Crop. Res.* **2018**, *219*, 106–112. [CrossRef]
39. Fullmer, D.; Chetty, V.; Warnick, S. How good is bad weather? In Proceedings of the 2014 American Control Conference, Portland, OR, USA, 4–6 June 2014; pp. 2711–2716.
40. Sadler, E.J.; Evans, R.G.; Stone, K.C.; Camp, C.R. Opportunities for conservation with precision irrigation. *J. Soil Water Conserv.* **2005**, *60*, 371–378.
41. Vellidis, G.; Snider, J.; Liakos, V.; Porter, W.; Perry, C. Dynamic Variable Rate Irrigation Management Using Soil Moisture and Canopy Temperature Sensors in the Southeastern USA. In Proceedings of the 2017 ASA, CSSA, and SSSA International Annual Meeting, Tampa, FL, USA, 22–25 October 2017.
42. Franzen, D.W. Profitable Use of Site-Specific Nutrient Management Technologies. In Proceedings of the 2017 ASA, CSSA, and SSSA International Annual Meeting, Tampa, FL, USA, 22–25 October 2017.
43. Scudiero, E.; Morari, F.; Skaggs, T.H.; Braga, F.; Teatini, P. Understanding spatiotemporal variability of soil–plant relationships in a heterogeneous coastal farmland in Northern Italy. In Proceedings of the 2016 ASA, CSSA, and SSSA International Annual Meeting, Phoenix, AZ, USA, 6–9 November 2016.
44. Liu, H.; Whiting, M.L.; Ustin, S.L.; Zarco-Tejada, P.J.; Huffman, T.; Zhang, X. Maximizing the relationship of yield to site-specific management zones with object-oriented segmentation of hyperspectral images. *Precis. Agric.* **2018**, *19*, 348–364. [CrossRef]
45. Myers, D.B. Measurements That Matter for Decision Agriculture. In Proceedings of the 2016 ASA, CSSA, and SSSA International Annual Meeting, Phoenix, AZ, USA, 6–9 November 2016.
46. Betzek, N.M.; Souza, E.G.D.; Bazzi, C.L.; Schenatto, K.; Gavioli, A. Rectification methods for optimization of management zones. *Comput. Electron. Agric.* **2018**, *146*, 1–11. [CrossRef]
47. Gavioli, A.; de Souza, E.G.; Bazzi, C.L.; Guedes, L.P.C.; Schenatto, K. Optimization of management zone delineation by using spatial principal components. *Comput. Electron. Agric.* **2016**, *127*, 302–310. [CrossRef]
48. Taylor, J.A.; McBratney, A.B.; Whelan, B.M. Establishing Management Classes for Broadacre Agricultural Production. *Agron. J.* **2007**, *99*, 1366–1376. [CrossRef]
49. Córdoba, M.A.; Bruno, C.I.; Costa, J.L.; Peralta, N.R.; Balzarini, M.G. Protocol for multivariate homogeneous zone delineation in precision agriculture. *Biosyst. Eng.* **2016**, *143*, 95–107. [CrossRef]
50. Corwin, D.L.; Lesch, S.M. Characterizing soil spatial variability with apparent soil electrical conductivity I. Survey protocols. *Comput. Electron. Agric.* **2005**, *46*, 103–133. [CrossRef]
51. Rossel, R.A.V.; Taylor, H.J.; McBratney, A.B. Multivariate calibration of hyperspectral γ-ray energy spectra for proximal soil sensing. *Eur. J. Soil Sci.* **2007**, *58*, 343–353. [CrossRef]
52. Minty, B.R.S. Fundamentals of airborne gamma-ray spectrometry. *AGSO J. Aust. Geol. Geophys.* **1997**, *17*, 39–50.

53. Priori, S.; Martini, E.; Andrenelli, M.C.; Magini, S.; Agnelli, A.E.; Bucelli, P.; Biagi, M.; Pellegrini, S.; Costantini, E.A.C. Improving Wine Quality through Harvest Zoning and Combined Use of Remote and Soil Proximal Sensing. *Soil Sci. Soc. Am. J.* **2013**, *77*, 1338–1348. [CrossRef]
54. Scudiero, E.; Corwin, D.L.; Wienhold, B.J.; Bosley, B.; Shanahan, J.F.; Johnson, C.K. Downscaling Landsat 7 canopy reflectance employing a multi-soil sensor platform. *Precis. Agric.* **2016**, *17*, 53–73. [CrossRef]
55. Lesch, S.M. Sensor-directed response surface sampling designs for characterizing spatial variation in soil properties. *Comput. Electron. Agric.* **2005**, *46*, 153–179. [CrossRef]
56. Van Groenigen, J.W.; Stein, A. Constrained optimization of spatial sampling using continuous simulated annealing. *J. Environ. Qual.* **1998**, *27*, 1078–1086. [CrossRef]
57. Hengl, T.; Heuvelink, G.B.M.; Stein, A. A generic framework for spatial prediction of soil variables based on regression-kriging. *Geoderma* **2004**, *120*, 75–93. [CrossRef]
58. Lesch, S.M.; Corwin, D.L. Prediction of spatial soil property information from ancillary sensor data using ordinary linear regression: Model derivations, residual assumptions and model validation tests. *Geoderma* **2008**, *148*, 130–140. [CrossRef]
59. Nelson, M.A.; Bishop, T.F.A.; Triantafilis, J.; Odeh, I.O.A. An error budget for different sources of error in digital soil mapping. *Eur. J. Soil Sci.* **2011**, *62*, 417–430. [CrossRef]
60. Gausman, H.W.; Allen, W.A. Optical parameters of leaves of 30 plant species. *Plant Physiol.* **1973**, *52*, 57–62. [CrossRef] [PubMed]
61. Huete, A.R.; Jackson, R.D.; Post, D.F. Spectral Response of a Plant Canopy with Different Soil Backgrounds. *Remote Sens. Environ.* **1985**, *17*, 37–53. [CrossRef]
62. Li, H.; Lascano, R.J.; Barnes, E.M.; Booker, J.; Wilson, L.T.; Bronson, K.F.; Segarra, E. Multispectral Reflectance of Cotton Related to Plant Growth, Soil Water and Texture, and Site Elevation. *Agron. J.* **2001**, *93*, 1327–1337. [CrossRef]
63. Brunsdon, C.; Fotheringham, S.; Charlton, M. Geographically weighted regression—Modelling spatial non-stationarity. *J. R. Stat. Soc. Ser. D-Stat.* **1998**, *47*, 431–443. [CrossRef]
64. Brunsdon, C.; Fotheringham, A.S.; Charlton, M.E. Geographically Weighted Regression: A Method for Exploring Spatial Nonstationarity. *Geogr. Anal.* **1996**, *28*, 281–298. [CrossRef]
65. Mitchell, A. *The ESRI Guide to GIS Analysis. Vol. II: Spatial Measurements and Statistics*; ESRI Press: Redlands, CA, USA, 2005.
66. Fotheringham, A.S.; Brunsdon, C.; Charlton, M.E. *Geographically Weighted Regression: The Analysis of Spatially Varying Relationships*; John Wiley: Hoboken, NJ, USA, 2002.
67. Odeh, I.O.A.; McBratney, A.B.; Chittleborough, D.J. Soil pattern recognition with fuzzy-c-means: Application to classification and soil-landform interrelationships. *Soil Sci. Soc. Am. J.* **1992**, *56*, 505–516. [CrossRef]
68. Venkatramanan, S.; Chung, S.Y.; Rajesh, R.; Lee, S.Y.; Ramkumar, T.; Prasanna, M.V. Comprehensive studies of hydrogeochemical processes and quality status of groundwater with tools of cluster, grouping analysis, and fuzzy set method using GIS platform: A case study of Dalcheon in Ulsan City, Korea. *Environ. Sci. Pollut. Res.* **2015**, *22*, 11209–11223. [CrossRef] [PubMed]
69. Fridgen, J.J.; Kitchen, N.R.; Sudduth, K.A.; Drummond, S.T.; Wiebold, W.J.; Fraisse, C.W. Management Zone Analyst (MZA): Software for subfield management zone delineation. *Agron. J.* **2004**, *96*, 100–108. [CrossRef]
70. Lowrance, C.; Fountas, S.; Liakos, V.; Vellidis, G. EZZone–An Online Tool for Delineating Management Zones. In Proceedings of the 13th International Conference on Precision Agriculture, St. Louis, MI, USA, 24–27 June 2018.
71. Caliński, T.; Harabasz, J. A dendrite method for cluster analysis. *Commun. Stat.* **1974**, *3*, 1–27. [CrossRef]
72. Bezdek, J.C. *Pattern Recognition with Fuzzy Objective Function Algorithms*; Plenum Press: New York, NY, USA, 1981.
73. Jenks, G.F. The data model concept in statistical mapping. *Int. Yearb. Cartogr.* **1967**, *7*, 186–190.
74. Kruskal, W.H.; Wallis, W.A. Use of ranks in one-criterion variance analysis. *J. Am. Stat. Assoc.* **1952**, *47*, 583–621. [CrossRef]
75. Rouse, J.; Haas, R.; Schell, J.; Deering, D. Monitoring vegetation systems in the Great Plains with ERTS. In Proceedings of the Third Earth Resources Technology Satellite Symposium, Washington, DC, USA, 10–14 December 1973; pp. 309–317.

76. Viscarra Rossel, R.A.; Adamchuk, V.I.; Sudduth, K.A.; McKenzie, N.J.; Lobsey, C. Chapter Five—Proximal Soil Sensing: An Effective Approach for Soil Measurements in Space and Time. In *Advances in Agronomy*; Sparks, D.L., Ed.; Academic Press: Cambridge, MA, USA, 2011; Volume 113, pp. 243–291.
77. McBratney, A.B.; Mendonça Santos, M.L.; Minasny, B. On digital soil mapping. *Geoderma* **2003**, *117*, 3–52. [CrossRef]
78. Heggemann, T.; Welp, G.; Amelung, W.; Angst, G.; Franz, S.O.; Koszinski, S.; Schmidt, K.; Pätzold, S. Proximal gamma-ray spectrometry for site-independent in situ prediction of soil texture on ten heterogeneous fields in Germany using support vector machines. *Soil Tillage Res.* **2017**, *168*, 99–109. [CrossRef]
79. Morari, F.; Castrignano, A.; Pagliarin, C. Application of multivariate geostatistics in delineating management zones within a gravelly vineyard using geo-electrical sensors. *Comput. Electron. Agric.* **2009**, *68*, 97–107. [CrossRef]
80. Weller, U.; Zipprich, M.; Sommer, M.; Castell, W.Z.; Wehrhan, M. Mapping Clay Content across Boundaries at the Landscape Scale with Electromagnetic Induction. *Soil Sci. Soc. Am. J.* **2007**, *71*, 1740–1747. [CrossRef]
81. Wardlow, B.D.; Egbert, S.L. Large-area crop mapping using time-series MODIS 250 m NDVI data: An assessment for the U.S. Central Great Plains. *Remote Sens. Environ.* **2008**, *112*, 1096–1116. [CrossRef]
82. Masialeti, I.; Egbert, S.; Wardlow, B.D. A Comparative Analysis of Phenological Curves for Major Crops in Kansas. *GIScience Remote Sens.* **2010**, *47*, 241–259. [CrossRef]
83. Xu, Y.; Smith, S.E.; Grunwald, S.; Abd-Elrahman, A.; Wani, S.P.; Nair, V.D. Estimating soil total nitrogen in smallholder farm settings using remote sensing spectral indices and regression kriging. *CATENA* **2018**, *163*, 111–122. [CrossRef]
84. Samuel-Rosa, A.; Heuvelink, G.B.M.; Vasques, G.M.; Anjos, L.H.C. Do more detailed environmental covariates deliver more accurate soil maps? *Geoderma* **2015**, *243*, 214–227. [CrossRef]
85. Hengl, T.; Heuvelink, G.B.M.; Rossiter, D.G. About regression-kriging: From equations to case studies. *Comput. Geosci.* **2007**, *33*, 1301–1315. [CrossRef]
86. Scudiero, E.; Teatini, P.; Corwin, D.L.; Deiana, R.; Berti, A.; Morari, F. Delineation of site-specific management units in a saline region at the Venice Lagoon margin, Italy, using soil reflectance and apparent electrical conductivity. *Comput. Electron. Agric.* **2013**, *99*, 54–64. [CrossRef]
87. Manoli, G.; Bonetti, S.; Scudiero, E.; Morari, F.; Putti, M.; Teatini, P. Modeling Soil–Plant Dynamics: Assessing Simulation Accuracy by Comparison with Spatially Distributed Crop Yield Measurements. *Vadose Zone J.* **2015**, *14*. [CrossRef]
88. Scudiero, E.; Teatini, P.; Corwin, D.L.; Dal Ferro, N.; Simonetti, G.; Morari, F. Spatiotemporal Response of Maize Yield to Edaphic and Meteorological Conditions in a Saline Farmland. *Agron. J.* **2014**, *106*, 2163–2174. [CrossRef]
89. Grosso, C.; Manoli, G.; Martello, M.; Chemin, Y.; Pons, D.; Teatini, P.; Piccoli, I.; Morari, F. Mapping Maize Evapotranspiration at Field Scale Using SEBAL: A Comparison with the FAO Method and Soil–Plant Model Simulations. *Remote Sens.* **2018**, *10*, 1452. [CrossRef]
90. FAO-UNESCO. *Soil Map of the World, Revised Legend*; FAO: Rome, Italy, 1989.
91. Rhoades, J.; Chanduvi, F.; Lesch, S.M. *Soil Salinity Assessment: Methods and Interpretation of Electrical Conductivity Measurements*; FAO: Rome, Italy, 1999; Volume 57.
92. Abdi, H.; Williams, L.J. Principal component analysis. *Wiley Interdiscip. Rev. Comput. Stat.* **2010**, *2*, 433–459. [CrossRef]
93. Vicente-Serrano, S.M.; Perez-Cabello, F.; Lasanta, T. Assessment of radiometric correction techniques in analyzing vegetation variability and change using time series of Landsat images. *Remote Sens. Environ.* **2008**, *112*, 3916–3934. [CrossRef]
94. Updike, T.; Comp, C. *Radiometric Use of WorldView-2 Imagery*; DigitalGlobe, Inc.: Longmont, CO, USA, 2010; pp. 1–17.
95. Vermote, E.F.; Tanre, D.; Deuze, J.L.; Herman, M.; Morcrette, J.J. Second Simulation of the Satellite Signal in the Solar Spectrum, 6S: An overview. *IEEE Trans. Geosci. Remote Sens.* **1997**, *35*, 675–686. [CrossRef]
96. Holben, B.N.; Eck, T.F.; Slutsker, I.; Tanré, D.; Buis, J.P.; Setzer, A.; Vermote, E.; Reagan, J.A.; Kaufman, Y.J.; Nakajima, T.; et al. AERONET—A Federated Instrument Network and Data Archive for Aerosol Characterization. *Remote Sens. Environ.* **1998**, *66*, 1–16. [CrossRef]

97. Xia, T.; Miao, Y.; Wu, D.; Shao, H.; Khosla, R.; Mi, G. Active Optical Sensing of Spring Maize for In-Season Diagnosis of Nitrogen Status Based on Nitrogen Nutrition Index. *Remote Sens.* **2016**, *8*, 605. [CrossRef]
98. Peng, Y.; Nguy-Robertson, A.; Arkebauer, T.; Gitelson, A. Assessment of Canopy Chlorophyll Content Retrieval in Maize and Soybean: Implications of Hysteresis on the Development of Generic Algorithms. *Remote Sens.* **2017**, *9*, 226. [CrossRef]
99. Gomez, C.; Oltra-Carrio, R.; Bacha, S.; Lagacherie, P.; Briottet, X. Evaluating the sensitivity of clay content prediction to atmospheric effects and degradation of image spatial resolution using Hyperspectral VNIR/SWIR imagery. *Remote Sens. Environ.* **2015**, *164*, 1–15. [CrossRef]
100. Gomez, C.; Viscarra Rossel, R.A.; McBratney, A.B. Soil organic carbon prediction by hyperspectral remote sensing and field vis-NIR spectroscopy: An Australian case study. *Geoderma* **2008**, *146*, 403–411. [CrossRef]
101. Corwin, D.L.; Lesch, S.M. Protocols and Guidelines for Field-scale Measurement of Soil Salinity Distribution with ECa-Directed Soil Sampling. *J. Environ. Eng. Geophys.* **2013**, *18*, 1–25. [CrossRef]
102. Corwin, D.L. Field-scale monitoring of the long-term impact and sustainability of drainage water reuse on the west side of California's San Joaquin Valley. *J. Environ. Monit.* **2012**, *14*, 1576–1596. [CrossRef] [PubMed]
103. Ransom, J.; Endres, G.J.; McWilliams, D.A. Corn Growth and Management Quick Guide A1173. Available online: www.ag.ndsu.edu/pubs/plantsci/crops/a1173.pdf (accessed on 20 October 2018).
104. Nielsen, R.L. Effects of Severe Stress During Grain Filling in Corn. Available online: http://www.kingcorn.org/news/timeless/GrainFillStress.html (accessed on 22 October 2018).
105. Lobell, D.B.; Lesch, S.M.; Corwin, D.L.; Ulmer, M.G.; Anderson, K.A.; Potts, D.J.; Doolittle, J.A.; Matos, M.R.; Baltes, M.J. Regional-scale Assessment of Soil Salinity in the Red River Valley Using Multi-year MODIS EVI and NDVI. *J. Environ. Qual.* **2010**, *39*, 35–41. [CrossRef] [PubMed]
106. Lobell, D.B.; Ortiz-Monasterio, J.I.; Gurrola, F.C.; Valenzuela, L. Identification of saline soils with multiyear remote sensing of crop yields. *Soil Sci. Soc. Am. J.* **2007**, *71*, 777–783. [CrossRef]
107. Madrigal, L.P.; Wiegand, C.L.; Meraz, J.G.; Rubio, B.D.R.; Estrada, X.C.; Ramirez, O.L. Soil salinity and its effect on crop yield—A study using satellite imagery in three irrigation districts. *Ing. Hidraul. En Mex.* **2003**, *18*, 83–97.

4

Wavelength Selection Method based on Partial Least Square from Hyperspectral Unmanned Aerial Vehicle Orthomosaic of Irrigated Olive Orchards

Antonio Santos-Rufo [1,*], Francisco-Javier Mesas-Carrascosa [2], Alfonso García-Ferrer [2] and Jose Emilio Meroño-Larriva [2]

1 Department of Agronomy, University of Cordoba, Campus de Rabanales, 14071 Córdoba, Spain
2 Department of Graphic Engineering and Geomatics, University of Cordoba, Campus de Rabanales, 14071 Córdoba, Spain; fjmesas@uco.es (F.-J.M.-C.); agferrer@uco.es (A.G.-F.); ir1melaj@uco.es (J.E.M.-L.)
* Correspondence: g02sarua@uco.es

Abstract: Identifying and mapping irrigated areas is essential for a variety of applications such as agricultural planning and water resource management. Irrigated plots are mainly identified using supervised classification of multispectral images from satellite or manned aerial platforms. Recently, hyperspectral sensors on-board Unmanned Aerial Vehicles (UAV) have proven to be useful analytical tools in agriculture due to their high spectral resolution. However, few efforts have been made to identify which wavelengths could be applied to provide relevant information in specific scenarios. In this study, hyperspectral reflectance data from UAV were used to compare the performance of several wavelength selection methods based on Partial Least Square (PLS) regression with the purpose of discriminating two systems of irrigation commonly used in olive orchards. The tested PLS methods include filter methods (Loading Weights, Regression Coefficient and Variable Importance in Projection); Wrapper methods (Genetic Algorithm-PLS, Uninformative Variable Elimination-PLS, Backward Variable Elimination-PLS, Sub-window Permutation Analysis-PLS, Iterative Predictive Weighting-PLS, Regularized Elimination Procedure-PLS, Backward Interval-PLS, Forward Interval-PLS and Competitive Adaptive Reweighted Sampling-PLS); and an Embedded method (Sparse-PLS). In addition, two non-PLS based methods, Lasso and Boruta, were also used. Linear Discriminant Analysis and nonlinear K-Nearest Neighbors techniques were established for identification and assessment. The results indicate that wavelength selection methods, commonly used in other disciplines, provide utility in remote sensing for agronomical purposes, the identification of irrigation techniques being one such example. In addition to the aforementioned, these PLS and non-PLS based methods can play an important role in multivariate analysis, which can be used for subsequent model analysis. Of all the methods evaluated, Genetic Algorithm-PLS and Boruta eliminated nearly 90% of the original spectral wavelengths acquired from a hyperspectral sensor onboard a UAV while increasing the identification accuracy of the classification.

Keywords: olive tree; UAV; hyperspectral; classification; irrigation technique; PLS; wavelength selection

1. Introduction

The intensification of agricultural practices, including better seeds, extensive fertilizer use and irrigation techniques, has altered the dynamics between humans and environmental systems across the world [1]. Although these agricultural practices have allowed for increased food production, they have also caused significant environmental impact on many regions. Consequently, accurate and precise information is in high demand from Earth System Science and global change research [1].

Today, irrigated agriculture is one of the most significant contributors of water consumption [2], necessitating modeling water exchange between land surface and atmosphere [3], managing water resources [4] and analyzing the variability of irrigation water requirements and supply [5]. As a result, estimations of water demand do not consider the spatial variability in irrigation practices and do not reflect the characteristics of irrigation techniques [6].

Irrigation is currently used in most new olive orchards, which suggests that the percentage of irrigated olive trees is very high throughout the world and increasing [7]. Since water is the largest agricultural input in many cultivation systems in Mediterranean areas, with water availability being one of the main limiting factors for crop yield, the use of more efficient irrigation techniques in olive orchards has become essential [8]. Water Use Efficiency (WUE) is a term that was coined more than 100 years ago [9] and it functions as an indicator of the balance between productivity and water availability. WUE is an essential parameter nowadays due to the great pressure of increasingly intense and frequent droughts associated with climate change effects on agricultural water availability and crop yields worldwide [10]. As WUE is a measurement of yield or biomass produced per unit of water [11], it is therefore particularly useful when trying to compare the efficiency of different irrigation systems. In this context, many research projects demonstrate that productivity could be increased with no change in the rate of water use resulting in greater WUE [12–14]. For the olive grove, Martínez and Reca (2014) [8] compared Subsurface-Drip Irrigation (SDI) and surface Drip Irrigation (DI). Both yield and WUE for SDI outperformed DI, with a water savings of up to 20% for the former. As such, evaluating crop productivity and means of irrigation in relationship to WUE would allow a better management of water resources [10].

Remote sensing has been demonstrated to be an effective tool for locating, mapping and monitoring irrigation techniques by providing data in several regions of the electromagnetic spectrum and with a variety of spatial and temporal resolutions to assess crop growth, maturity and yield [15–18]. Images are obtained remotely via a broad range of sensors on-board three main types of platforms: satellite, manned aerial and unmanned [19]. Each of these approaches have pros and cons that involve economic, operational, and technological factors [19]. Satellite imagery covers extensive areas, and some Earth observation programs provide free low spatial-resolution datasets, e.g., Moderate Resolution Imaging Spectroradiometer (MODIS), or medium resolution datasets, e.g., Sentinel-1 and -2 from the Copernicus program [20,21]. The effectiveness of satellite imagery applied to arable crops, forests and extensive plantations has been demonstrated in many studies [1,21]. However, satellite imagery may suffer from cloud cover and constrain image timing for specific phenological characteristics due to the limits of temporal resolution [22]. Remote sensing becomes more challenging when considering crops with discontinuous layouts, such as olive trees, vineyards or orchards [23]. The presence of inter-row paths may deeply affect the overall computation of spectral indices, leading to an inadequate assessment of crop status [23]. On the other hand, manned aircraft surveys offer more operational flexibility, providing spatial resolutions in the range of centimeters, but comes with high operational and logistic costs [24,25], making it difficult to perform frequent flights in phenological studies [26]. UAV platforms offer greater flexibility still [27], allowing for the possibility to differentiate pure vegetation pixels in images over woody crops. However, UAV platforms are limited both in regard to payload and flight time [28]. Despite these limitations, UAV platforms have been shown to be very useful tools in the mapping of irrigated areas when working in concordance with traditional platforms [29]. In the context of this research, UAV platforms have been used in post-classification correction of traditional platforms to identify anomalies in the mapped irrigated plots and improve classification accuracy [29].

The UAV payload included a variety of sensors, including RGB, multispectral, hyperspectral, thermal and LiDAR. Multi- and hyper-spectral sensors have been successfully used in many applications which require accurate spectral information [30,31]. The main difference between both type of sensors is based on the number of spectral bands. While multispectral images generally range from 4 to 12 spectral bands that are represented in each pixel, hyperspectral images consist of hundreds of spectral bands arranged in a very narrow bandwidth [32]. The high spectral resolution in hyperspectral images allow the detection of spectral details that can be imperceptible in multispectral images due to their discrete spectral nature [32]. UAV-based hyperspectral imaging in agriculture has been successfully used in chlorophyll [33], biomass [34], nitrogen [35] or water [36] content estimation; the detection of diseases [37]; weed classification [38]; the evaluation and classification of crop water status [39]; etc. Therefore, hyperspectral remote sensing technologies have improved our capability for understanding the processes of biophysical and biochemical properties of vegetation [40].

As a result of the high number of spectral bands, many of them are highly correlated and therefore a dimension reduction or wavelength selection method is essential to apply in pre-processing of the hyperspectral image to improve its usability [41]. These selection methods can be grouped in three categories: (i) wave band features [42,43]; (ii) spectral position features [44,45]; and (iii) vegetation indices [45,46]. These methods are performed through a variety of techniques such as Principal Components Analysis (PCA) [47,48], Minimum Noise Fraction [49,50], Singular Value Decomposition [51] or Partial Least Square (PLS) [52,53] among others. As a result, these techniques reduce the data size by selecting those wavelengths sensitive to the object of interest [54]. Specifically, PLS regression is a nonparametric and supervised technique particularly useful to achieve the "large p -small n" problem in high-dimensional datasets [55]. This technique combines features from PCA and multiple regression to predict dependent variables from a set of orthogonal factors (latent variables) extracted from predictors with the best predictive power [56]. To achieve this, a simultaneous decomposition of predictors and dependent variables are performed with the constraint that these components explain the covariance as much as possible. Afterwards, a regression step is performed where the decomposition of predictors is used to predict dependent variables [56].

Previous studies have identified wavelengths that are sensitive to crop properties such as chlorophyll content, nitrogen status and water content or estimation of biomass [57–60]. PLS methods have proven to be very versatile for multivariate data analysis in applications related to bioinformatics [61,62] or chemometrics [63,64], as well as remote sensing [65,66]. Initially, in these studies, PLS was not implemented to select variables, since the objective was to find a relevant linear subspace of the explanatory variables, but, eventually, several PLS selection methods for variable selection were finally proposed [67]. These methods can be categorized in three types: filter, wrapper and embedded methods. Filter techniques evaluate the relevance of the characteristics by only looking at the intrinsic properties of the data. In most cases, a feature relevance score is calculated, and low scoring features are removed. Subsequently, this subset of characteristics is presented as input to the classification algorithm [67]. These methods require some type of filter measure (loading weights, regression coefficients or importance of the variable in projection) that represents the response relationship with the respective variable, and, for this, a threshold is required to classify the variables as selected or not [55]. For Loading Weights Method (LW-PLS), the peaks or valleys with the maximum absolute load weights from the first major factor to the optimal principal factor are selected as sensible wavelengths [68]. For the Regression Coefficient Method (RC-PLS), the sensitive wavelengths are generally selected according to the regression coefficient of the optimal PLS model [67]. In general, the peaks or bands are selected as the sensible wavelength or waveband when the absolute value of the regression coefficient is greater than the threshold [69]. On the other hand, the basis of the Variable Importance in Projection Method (VIP-PLS) is to accumulate the importance of each variable, this being reflected by the load weights from each component [70]. While filter techniques address the problem of finding a good subset of features regardless of the model selection step, wrapper methods incorporate the model hypothesis search into the feature subset search [67]. The methods are mainly distinguished by the choice of the underlying filter method and how the wrapper is implemented, and they are primarily based on procedures that iterate between model fitting and variable selection [55]. These procedures, for example, a Genetic Algorithm integrated with the PLS regression Method (GA-PLS), combine the advantage of GA and PLS and return a vector of variable numbers that corresponds to the model that has the lowest prediction error [71]. For Uninformative Variable Elimination PLS (UVE-PLS), artificial noise variables are added to the predictor set before the PLS model is fitted and all original variables that are less important than the artificial noise variables are removed before the procedure is repeated until a stop criterion is reached [72]. In the general procedure for Backward Variable Elimination PLS (BVE-PLS), variables are ordered first with respect to some measure of importance, and one of the filter measures described above is generally used [73]. Secondly, a threshold is used to eliminate a subset of the least informative variables, and then a model is fitted again to the remaining variables and performance is measured. The procedure is repeated until the maximum performance of the model is achieved. Other wrapper methods are Subwindow Permutation Analysis coupled with PLS (SwPA-PLS), which provides the influence of each variable without considering the influence of the rest of the variables [74]; Iterative Predictive Weighting PLS (IPW-PLS), which is an iterative elimination procedure where a measure of predictor importance is computed after fitting a PLS model [75]; Regularized Elimination Procedure in PLS (REP-PLS), where a stability-based variable selection procedure is adopted [67]; Backward [76] and Forward [77] Interval PLS (BiPLS and FiPLS, respectively), where the dataset is divided into a given number of intervals, and the PLS models are then calculated with each interval left in a sequence, giving the first omitted interval the worst performance model with respect to the mean squared error of cross-validation (MSECV); and Competitive Adaptive Reweighted Sampling (CARS-PLS), which is a function variable selection method that combines Monte Carlo sampling with the PLS regression coefficient [78]. For embedded methods, the search for an optimal subset of features is built into the classifier construction and can be seen as a search in the combined space of feature subsets and hypotheses [79]. Similar to wrapper approaches, embedded approaches are thus specific to a given learning algorithm. One of these methods is Sparse-PLS (S-PLS), which is a version of PLS that aims to combine selection and modeling in a one-step procedure [80].

On the other hand, machine learning algorithms, including Linear Discriminant Analysis (LDA) and K-Nearest Neighbors (KNN), are powerful tools for analyzing hyperspectral information since they can process a large number of variables efficiently [81,82]. LDA is a subspace technique that seeks to find the maximum Fisher's ratio [83] while KNN is a non-parametric learning algorithm since there is no assumption for the underlying data distribution. KNN also utilizes lazy learning, meaning that it does not need any training data point for the generation of the model [84]. Specifically, these machine learning algorithms have been widely used in the remote sensing field for agricultural applications [37,85]. For instance, Suarez et al. [85] estimated phenoxy herbicide dosage in cotton crops through the analysis of hyperspectral data with LDA. In addition, Bohnenkamp et al. [37] utilized KNN to detect yellow rust in wheat with the application of hyperspectral imaging technology. In general, these models have shown to be effectives for investigating agricultural features using hyperspectral imagery [86].

Many authors have compared selection PLS methods to analyze which offers better performance [69,87]. Likely, there is no best variable selection method due to the interaction between method and properties of the analyzed data [55]. As per the above discussion, this article focuses on the use of wavelength selection methods in UAV hyperspectral images to compare two irrigation systems commonly used in olive orchards, SDI and DI. For this purpose, 16 methods (13 based on PLS) were evaluated, and the quality of the results were assessed by two linear and nonlinear classification techniques. This paper is organized as follows. Section 2 describe the materials and methods used. Section 3 shows the results. Section 4 includes the final concluding remarks.

2. Materials and Methods

The workflow for classifying olive tree crowns according to the irrigation technique used is summarized in Figure 1. Two UAV flights were performed using RGB and hyperspectral sensors, respectively. The RGB UAV flight was used to geometrically define each of the olive trees while the hyperspectral UAV flight was used to characterize them radiometrically. Then, different PLS- and non-PLS-based methods were used to select the most significant wavelengths to classify the two irrigation techniques methods of the study area. Finally, with the selected wavelengths, two classifications were performed using LDA and KNN, evaluating the quality and efficiency of the results.

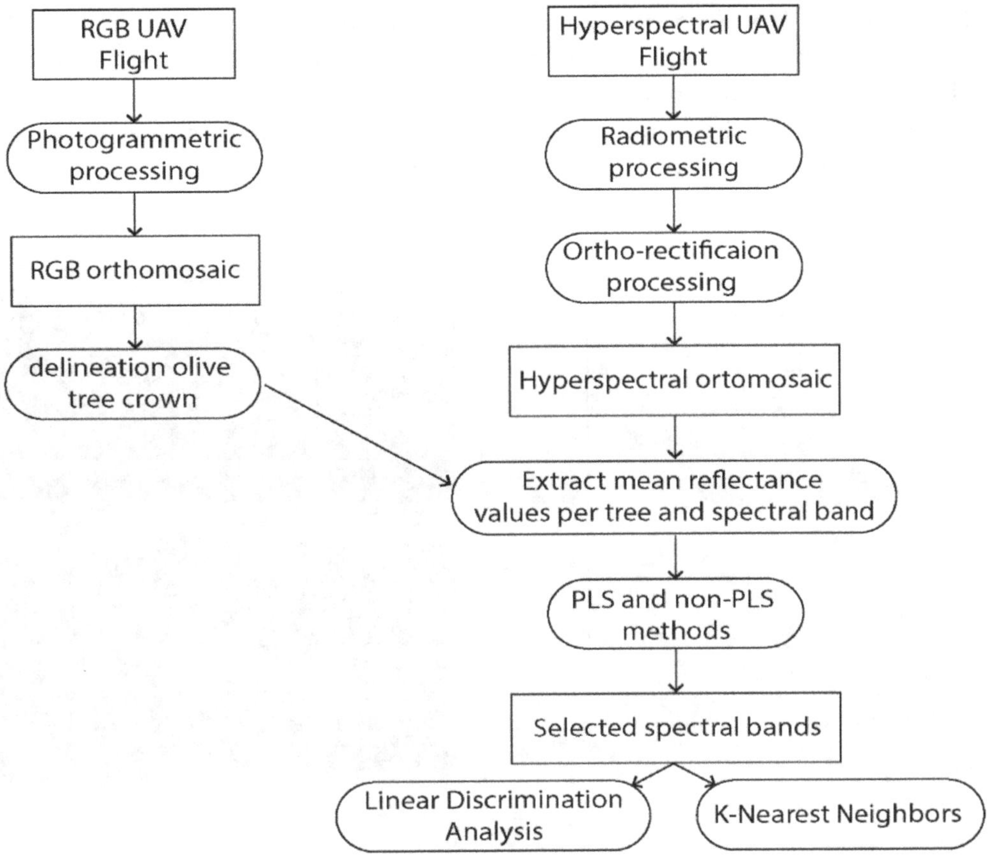

Figure 1. Flowchart used for the classification of irrigation techniques.

2.1. *Study Area and UAV Flights*

The study site was located in Marchena (Seville province, Spain) (37°24′52.7″ N, 5°27′44.9″ W, WGS84) (Figure 2a) in a 7-ha commercial orchard planted in 2015 with the olive cultivar Arbequina at a spacing of 3 m × 1.5 m (Figure 2b). The hedgerow orchard was adapted in equal parts to DI or SDI with carry self-compensating dripper pipes at 2.2 L/h spaced 0.5 m apart along the irrigation line and placed 0.5 m from the trunks and, in the case of SDI, buried 0.45 m deep. The trees were irrigated weekly with the seasonal water amount equivalent to 100% of ETc. Soil properties and agricultural practices were similar in both areas where the irrigation systems were implemented. The climate is Mediterranean with an average annual temperature and precipitation (concentrated mostly in late fall and winter) of 18.8 °C and 544 mm, respectively. A representative area including 413 olive trees (200 olives trees irrigated with DI and 213 with SDI) was used for data analyses.

2.2. *UAV Flights and Processing*

Two UAV flights, RGB and hyperspectral, were performed on 22 May 2020 at solar noon to take advantage of the position of the sun and minimize shadows in the images acquired. The RGB UAV flight was used to precisely delineate each of the olive trees in the study area while the hyperspectral UAV flight characterized each of them spectrally. A DJI Mavic Pro 2 (SZ DJI Technology Co., Shenzhen, China) was used to perform the RGB flight. As the payload, a Hasselblad L1D-20c (Hasselblad Group, Göteborg, Sweden) was used, which provides an image of 13.2 mm × 8.8 mm, a focal length of 10.3 mm and an image size of 5472 × 3648 pixels. This UAV was flown at an altitude of 65 m above ground level (AGL) and forward and side-lap were 80% and 70%, respectively. Ground Sample Distance (GSD) was 1.6 cm taking into account the characteristics of the sensor. For the hyperspectral flight, a DJI Matrice 600 Pro (SZ DJI Technology Co., Shenzhen, China) was used. This UAV platform was also equipped with a Nano Hyperspec sensor (Headwall Photonics Inc., Boston, MA, USA) with 270 spectral bands

(6-nm FWHM) both in the VNIR spectral range (400–1000 nm) with a sampling of 2.2 nm. Being a push-broom sensor, imagery is collected, line by line, at 12-bit radiometric resolution along the flight path where each line of pixels comprises 640 spatial pixels. The hyperspectral sensor was mounted on a Ronin-MX gimbal system (SZ DJI Technology Co., Shenzhen, China) to minimize external disturbances such as roll, pitch and yaw oscillations. This UAV platform was flown at an altitude of 100 m AGL, GSD being equal to 6 cm taking into account the characteristic of the sensor.

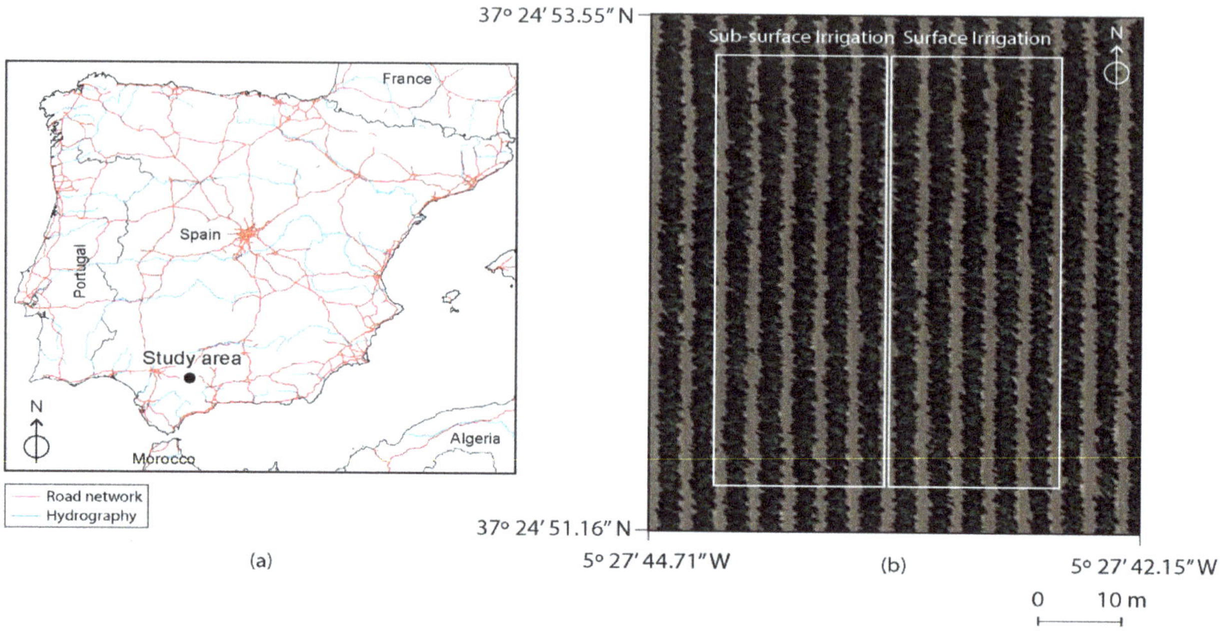

Figure 2. Study area: (**a**) Location; and (**b**) olive tree plantation details.

Prior to the UAV flights, five artificial targets were placed in the study area as Ground Control Points (GCPs), one in each corner and the other in the center. Each GCP was measured with the stop-and-go technique through relative positioning by means of the NTRIP protocol (The Radio Technical Commission for Maritime Services, RTCM, for Networked Transfer via Internet Protocol) using two GNSS (Global Navigation Satellite System) receivers. One of the receivers was a reference station for the GNSS Red Andaluza de Posicionamiento (RAP) network from the Institute of Statistics and Cartography of Andalusia, Spain, and the other, a Leica GS15 GNSS (Leica Geosystems AG, Heerbrugg, Switzerland), functioned as the rover receiver. In addition, a known reflectance value calibration tarp was placed in the center of the plot to subsequently correct the spectral cubes radiometrically. Before the hyperspectral UAV flight, the sensor capture mode was configured to set up the number of frames per second in accordance with the flight speed and AGL flight height. Likewise, the exposure level was established through the determination of the reflectance curve of a 90% target in order to avoid saturating the sensor in the recording of the spectral cubes. In addition, black body spectral information was taken by closing the aperture to later convert the digital levels to radiance values.

To produce an RGB orthomosaic from the RGB UAV flight, photogrammetric processing was divided into four stages: aerial triangulation, Digital Surface Model (DSM) generation, rectification of individual images and orthomosaicking. Aerial triangulation determined the individual external orientation of each image of the photogrammetric block. Afterwards, a dense point cloud was generated using Structure from Motion (SfM) techniques [88] to create a DSM. Finally, individual images were rectified and mosaicked to generate an RGB UAV orthomosaic of the study area. This methodology has been validated in previous research projects [19,89,90] and was performed using Pix4Dmapper (Pix4D S.A., Prilly, Switzerland). The hyperspectral data processing was divided into three stages [91]. Firstly, each hyperspectral cube was converted from digital number to radiance values, using dark reference captured prior to the UAV flight. Secondly, radiance values were converted to reflectance values using the values from the calibration tarp. Finally, all the reflectance hyperspectral cubes were orthorectified using data from the GNSS receivers, the Inertial Measurement Unit (IMU) and the Shuttle Radar Topography Mission (SRTM) Digital Elevation Model [92]. The accuracy and spatial resolution of the SRTM model was adequate due to the smooth relief of the study area. This processing was performed using SpectralView software (Headwall Photonics Inc., Boston, MA, USA).

Due to the difficulty in segmenting individual trees, an RGB orthomosaic was used to manually digitize polygons of all the olive trees, using the QGIS desktop Geographic Information System. As an example, a partial view of the hedgerow in the RGB and hyperspectral orthomosaic is shown in Figure 3. Then, for each olive tree, the mean spectral reflectance value for each of the 270 spectral bands was calculated, generating a spectral curve at the canopy level. This step was automated using a script developed in the Python programming language [93].

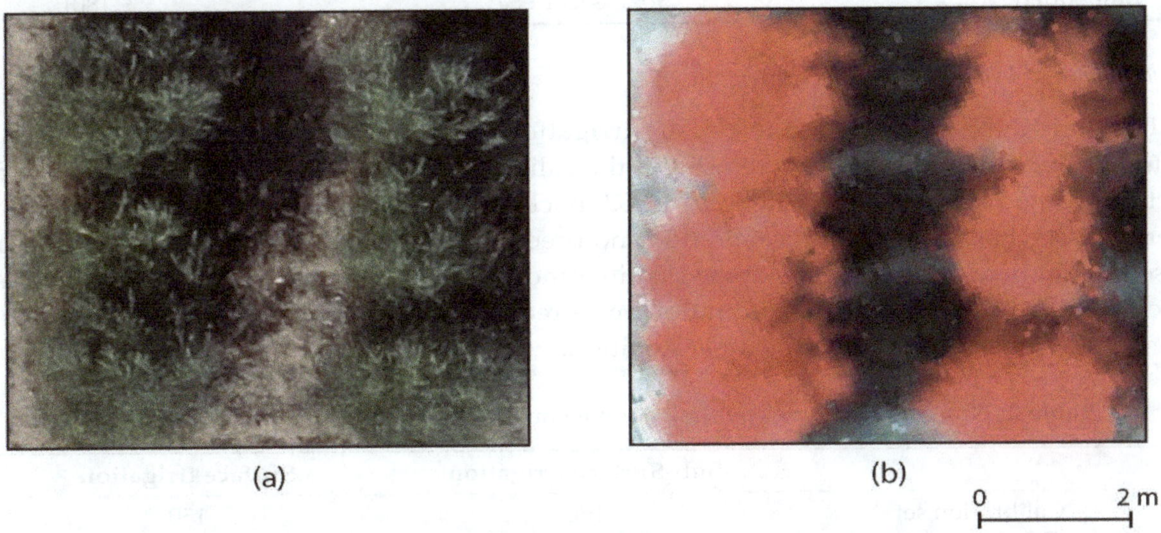

Figure 3. Partial view of (**a**) RGB and (**b**) hyperspectral orthomosaic of the study area.

2.3. *Wavelength Selection Methods Used and Evaluation*

Although there are other methods, this project focused on applying those methods that use the output of a PLS algorithm to identify a subset of important variables. As such, the variables analyzed in this project were the wavelengths registered by the hyperspectral sensor. The PLS selection methods used in this study are grouped by category (Table 1). More information on the PLS methods used in this study can be found in the work by Mehmood et al. [67].

Other wavelength selection methods, such as the Least Absolute Shrinkage and Selection Operator (Lasso) and Boruta algorithms, were also included in this study [94,95]. Lasso can handle ill-posed problems (i.e., a large number of correlated variables compared to sample size) and works by penalizing the magnitude of the characteristic coefficients along with minimizing the error between the predicted and actual observations [96]. On the other hand, the Boruta algorithm was developed to identify all relevant variables within a classification framework [97]. This method searches for relevant characteristics by comparing the importance of the original attributes with the importance that can be obtained at random, estimated using their permuted copies, and progressively eliminating irrelevant characteristics to stabilize the test. First, the dataset is duplicated, and the values are shuffled in each column; these are called shadow characteristics. Then, a classifier is trained on the dataset to extract the importance of each characteristic. Random Forest is one of the most widely used classifiers for this purpose [97]. As a final step, a combined analysis was performed considering the individual results from all the PLS, Lasso and Boruta methods used, which has been termed the All-together method.

Table 1. Evaluated wavelength selection PLS methods.

Category	PLS Method	Reference
Filter	Loading Weights (LW-PLS)	[68]
	Regression Coefficient (RC-PLS)	[69]
	Variable Importance in Projection (VIP-PLS)	[70]
Wrapper	Genetic Algorithm (GA-PLS)	[71]
	Uninformative Variable Elimination (UVE-PLS)	[72]
	Backward Variable Elimination (BVE-PLS)	[73]
	Subwindow Permutation Analysis (SwPA-PLS)	[74]
	Iterative Predictive Weighting (IPW-PLS)	[75]
	Regularized Elimination Procedure (REP-PLS)	[67]
	Backward Interval (BiPLS)	[76]
	Forward Interval (FiPLS)	[77]
	Competitive Adaptive Reweighted Sampling (CARS-PLS)	[78]
Embedded	Sparse (S-PLS)	[80]

2.4. Evaluation of Wavelength Selection Methods

LDA and KNN were applied to classify irrigation techniques used in each olive tree. Of the total number of olive trees, 75% were used in the calibration phase (explained below) and 25% for prediction (Table 2). Before applying LDA or KNN classification, the dataset was divided into two smaller datasets that were used for calibration and prediction purposes. The calibration subset of data was used to estimate the parameters of the classifier model and the prediction subset of data was used to check the results of the model. Calibration and prediction were performed following an iterative process where subsets of data changed per iteration.

Table 2. Number of samples per irrigation technique for calibration and prediction set.

	Sub-Surface Irrigation	Surface Irrigation
Calibration set	160	150
Prediction set	53	50
Total	213	200

Overall accuracy (OA) of LDA and KNN results were calculated by summing the number of correctly classified olive trees and dividing by the total number of trees. Moreover, the accuracy of each irrigation technique was evaluated [98]. In addition, efficiency of methods applied were calculated in the prediction stage. According to Xia et al. [87], the efficiency of a wavelength selection method is based on the prediction rate and the number of variables, being calculated as follows:

$$E = \frac{\left(D_s - D_f\right) \times \left(N_f - N_s\right)}{N_f} \times 100 \tag{1}$$

where E is the efficiency of the wavelength selection method evaluated, D_s is the OA obtained by a PLS method, D_f is the OA obtained using all wavelengths registered by the UAV hyperspectral sensor, N_f is the total number of wavelengths and N_s is the number of wavelengths selected by the method used. If E is higher than or equal to 0.5, the selection method is highly efficient. On the other hand, if E ranges from −0.5 to 0.5 the method is shown to be efficient, except if E is equal to 0 and $\left(N_f - N_s\right)/N_f$ is greater than or equal to 0.8, indicating the method to be highly efficient. Finally, if E is less than or equal to −0.5 the method is of low efficiency.

2.5. Software

LW-PLS, RC-PLS, VIP-PLS, AG-PLS, UVE-PLS, BVE-PLS, SwPA-PLS, IPW-PLS, REP-PLS, BiPLS, FiPLS, CARS-PLS, S-PLS, Lasso and Boruta were conducted within the R software environment [99] using RStudio [100], with the packages shown in Table 3. For the evaluation of the identification, linear and nonlinear techniques were established through LDA and KNN, respectively. The implementation of these algorithms was performed in Python [93] using the Scikit-learn library [84].

Table 3. R packages used for wavelength selection methods.

PLS Method	R Packages	Reference
Loading Weights (LW-PLS)	plsVarSel, pls	[67,101]
Regression Coefficient (RC-PLS)	plsVarSel, pls, threshr	[67,101,102]
Variable Importance in Projection (VIP-PLS)	plsVarSel, pls	[67,101]
Genetic Algorithm (GA-PLS)	plsVarSel	[67]
Uninformative Variable Elimination (UVE-PLS)	plsVarSel	[67]
Backward Variable Elimination (BVE-PLS)	plsVarSel	[67]
Subwindow Permutation Analysis (SwPA-PLS)	plsVarSel	[67]
Iterative Predictive Weighting (IPW-PLS)	plsVarSel	[67]
Regularized Elimination Procedure (REP-PLS)	plsVarSel	[67]
Backward Interval (BiPLS)	mdatools	[103]
Forward Interval (FiPLS)	mdatools	[103]
Competitive Adaptive Reweighted Sampling (CARS-PLS)	libPLSn	[104]
Sparse (S-PLS)	spls	[105]
Lasso	glmnet	[106]
Boruta	Boruta	[97]

3. Results and Discussion

3.1. Spectral Reflectance Data

The mean spectral reflectance curves of olive trees, being irrigated with SDI or DI systems, were similar to the results obtained in related studies [107] (Figure 4). No differences were observed in the 400–550 and 880–947 nm ranges. However, different magnitudes of spectral reflectance were found in the 550–880 nm range. Therefore, within that range, there are 165 possible wavelengths to differentiate between both irrigation techniques. Given this high number, different wavelength selection methods, detailed above, were applied to identify those that best classify both types of irrigation.

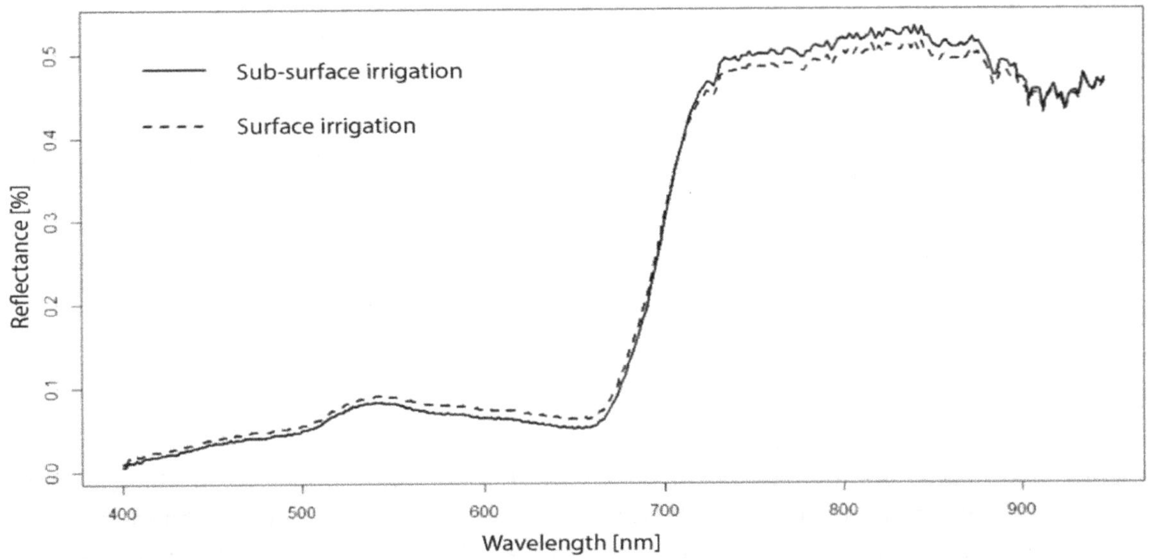

Figure 4. Average raw spectra reflectance curves of olive canopies irrigated with SDI or DI.

3.2. Wavelength Selection Results

The results of the application of the methods used, including the number of the selected wavelengths as well as their wavelengths, are shown in Table 4. Depending on the method employed, the number of wavelengths varied considerably. The method that selected the least number of wavelengths was CARS, followed by LW-PLS, neither exceeding five wavelengths, less than 1% of the original wavelengths registered. A set of methods (RC-PLS, VIP-PLS and IPW-PLS) selected a range of 10 wavelengths, equal to 3.6% of the original wavelength number, and others selected 17–77 (6.2–28.1%) wavelengths (GA-PLS, BVE-PLS, SwPA-PLS, REP_PLS, FiPLS, Lasso and Boruta). Finally, the S-PLS and BiPLS methods were the ones that selected the largest number of wavelengths, 192 (70%) and 265 (96.7%), respectively. As such, the reduction in the number of wavelengths of interest will be higher or lower depending on the method used.

Table 4. Wavelengths selected by different methods.

Method	Number of Wavelengths	Wavelengths [nm]
LW-PLS	5	882, 884, 890, 934, 942
RC-PLS	12	726, 728, 888, 904–908, 914, 920, 924, 928, 930–942
VIP-PLS	10	726, 888, 904, 906, 914, 924, 928, 936, 938, 942
GA-PLS	31	424, 428, 436, 442, 444, 458, 460, 522, 588, 612, 630, 640, 662, 698, 714, 716, 744, 758, 770, 780, 826, 846, 854, 860, 870, 878, 888, 912, 918, 920, 938
UVE-PLS	10	428, 696, 698, 700–704, 734, 792, 812, 856
BVE-PLS	69	686–734, 746, 776, 790, 840, 846, 848, 858, 860, 864–870, 880–890, 894–946
SwPA-PLS	77	410, 414, 418, 424, 436, 450, 466, 476, 490, 494, 496, 500, 506, 508, 518, 530–538, 542, 558, 562, 568, 576–580, 584, 588, 596, 612, 622–630, 640, 646, 648, 660, 668, 690, 702–706, 728, 738, 746, 752, 756, 768, 774, 782, 786, 804, 812, 822, 828, 836, 838, 848, 850, 856, 858, 860, 862, 874, 878, 882, 888, 896, 900, 904, 910, 914, 926, 936, 940
IPW-PLS	12	710, 790, 832, 846, 888, 914, 920–924, 936–940
REP-PLS	31	726, 728, 882, 884, 888, 890, 894, 898, 902–946
BiPLS	265	400–604, 624–946
FiPLS	54	660–676, 696–730, 768–784, 858–874, 912–928
CARS	2	436, 790
S-PLS	192	560–714, 720–946
Lasso	17	404, 662, 698, 704, 788, 844, 846, 858, 886, 904, 912, 918–922, 934–938
Boruta	29	430, 456, 464, 474, 636, 638, 644, 646, 650–660, 698, 774, 776, 780, 792, 794, 810, 816, 818, 838, 866, 868, 914, 930, 936
All-together	18	790, 814, 846, 860, 866, 868, 870, 882, 888, 892, 898, 902, 904, 920, 928, 934, 936, 940

The total number of times a wavelength was selected by the total number of methods simultaneously employed is shown in Figure 5. Although there was a difference in the spectrum of SDI and DI in the wavelengths of around 600–650 nm (Figure 4), the frequency of selection by the selection methods employed in this range was low (Figure 5). Of all the methods applied, 18 wavelengths were selected by almost half the methods (All-together method; Table 4). All within the infrared region and for a total of nine times, the wavelength 920 nm was selected the most to predict the irrigation method used in this study. The utility of the All-together method was demonstrated for investigating agricultural features using hyperspectral imagery.

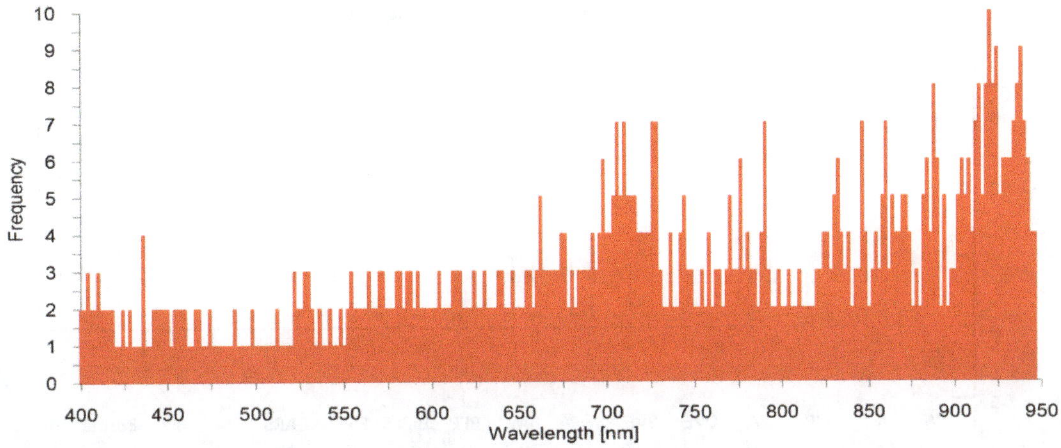

Figure 5. Number of times a wavelength has been selected when performing the All-together method.

3.3. *Quality and Efficiency of Classification with Each Selection Method*

The accuracy and efficiency results of the irrigation technique rating using LDA and KNN methods and their efficiency are shown in Table 5, while Figure 6 shows their efficiency. In general terms, LDA classifications showed better overall accuracy than KNN, in both calibration and prediction. Considering LDA, overall accuracy prediction (OAP) using all wavelengths was equal to 65%. Using only those selected wavelengths from a selection method, OAP improved, except in LW-PLS, the mean OPA being equal to 75%. The highest percentage of improvement was offered by GA-PLS, at 20%, while the lowest was offered by CARS, at 1%. In addition, ten of the methods evaluated improved the predictive quality of the irrigation technique by more than 10% (RC-PLS, GA-PLS, BVE-PLS, SwPA-PLS, IPW-PLS, BiPLS, S-PLS, Lasso, Boruta and All-together). On the other hand, the OAP applying KNN using all wavelengths was equal to 68.1% while just using the wavelengths selected by a PLS method generally resulted in a slightly lower OAP, with a mean value equal to 66.4%. Boruta was the method that showed the highest percentage of improvement, with an OAP of 74%. In total, seven of the methods offered an improvement in irrigation system prediction (GA-PLS, IPW-PLS, BiPLS, S-PLS, Lasso, Boruta and All-together). In addition, while LDA showed similar accuracy classifying both irrigation techniques, KNN offered worse results with the SDI technique.

Table 5. Overall accuracy (OA), Accuracy of Sub-Surface drip irrigation (A SDI), Accuracy of Surface drip irrigation (A DI) and Efficiency (E) results of Linear Discriminant Analysis (LDA) and K-Nearest Neighbors (KNN) using different selection methods.

	LDA				KNN			
Method	**OA (%)**	**A SDI (%)**	**A DI (%)**	**E**	**OA (%)**	**A SDI (%)**	**A DI (%)**	**E**
All bands	65.0	62.5	68.7	-	68.1	64.6	73.6	-
LW-PLS	62.5	57.5	70.7	−2.9	54.8	51.7	61.1	−13.1
RC-PLS	81.3	80.7	81.1	15.3	68.3	64.4	73.3	0.2
VIP-PLS	72.2	69.5	73.3	6.7	60.6	55.1	71.9	−7.2
GA-PLS	85.2	83.6	87.6	17.7	69.2	65.4	73.0	1.0
UVE-PLS	66.5	61.9	72.8	1.0	63.5	59.9	70.1	−4.4
BVE-PLS	79.0	78.5	79.5	10.5	65.4	61.7	71.1	−2.0
SwPA-PLS	76.3	76.2	76.4	7.9	68.3	66.6	70.0	0.1
IPW-PLS	75.5	71.5	79.5	9.6	71.2	67.4	75.4	3.0
REP-PLS	69.3	67.9	71.3	3.5	59.6	54.5	71.1	−7.5
BiPLS	70.2	68.3	72.1	0.2	69.2	64.2	75.2	0.0
FiPLS	80.1	81.6	79.0	12.0	66.3	63.5	70.1	−1.4
CARS	66.0	60.1	76.1	1.0	63.5	60.1	67.9	−4.6
S-PLS	75.3	75.3	75.3	3.0	69.2	66.2	72.2	0.5
Lasso	84.0	86.4	82.3	17.8	71.2	72.3	72.2	2.9
Boruta	78.0	76.4	80.2	11.6	74.0	73.5	75.6	5.3
All-together	82.5	78.6	85.3	15.9	69.2	65.3	74.9	1.0

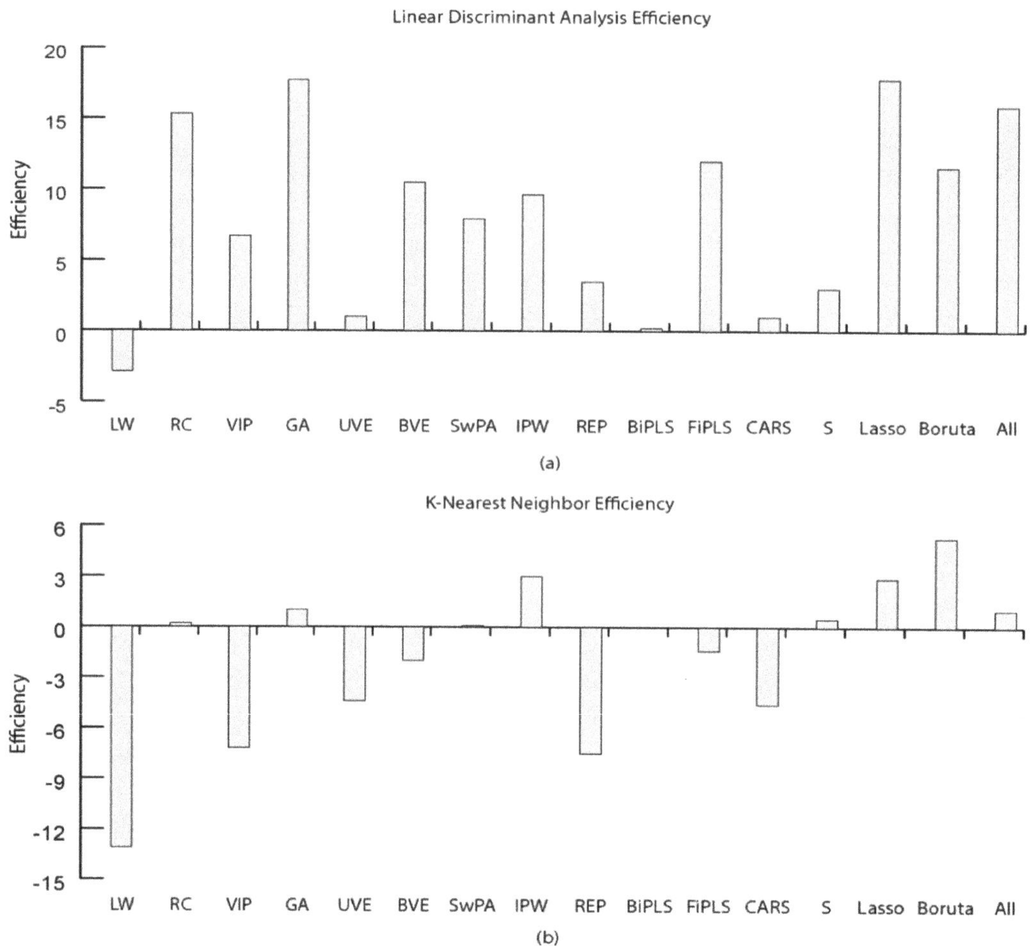

Figure 6. Efficiency of PLS method using (**a**) Linear Discriminant Analysis and (**b**) K-Nearest Neighbor.

While LDA offered better results than KNN, all the methods used were highly efficient except for LW-PLS (Figure 6). GA-PLS and Lasso scored the highest with an efficiency equal to 17.7 and 17.8, respectively (Figure 6a). In contrast, only six of the sixteen methods were highly efficient using KNN, Boruta being the most efficient with a value equal to 5.3 (Figure 6b). In addition, the irrigation classification maps of some of the selection methods assessed are shown in Figure 7. From a visual analysis, no spatial correlation was detected in the errors obtained as well as the presence of a higher concentration of errors in the perimeter of each type of irrigation.

Although combining bands when using hyperspectral data for olive trees is common [108], this study shows that the use of individual wavelengths could be an interesting and more accessible way to manage hyperspectral data. The results obtained from each selection method were different, showing methods that improved overall accuracy and efficiency in the classification of irrigation systems and methods that did not. This variation in the results obtained by selection methods were similar to those obtained by other authors [87]. In this study, according to the LDA and KNN results, GA-PLS, IPW-PLS, Lasso, Boruta and All-together methods were those which improved OAP while being highly efficient at the same time. In addition, RC-PLS and FiPLS methods showed a high overall accuracy using LDA. The GA-PLS method was also one of the most efficient, as seen by Xia et al. (2017) [87]. These authors found spectral differences between different samples of *Ophiopogon japonicus* from differing origins, using, in their case, an imaging spectrograph. In addition, although the Boruta method has not been widely used for the selection of variables in remote sensing or other disciplines, results obtained in this study show that it could be considered as a promising method for the selection of hyperspectral wavelengths. On the other hand, LW-PLS showed low efficiency and lower OAP. This can be explained because this method greatly reduces wavelengths and therefore eliminated useful information as indicated in other works [87]. The efficiency of RC-PLS, VIP-PLS, UVE-PLS, BVE-PLS, SwPA-PLS, REP-PLS, BiPLS, FiPLS, CARS-PLS and S-PLS varied depending on whether LDA or KNN was used. In general, LDA showed better classification results than KNN, which can be explained due to the fact that KNN needs to have large training data in order to achieve acceptable classification results, as indicated by Starzacher and Rinner (2008) [109].

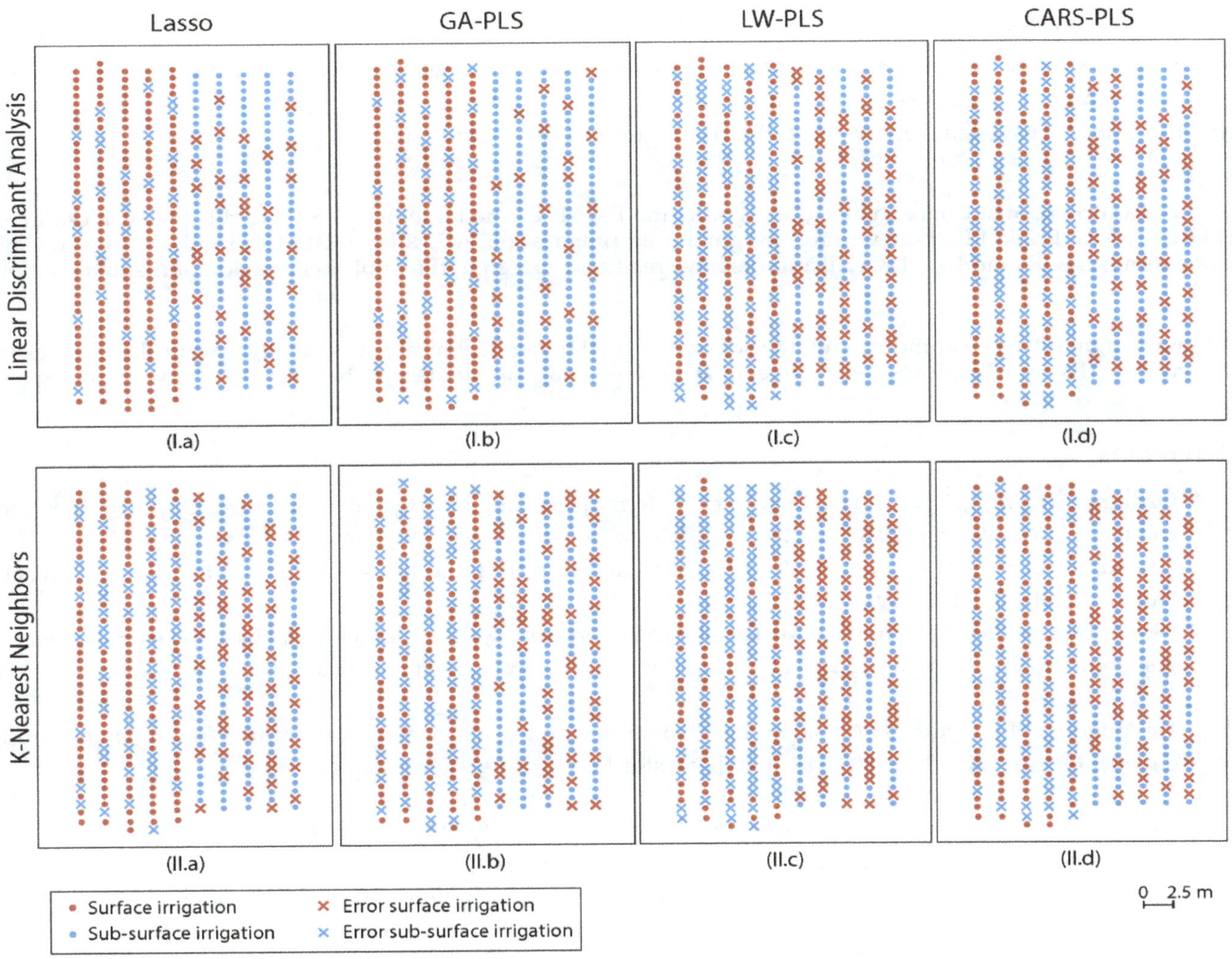

Figure 7. Classification results per classifier ((I) Linear Discriminant Analysis; and (II) K-Nearest Neighbors) and PLS method: (**a**) Lasso; (**b**) Genetic Algorithm; (**c**) Loading Weights; and (**d**) Competitive Adaptative Reweighted Sampling.

In general, these preliminary results show the need to analyze the relationships between wavelengths registered by a hyperspectral sensor and the object of study to optimize the number of wavelengths utilized. As a future line of research, the analysis of which combinations of wavelengths may be of interest to characterize irrigated areas in more detail is proposed. In addition, other features such as soil properties or olive cultivars along with other classifiers should be considered for improving the characterization of irrigation areas.

4. Conclusions

This study explored the use of UAV hyperspectral reflectance measurements of olive trees as a means for differentiating irrigation systems. Because of the high dimension and multicollinearity of the data, selection methods were found appropriate due to their capacity to extract useful wavelengths for analyzing categorial data, even when using individualized wavelengths. The results showed how the spectral response of olive trees is sensitive to the irrigation technique used, allowing improved information for the mapping of irrigated areas. Overall accuracy in the classification of irrigation systems in olives trees using LDA and KNN ranged from 54.8% and 85.2%. These variations showed the need to select the appropriate wavelength selection method. In addition, LDA offered more accurate results than KNN. In our study, GA-PLS, RC-PLS, Lasso, FiPLS, Boruta and All-together showed an overall accuracy of 75% or higher. They were all highly efficient methods and resulted in an improved classification.

The study has shown how the use of hyperspectral UAV data allows the irrigated areas of olive groves to be characterized in greater detail. This will generate information of interest for decision-making processes in the context of water use policies, enabling better understanding of irrigated olive groves and improving the management of water resources.

Author Contributions: Conceptualization, A.S.-R. and F.-J.M.-C.; methodology, A.S.-R. and F.-J.M.-C.; resources, F.-J.M.-C., A.G.-F. and J.E.M.-L.; writing—original draft preparation, A.S.-R. and F.-J.M.-C.; and writing—review and editing, A.S.-R. and F.-J.M.-C. All authors have read and agreed to the published version of the manuscript.

Acknowledgments: The authors thank the support of the Higher Technical School of Agricultural and Forestry Engineering (Escuela Técnica Superior Ingeniería Agronómica y de Montes) of the University of Córdoba, Spain.

References

1. Ozdogan, M.; Yang, Y.; Allez, G.; Cervantes, C. Remote sensing of irrigated agriculture: Opportunities and challenges. *Remote Sens.* **2010**, *2*, 2274–2304. [CrossRef]
2. Cai, X.; Rosegrant, M.W. Global Water Demand and Supply Projections: Part 1. A Modeling Approach. *Water Int.* **2002**, *27*, 159–169. [CrossRef]
3. Wisser, D.; Frolking, S.; Douglas, E.M.; Fekete, B.M.; Vörösmarty, C.J.; Schumann, A.H. Global irrigation water demand: Variability and uncertainties arising from agricultural and climate data sets. *Geophys. Res. Lett.* **2008**, *35*, 1–5. [CrossRef]
4. Kueppers, L.M.; Snyder, M.A.; Sloan, L.C. Irrigation cooling effect: Regional climate forcing by land-use change. *Geophys. Res. Lett.* **2007**, *34*, 1–5. [CrossRef]
5. Droogers, P.; Aerts, J. Adaptation strategies to climate change and climate variability: A comparative study between seven contrasting river basins. *Phys. Chem. Earth* **2005**, *30*, 339–346. [CrossRef]
6. Thenkabail, P.S.; Dheeravath, V.; Biradar, C.M.; Gangalakunta, O.R.P.; Noojipady, P.; Gurappa, C.; Velpuri, M.; Gumma, M.; Li, Y. Irrigated area maps and statistics of India using remote sensing and national statistics. *Remote Sens.* **2009**, *1*, 50–67. [CrossRef]
7. Escobar, R.F.; de la Rosa, R.; Leon, L.; Gomez, J.A.; Testi, F.; Orgaz, M.; Gil-Ribes, J.A.; Quesada-Moraga, E.; Trapero, A. Evolution and sustainability of the olive production systems. In *Present and Future of the Mediterranean Olive Sector*; CIHEAM/IOC: Zaragoza, Spain, 2013; Volume 106, pp. 11–41. ISBN 2-85352-512-0.
8. Martínez, J.; Reca, J. Water use efficiency of surface drip irrigation versus an alternative subsurface drip irrigation method. *J. Irrig. Drain. Eng.* **2014**, *140*, 1–9. [CrossRef]
9. Briggs, L.J.; Shantz, H.L. The water requirement of plants. In *Bureau of Plant Industry Bulletin*; Wiley: Akron, OH, USA, 1913; p. 96.
10. Yu, L.; Gao, X.; Zhao, X. Global synthesis of the impact of droughts on crops' water-use efficiency (WUE): Towards both high WUE and productivity. *Agric. Syst.* **2020**, *177*, 102723. [CrossRef]
11. Tolk, J.; Howell, T.; Evett, S. Effect of mulch, irrigation, and soil type on water use and yield of maize. *Soil Tillage Res.* **1999**, *50*, 137–147. [CrossRef]
12. Basso, B.; Ritchie, J.T. Evapotranspiration in High-Yielding Maize and under Increased Vapor Pressure Deficit in the US Midwest. *Agric. Environ. Lett.* **2018**, *3*, 170039. [CrossRef]
13. Bota, J.; Tomás, M.; Flexas, J.; Medrano, H.; Escalona, J.M. Differences among grapevine cultivars in their stomatal behavior and water use efficiency under progressive water stress. *Agric. Water Manag.* **2016**, *164*, 91–99. [CrossRef]
14. Michelon, N.; Pennisi, G.; Myint, N.O.; Orsini, F.; Gianquinto, G. Strategies for improved Water Use Efficiency (WUE) of field-grown lettuce (*Lactuca sativa* L.) under a semi-arid climate. *Agronomy* **2020**, *10*, 668. [CrossRef]

15. Thenkabail, P.S.; Schull, M.; Turral, H. Ganges and Indus river basin land use/land cover (LULC) and irrigated area mapping using continuous streams of MODIS data. *Remote Sens. Environ.* **2005**, *95*, 317–341. [CrossRef]
16. Toureiro, C.; Serralheiro, R.; Shahidian, S.; Sousa, A. Irrigation management with remote sensing: Evaluating irrigation requirement for maize under Mediterranean climate condition. *Agric. Water Manag.* **2017**, *184*, 211–220. [CrossRef]
17. Gao, Q.; Zribi, M.; Escorihuela, M.J.; Baghdadi, N.; Segui, P.Q. Irrigation mapping using Sentinel-1 time series at field scale. *Remote Sens.* **2018**, *10*, 1495. [CrossRef]
18. Jalilvand, E.; Tajrishy, M.; Hashemi, S.A.G.Z.; Brocca, L. Quantification of irrigation water using remote sensing of soil moisture in a semi-arid region. *Remote Sens. Environ.* **2019**, *231*, 111226. [CrossRef]
19. Mesas-Carrascosa, F.-J.; Torres-Sánchez, J.; Clavero-Rumbao, I.; García-Ferrer, A.; Peña, J.-M.; Borra-Serrano, I.; López-Granados, F. Assessing Optimal Flight Parameters for Generating Accurate Multispectral Orthomosaicks by UAV to Support Site-Specific Crop Management. *Remote Sens.* **2015**, *7*, 12793–12814. [CrossRef]
20. Pageot, Y.; Baup, F.; Inglada, J.; Baghdadi, N.; Demarez, V. Detection of Irrigated and Rainfed Crops in Temperate Areas Using Sentinel-1 and Sentinel-2 Time Series. *Remote Sens.* **2020**, *12*, 3044. [CrossRef]
21. Velpuri, N.M.; Thenkabail, P.S.; Gumma, M.K.; Biradar, C.; Dheeravath, V.; Noojipady, P.; Yuanjie, L. Influence of resolution in irrigated area mapping and area estimation. *Photogramm. Eng. Remote Sens.* **2009**, *75*, 1383–1395. [CrossRef]
22. Matese, A.; Toscano, P.; Di Gennaro, S.F.; Genesio, L.; Vaccari, F.P.; Primicerio, J.; Belli, C.; Zaldei, A.; Bianconi, R.; Gioli, B. Intercomparison of UAV, aircraft and satellite remote sensing platforms for precision viticulture. *Remote Sens.* **2015**, *7*, 2971–2990. [CrossRef]
23. Eastman, J.R.; Sangermano, F.; Machado, E.A.; Rogan, J.; Anyamba, A. Global trends in seasonality of Normalized Difference Vegetation Index (NDVI), 1982–2011. *Remote Sens.* **2013**, *5*, 4799–4818. [CrossRef]
24. Berni, J.A.J.; Zarco-Tejada, P.J.; Suárez, L.; Fereres, E. Thermal and Narrowband Multispectral Remote Sensing for Vegetation Monitoring from an Unmanned Aerial Vehicle Improved Evapotranspiration using Unmanned Aerial Vehicles View project High throughput and remote trait measurement View project Thermal and Nar. *IEEE Trans. Geosci. Remote Sens.* **2009**, *47*, 722–738. [CrossRef]
25. Anderson, K.; Gaston, K.J. Lightweight unmanned aerial vehicles will revolutionize spatial ecology. *Front. Ecol. Environ.* **2013**, *11*, 138–146. [CrossRef]
26. Berra, E.F.; Gaulton, R.; Barr, S. Assessing spring phenology of a temperate woodland: A multiscale comparison of ground, unmanned aerial vehicle and Landsat satellite observations. *Remote Sens. Environ.* **2019**, *223*, 229–242. [CrossRef]
27. Jay, S.; Baret, F.; Dutartre, D.; Malatesta, G.; Héno, S.; Comar, A.; Weiss, M.; Maupas, F. Exploiting the centimeter resolution of UAV multispectral imagery to improve remote-sensing estimates of canopy structure and biochemistry in sugar beet crops. *Remote Sens. Environ.* **2019**, *231*, 110898. [CrossRef]
28. Radoglou-Grammatikis, P.; Sarigiannidis, P.; Lagkas, T.; Moscholios, I. A compilation of UAV applications for precision agriculture. *Comput. Netw.* **2020**, *172*, 107148. [CrossRef]
29. Nhamo, L.; van Dijk, R.; Magidi, J.; Wiberg, D.; Tshikolomo, K. Improving the accuracy of remotely sensed irrigated areas using post-classification enhancement through UAV capability. *Remote Sens.* **2018**, *10*, 712. [CrossRef]
30. Qin, J.; Chao, K.; Kim, M.S.; Lu, R.; Burks, T.F. Hyperspectral and multispectral imaging for evaluating food safety and quality. *J. Food Eng.* **2013**, *118*, 157–171. [CrossRef]
31. Park, B.; Lu, R. (Eds.) *Hyperspectral Imaging Technology in Food and Agriculture*; Springer US: New York, NY, USA, 2015; ISBN 9781493928354.
32. Adão, T.; Hruška, J.; Pádua, L.; Bessa, J.; Peres, E.; Morais, R.; Sousa, J.J. Hyperspectral imaging: A review on UAV-based sensors, data processing and applications for agriculture and forestry. *Remote Sens.* **2017**, *9*, 1110. [CrossRef]
33. Yu, F.H.; Xu, T.Y.; Du, W.; Ma, H.; Zhang, G.S.; Chen, C.L. Radiative transfer models (RTMs) for field phenotyping inversion of rice based on UAV hyperspectral remote sensing. *Int. J. Agric. Biol. Eng.* **2017**, *10*, 150–157. [CrossRef]
34. Yue, J.; Yang, G.; Li, C.; Li, Z.; Wang, Y.; Feng, H.; Xu, B. Estimation of Winter Wheat Above-Ground Biomass Using Unmanned Aerial Vehicle-Based Snapshot Hyperspectral Sensor and Crop Height Improved Models. *Remote Sens.* **2017**, *9*, 708. [CrossRef]

35. Akhtman, Y.; Golubeva, E.I.; Tutubalina, O.V.; Zimin, M. Application of hyperspectural images and ground data for precision farming. *Geogr. Environ. Sustain.* **2017**, *10*, 117–128. [CrossRef]
36. Bauer, M.E.; Daughtry, C.S.T.; Vanderbilt, V.C. Spectral-agronomic relationships of corn, soybean and wheat canopies. In Proceedings of the Signatures Spectrales D'objets En Teledetection, Avignon, France, 18–22 January 1981; pp. 261–272.
37. Bohnenkamp, D.; Behmann, J.; Mahlein, A.K. In-field detection of Yellow Rust in Wheat on the Ground Canopy and UAV Scale. *Remote Sens.* **2019**, *11*, 2495. [CrossRef]
38. Scherrer, B.; Sheppard, J.; Jha, P.; Shaw, J.A. Hyperspectral imaging and neural networks to classify herbicide-resistant weeds. *J. Appl. Remote Sens.* **2019**, *13*, 044516. [CrossRef]
39. Rinaldi, M.; Castrignanò, A.; De Benedetto, D.; Sollitto, D.; Ruggieri, S.; Garofalo, P.; Santoro, F.; Figorito, B.; Gualano, S.; Tamborrino, R. Discrimination of tomato plants under different irrigation regimes: Analysis of hyperspectral sensor data. *Environmetrics* **2015**, *26*, 77–88. [CrossRef]
40. Renzullo, L.J.; Blanchfield, A.L.; Powell, K.S. A method of wavelength selection and spectral discrimination of hyperspectral reflectance spectrometry. *IEEE Trans. Geosci. Remote Sens.* **2006**, *44*, 1986–1994. [CrossRef]
41. Lu, B.; Dao, P.D.; Liu, J.; He, Y.; Shang, J. Recent Advances of Hyperspectral Imaging Technology and Applications in Agriculture. *Remote Sens.* **2020**, *12*, 2659. [CrossRef]
42. Thenkabail, P.S.; Enclona, E.A.; Ashton, M.S.; Van Der Meer, B. Accuracy assessments of hyperspectral waveband performance for vegetation analysis applications. *Remote Sens. Environ.* **2004**, *91*, 354–376. [CrossRef]
43. Becker, B.L.; Lusch, D.P.; Qi, J. Identifying optimal spectral bands from in situ measurements of Great Lakes coastal wetlands using second-derivative analysis. *Remote Sens. Environ.* **2005**, *97*, 238–248. [CrossRef]
44. Pu, R.; Gong, P.; Biging, G.S.; Larrieu, M.R. Extraction of red edge optical parameters from hyperion data for estimation of forest leaf area index. *IEEE Trans. Geosci. Remote Sens.* **2003**, *41*, 916–921. [CrossRef]
45. Darvishzadeh, R.; Atzberger, C.; Skidmore, A.K.; Abkar, A.A. Leaf Area Index derivation from hyperspectral vegetation indicesand the red edge position. *Int. J. Remote Sens.* **2009**, *30*, 6199–6218. [CrossRef]
46. Haboudane, D.; Miller, J.R.; Pattey, E.; Zarco-Tejada, P.J.; Strachan, I.B. Hyperspectral vegetation indices and novel algorithms for predicting green LAI of crop canopies: Modeling and validation in the context of precision agriculture. *Remote Sens. Environ.* **2004**, *90*, 337–352. [CrossRef]
47. Farrell, M.D.; Mersereau, R.M. On the impact of PCA dimension reduction for hyperspectral detection of difficult targets. *IEEE Geosci. Remote Sens. Lett.* **2005**, *2*, 192–195. [CrossRef]
48. Koonsanit, K.; Jaruskulchai, C.; Eiumnoh, A. Band Selection for Dimension Reduction in Hyper Spectral Image Using Integrated InformationGain and Principal Components Analysis Technique. *Int. J. Mach. Learn. Comput.* **2012**, *2*, 248–251. [CrossRef]
49. Burger, J.; Gowen, A. Data handling in hyperspectral image analysis. *Chemom. Intell. Lab. Syst.* **2011**, *108*, 13–22. [CrossRef]
50. Gao, L.; Zhao, B.; Jia, X.; Liao, W.; Zhang, B. Optimized kernel minimum noise fraction transformation for hyperspectral image classification. *Remote Sens.* **2017**, *9*, 548. [CrossRef]
51. Menon, V.; Du, Q.; Fowler, J.E. Fast SVD with Random Hadamard Projection for Hyperspectral Dimensionality Reduction. *IEEE Geosci. Remote Sens. Lett.* **2016**, *13*, 1275–1279. [CrossRef]
52. Fordellone, M.; Bellincontro, A.; Mencarelli, F. Partial least squares discriminant analysis: A dimensionality reduction method to classify hyperspectral data. *Stat. Appl. Ital. J. Appl. Stat.* **2018**, *31*, 181–200. [CrossRef]
53. Krishnan, A.; Williams, L.J.; McIntosh, A.R.; Abdi, H. Partial Least Squares (PLS) methods for neuroimaging: A tutorial and review. *Neuroimage* **2011**, *56*, 455–475. [CrossRef]
54. Ray, S.S.; Das, G.; Singh, J.P.; Panigrahy, S. Evaluation of hyperspectral indices for LAI estimation and discrimination of potato crop under different irrigation treatments. *Int. J. Remote Sens.* **2006**, *27*, 5373–5387. [CrossRef]
55. Mehmood, T.; Sæbø, S.; Liland, K.H. Comparison of variable selection methods in partial least squares regression. *J. Chemom.* **2020**, *34*. [CrossRef]
56. Abdi, H. Partial least squares regression and projection on latent structure regression (PLS Regression). *WIREs Comput. Stat.* **2010**, *2*, 97–106. [CrossRef]
57. Lu, B.; He, Y. Evaluating empirical regression, machine learning, and radiative transfer modelling for estimating vegetation chlorophyll content using bi-seasonal hyperspectral images. *Remote Sens.* **2019**, *11*, 1979. [CrossRef]

58. Thenkabail, P.S.; Mariotto, I.; Gumma, M.K.; Middleton, E.M.; Landis, D.R.; Huemmrich, K.F. Selection of hyperspectral narrowbands (hnbs) and composition of hyperspectral twoband vegetation indices (HVIS) for biophysical characterization and discrimination of crop types using field reflectance and hyperion/EO-1 data. *IEEE J. Sel. Top. Appl. Earth Obs. Remote Sens.* **2013**, *6*, 427–439. [CrossRef]
59. Thenkabail, P.S.; Smith, R.B.; De Pauw, E. Hyperspectral Vegetation Indices and Their Relationships with Agricultural Crop Characteristics. *Remote Sens. Environ.* **2000**, *71*, 158–182. [CrossRef]
60. Thenkabail, P.; Lyon, J. (Eds.) *Hyperspectral Remote Sensing of Vegetation*; CRC Press: Boca Raton, FL, USA, 2012.
61. Durif, G.; Modolo, L.; Michaelsson, J.; Mold, J.E.; Lambert-Lacroix, S.; Picard, F. High dimensional classification with combined adaptive sparse PLS and logistic regression. *Bioinformatics* **2018**, *34*, 485–493. [CrossRef]
62. Deb, D.; Mackey, D.; Opiyo, S.O.; McDowell, J.M. Application of alignment-free bioinformatics methods to identify an oomycete protein with structural and functional similarity to the bacterial AvrE effector protein. *PLoS ONE* **2018**, *13*, 1–19. [CrossRef]
63. Sampaio, P.S.; Soares, A.; Castanho, A.; Almeida, A.S.; Oliveira, J.; Brites, C. Optimization of rice amylose determination by NIR-spectroscopy using PLS chemometrics algorithms. *Food Chem.* **2018**, *242*, 196–204. [CrossRef]
64. Li, J.; Tong, Y.; Guan, L.; Wu, S.; Li, D. Optimization of COD determination by UV–vis spectroscopy using PLS chemometrics algorithms. *Optik* **2018**, *174*, 591–599. [CrossRef]
65. Luedeling, E.; Gassner, A. Partial Least Squares Regression for analyzing walnut phenology in California. *Agric. For. Meteorol.* **2012**, *158–159*, 43–52. [CrossRef]
66. Song, S.; Gong, W.; Zhu, B.; Huang, X. Wavelength selection and spectral discrimination for paddy rice, with laboratory measurements of hyperspectral leaf reflectance Shalei. *ISPRS J. Photogramm. Remote Sens.* **2011**, *66*, 672–682. [CrossRef]
67. Mehmood, T.; Liland, K.H.; Snipen, L.; Sæbø, S. A review of variable selection methods in Partial Least Squares Regression. *Chemom. Intell. Lab. Syst.* **2012**, *118*, 62–69. [CrossRef]
68. Wang, Y.; Gao, Y.; Yu, X.; Wang, Y.; Deng, S.; Gao, J.-M. Rapid Determination of Lycium Barbarum Polysaccharide with Effective Wavelength Selection Using Near-Infrared Diffuse Reflectance Spectroscopy. *Food Anal. Methods* **2016**, *9*, 131–138. [CrossRef]
69. Chen, H.; Xia, Z. Determination of total flavonoids in propolis based on NIR spectroscopy technology. *Chin. J. Pharm. Anal.* **2014**, *34*, 1868–1873.
70. Eriksson, I.; Johansson, E.; Kettaneh-Wold, N.; Wold, S. *Multi- and Megavariate Data Analysis. Principles and Applications*; MKS Umetrics: Malmö, Sweden, 2002; Volume 16.
71. Ding, X.-B.; Zhang, C.; Liu, F.; Song, X.L.; Kong, W.W.; He, Y. Determination of soluble solid content in strawberry using hyperspectral imaging combined with feature extraction methods. *Spectroscopy Spectr. Anal.* **2015**, *35*, 1020–1024.
72. Pan, L.; Lu, R.; Zhu, Q.; Tu, K.; Cen, H. Predict compositions and mechanical properties of sugar beet using hyperspectral scattering. *Food Bioprocess Technol.* **2016**, *9*, 1177–1186. [CrossRef]
73. Guzmán, E.; Baeten, V.; Pierna, J.A.F.; García-Mesa, J.A. Application of low-resolution Raman spectroscopy for the analysis of oxidized olive oil. *Food Control.* **2011**, *22*, 2036–2040. [CrossRef]
74. Li, H.D.; Zeng, M.M.; Tan, B.-B.; Liang, Y.Z.; Xu, Q.S.; Cao, D.S. Recipe for revealing informative metabolites based on model population analysis. *Metabolomics* **2010**, *6*, 353–361. [CrossRef]
75. Pan, T.; Chen, W.; Chen, Z.; Xie, J. Wavelength selection for NIR spectroscopic analysis of chemical oxygen demand based on different partial least squares models. *Key Eng. Mater.* **2011**, *480–481*, 393–396. [CrossRef]
76. Zou, X.; Zhao, J.; Li, Y. Selection of the efficient wavelength regions in FT-NIR spectroscopy for determination of SSC of "Fuji" apple based on BiPLS and FiPLS models. *Vib. Spectrosc.* **2007**, *44*, 220–227. [CrossRef]
77. Yang, Q.; Zhu, G.; Ren, P.; Long, S.; Yang, J. Wavelength selection for NIR spectroscopic analysis of chemical oxygen demand based on different partial least squares models. *J. Anal. Sci.* **2016**, *32*, 485–489.
78. Fan, S.; Zhang, B.; Li, J.; Huang, W.; Wang, C. Effect of spectrum measurement position variation on the robustness of NIR spectroscopy models for soluble solids content of apple. *Biosyst. Eng.* **2016**, *143*, 9–19. [CrossRef]
79. Saeys, Y.; Inza, I.; Larrañaga, P. A review of feature selection techniques in bioinformatics. *Bioinformatics* **2007**, *23*, 2507–2517. [CrossRef]
80. Lê Cao, K.A.; Rossouw, D.; Robert-Granié, C.; Besse, P. A sparse PLS for variable selection when integrating omics data. *Stat. Appl. Genet. Mol. Biol.* **2008**, *7*. [CrossRef] [PubMed]
81. Bax, E.; Weng, L.; Tian, X. Speculate-correct error bounds for k-nearest neighbor classifiers. *Mach. Learn.* **2019**, *108*, 2087–2111. [CrossRef]

82. Grove, A.J. General Convergence Results for Linear Discriminant Updates. *Mach. Learn.* **2001**, *43*, 173–210. [CrossRef]
83. AbuZeina, D.; Al-Anzi, F.S. Employing fisher discriminant analysis for Arabic text classification. *Comput. Electr. Eng.* **2018**, *66*, 474–486. [CrossRef]
84. Pedregosa, F.; Varoquaux, G.; Gramfort, A.; Michel, V.; Thirion, B.; Grisel, O.; Blondel, M.; Prettenhofer, P.; Weiss, R.; Dubourg, V.; et al. Scikit-learn: Machine Learning in Python. *J. Mach. Learn. Res.* **2011**, *12*, 2825–2830.
85. Suarez, L.A.; Apan, A.; Werth, J. Detection of phenoxy herbicide dosage in cotton crops through the analysis of hyperspectral data. *Int. J. Remote Sens.* **2017**, *38*, 6528–6553. [CrossRef]
86. Zhou, Q.; Huang, W.; Fan, S.; Zhao, F.; Liang, D.; Tian, X. Non-destructive discrimination of the variety of sweet maize seeds based on hyperspectral image coupled with wavelength selection algorithm. *Infrared Phys. Technol.* **2020**, *109*, 103418. [CrossRef]
87. Xia, Z.; Zhang, C.; Weng, H.; Nie, P.; He, Y. Sensitive Wavelengths Selection in Identification of Ophiopogon japonicus Based on Near-Infrared Hyperspectral Imaging Technology. *Int. J. Anal. Chem.* **2017**, *2017*, 1–11. [CrossRef]
88. Snavely, N.; Seitz, S.M.; Szeliski, R. Modeling the world from Internet photo collections. *Int. J. Comput. Vis.* **2008**, *80*, 189–210. [CrossRef]
89. Mesas-Carrascosa, F.J.; de Castro, A.I.; Torres-Sánchez, J.; Triviño-Tarradas, P.; Jiménez-Brenes, F.M.; García-Ferrer, A.; López-Granados, F. Classification of 3D point clouds using color vegetation indices for precision viticulture and digitizing applications. *Remote Sens.* **2020**, *12*, 317. [CrossRef]
90. Mesas-carrascosa, F.J.; Rumbao, I.C.; Alberto, J.; Berrocal, B.; Porras, A.G. Positional Quality Assessment of Orthophotos Obtained from Sensors Onboard Multi-Rotor UAV Platforms. *Sensors* **2014**, *14*, 22394–22407. [CrossRef] [PubMed]
91. Barreto, M.A.P.; Johansen, K.; Angel, Y.; McCabe, M.F. Radiometric assessment of a UAV-based push-broom hyperspectral camera. *Sensors* **2019**, *19*, 4699. [CrossRef] [PubMed]
92. van Zyl, J.J. The Shuttle Radar Topography Mission (SRTM): A breakthrough in remote sensing of topography. *Acta Astronaut.* **2001**, *48*, 559–565. [CrossRef]
93. Van Rossum, G.; Drake, F.L. *The Python Language Reference*; Python Software Foundation: Scotts Valley, CA, USA, 2011; p. 109.
94. Degenhardt, F.; Seifert, S.; Szymczak, S. Evaluation of variable selection methods for random forests and omics data sets. *Brief. Bioinform.* **2019**, *20*, 492–503. [CrossRef]
95. Tibshirani, R. Regression Shrinkage and Selection via the Lasso. *J. R. Stat. Soc.* **1996**, *58*, 267–288. [CrossRef]
96. Jain, A. A Complete Tutorial on Ridge and Lasso Regression in Python. Available online: https://www.analyticsvidhya.com/blog/2016/01/complete-tutorial-ridge-lasso-regression-python/ (accessed on 5 August 2020).
97. Kursa, M.B.; Rudnicki, W.R. Feature selection with the boruta package. *J. Stat. Softw.* **2010**, *36*, 1–13. [CrossRef]
98. Alba-Fernández, M.V.; Ariza-López, F.J.; Rodríguez-Avi, J.; García-Balboa, J.L. Statistical methods for thematic-accuracy quality control based on an accurate reference sample. *Remote Sens.* **2020**, *12*, 816. [CrossRef]
99. R Core Team. *R: A Language and Environment for Statistical Computing*; R Core Team: Vienna, Austria, 2018.
100. RStudio Team. *RStudio: Integrated Development for R*; RStudio: Boston, MA, USA, 2017.
101. Mevik, B.H.; Wehrens, R. The pls package: Principal component and partial least squares regression in R. *J. Stat. Softw.* **2007**, *18*, 1–23. [CrossRef]
102. Northrop, P.J.; Attalides, N.; Jonathan, P. Cross-validatory extreme value threshold selection and uncertainty with application to ocean storm severity. *J. R. Stat. Soc. Ser. C Appl. Stat.* **2017**, *66*, 93–120. [CrossRef]
103. Kucheryavskiy, S. Mdatools—R package for chemometrics. *Chemom. Intell. Lab. Syst.* **2020**, *198*. [CrossRef]
104. Li, H.D.; Xu, Q.S.; Liang, Y.Z. LibPLS: An integrated library for partial least squares regression and linear discriminant analysis. *Chemom. Intell. Lab. Syst.* **2018**, *176*, 34–43. [CrossRef]
105. Chun, H.; Keleş, S. Sparse partial least squares regression for simultaneous dimension reduction and variable selection. *J. R. Stat. Soc. Ser. B Stat. Methodol.* **2010**, *72*, 3–25. [CrossRef]
106. Friedman, J.; Trevor, H.; Tibshirani, R. Regularization paths for generalized linear models via coordinate descent. *J. Stat. Softw.* **2010**, *33*, 1–22. [CrossRef]
107. Calderón, R.; Navas-Cortés, J.A.; Lucena, C.; Zarco-Tejada, P.J. High-resolution airborne hyperspectral and thermal imagery for early detection of Verticillium wilt of olive using fluorescence, temperature and narrow-band spectral indices. *Remote Sens. Environ.* **2013**, *139*, 231–245. [CrossRef]

108. Zarco-Tejada, P.J.; Camino, C.; Beck, P.S.A.; Calderon, R.; Hornero, A.; Hernández-Clemente, R.; Kattenborn, T.; Montes-Borrego, M.; Susca, L.; Morelli, M.; et al. Previsual symptoms of Xylella fastidiosa infection revealed in spectral plant-trait alterations. *Nat. Plants* **2018**, *4*, 432–439. [CrossRef]
109. Starzacher, A.; Rinner, B. Evaluating KNN, LDA and QDA classification for embedded online feature fusion. In Proceedings of the 2008 International Conference on Intelligent Sensors, Sensor Networks and Information Processing, Sydney, NSW, Australia, 15–18 December 2008; IEEE: Piscataway Township, NJ, USA, 2008; pp. 85–90.

5

A Precision Agriculture Approach for Durum Wheat Yield Assessment using Remote Sensing Data and Yield Mapping

Piero Toscano [1], Annamaria Castrignanò [2], Salvatore Filippo Di Gennaro [1,*], Alessandro Vittorio Vonella [2], Domenico Ventrella [2] and Alessandro Matese [1]

1 Institute of BioEconomy (IBE), National Research Council (CNR), Via Caproni 8, 50145 Florence, Italy
2 Council for Agricultural Research and Economics, Research Centre for Agriculture and Environment (CREA-AA), Via Celso Ulpiani 5, 70125 Bari, Italy
* Correspondence: salvatorefilippo.digennaro@cnr.it

Abstract: The availability of big data in agriculture, enhanced by free remote sensing data and on-board sensor-based data, provides an opportunity to understand within-field and year-to-year variability and promote precision farming practices for site-specific management. This paper explores the performance in durum wheat yield estimation using different technologies and data processing methods. A state-of-the-art data cleaning technique has been applied to data from a yield monitoring system, giving a good agreement between yield monitoring data and hand sampled data. The potential use of Sentinel-2 and Landsat-8 images in precision agriculture for within-field production variability is then assessed, and the optimal time for remote sensing to relate to durum wheat yield is also explored. Comparison of the Normalized Difference Vegetation Index(NDVI) with yield monitoring data reveals significant and highly positive linear relationships (r ranging from 0.54 to 0.74) explaining most within-field variability for all the images acquired between March and April. Remote sensing data analyzed with these methods could be used to assess durum wheat yield and above all to depict spatial variability in order to adopt site-specific management and improve productivity, save time and provide a potential alternative to traditional farming practices.

Keywords: yield mapping; remote sensing; durum wheat; precision agriculture

---

1. Introduction

Durum wheat (*Triticum durum*, Desf.), although it represents only 8% of global wheat production, is one of the most common cereal crops in the Mediterranean basin, traditionally grown under rainfed conditions using conventional tillage [1–3]. Climate variability, price volatility and socio-economic factors are the main sources of uncertainty and concern for farmers in durum wheat cultivation [4,5]. For climate variability, it was shown how, in rainfed conditions, these affected both the quality and quantity of durum wheat production [6]. For other crops, evidence was given by Bowman and Zilberman [7] of how both price volatility and socio-economic factors might influence the agronomic techniques adopted. In light of this, it is increasingly urgent to provide information to optimize crop management and estimate crop yields before harvest for a sustainable agricultural income and ensuring food security. Precision agriculture (PA) has been used for ≥25 years to optimize the use of farm inputs such as fertilizers and herbicides [8,9], and thus maximize profit and minimize negative environmental impacts [10] by addressing spatial variability.

Yield mapping is one of the most widely-used precision agriculture techniques [11–13]. Most of these datasets are characterized by a non-normal distribution due to the presence of errors and outliers and can be misleading if used for decision making processes. Over the last 25 years, several studies have analyzed the sources of errors that cause this non-normality and proposed different processing techniques to reduce their effect [11,14,15]. However, among the studies highlighting accuracy issues associated with the use of the yield monitoring systems, only one compared yield monitoring with hand sampled data [16].

During the same period many studies have been published using satellite imagery to estimate crop parameters and yields [17–19], many of these using empirical relationships between yields and various vegetation indices (VIs) with limited applicability to different areas or years [9], especially in the recent era of prolific satellite data availability [20].

In the last years many studies have focused on the use of medium resolution satellite data (10–100 m) for yield estimation at broader spatial resolution (local, regional, country scales) even for long-term yield series analysis [20–26]. While studies are conducted by the use of very high resolution imagery [27–30] to identify within-field variability of crop growth and yield and for the definition of management zones, few [31] have used Sentinel-2 to provide an insight into field productivity variation for better future management [32–34]. The objectives of this study were to:

- Evaluate the correctness of yield monitoring maps comparing them with hand sampled yield data;
- Evaluate the ability of the most commonly used VI (NDVI) calculated from Landsat-8 and Sentinel-2 satellite platforms to understand within-field variability;
- Understand the optimal time for NDVI acquisition for better yield evaluation;
- Evaluate the relations between NDVI and yield for four durum wheat crop seasons with different climatic conditions and yield performance.

For the last two points, given the scarcity of satellite images, a durum wheat simulation model [5] was used to reconstruct the crop growth variables and analyze in detail the differences that emerged in the yield-NDVI correlation for the four crop seasons.

2. Materials and Methods

2.1. *Study Site and Field Trial*

The research was performed at the Menichella Experimental Farm of CREA-AA (Council for Agricultural Research and Economics—Research Centre for Agriculture and Environment), located in the Foggia countryside (Southern Italy, 41°27′05.9″ N, 15°30′43.6″ E; 88 m a.s.l.), within the study area of the JECAM site (http://jecam.org/studysite/italy-apulian-tavoliere/), during the 2013–2014, 2014–2015, 2015–2016 and 2016–2017 crop seasons. This study was conducted on a 5 ha field cropped with rainfed durum wheat (*Triticum durum*, Desf., cv Claudio) under conventional management and continuous cultivation.

The field is in a flat area called 'Apulian Tavoliere' and the soil is silty-clay Vertisol of alluvial origin classified as Fine Mesic Typic Cromoxerert by Soil Taxonomy USDA [35].

The soil in the upper 60 cm layer has a good availability of total nitrogen (0.12 g 100 g^{-1}), organic matter (2.07 g 100 g^{-1}) and 41 mg kg^{-1} of available phosphorus (P_2O_5). In summer 4–5 cm wide cracks frequently appear from the surface to about 50 cm depth.

The climate is classified as Mediterranean subtropical with a thermic soil temperature regime. Rainfall, unevenly distributed throughout the seasons and with a long-term annual average of 550 mm, is mostly concentrated in the winter months, while the dry period is from May to September [36]. Daily weather parameters (air temperature, relative humidity, global solar radiation, rainfall and wind speed), were recorded at the agro-meteorological station of the CNR-IBE weather station network (Foggia, 41°30′00.4″ N, 15°30′46.4″ E, 69 m a.s.l.). The field has been cultivated with a common agronomic management, applying 36 kg ha^{-1} of N as diammonium phosphate (18–46) before sowing and 68.4 kg ha^{-1} as ammonium nitrate (34.2) as top dressing at the end of tillering stage.

Sowing date varied between November and December due to the weather conditions (12 December 2013, 20 November 2014, 19 November 2015 and 29 November 2016). In each cropping season, the sowing density was of 350 germinable seeds/m^2 with 15 cm row spacing.

2.2. Hand Yield Samplings and Yield Map Monitoring

At maturity stage of each season aboveground biomass was collected over 1 m^2 areas in proximity to the 104 sampling points at the nodes of a 20 × 20 m cell-grid (Figure 1). The points were georeferenced in UTM coordinates WGS 84 using a TOPCON GPS, differentially corrected with an accuracy of less than one meter. All measurements were repeated at the same points over the years.

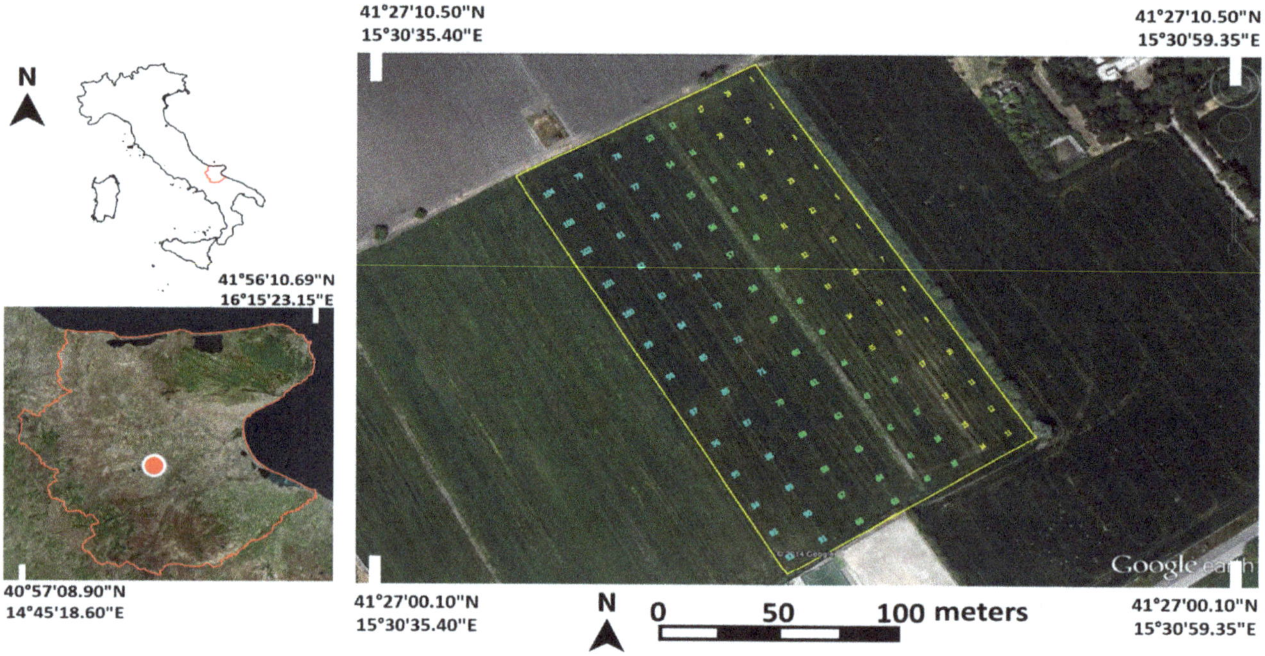

Figure 1. Location of study site and 104 sampling points.

At harvesting in 14 July 2014, 29 June 2015 and 12 July 2016, yield data were recorded by external services provider with a John Deere T670i combine (Deere & Company, Moline, USA) equipped with a yield monitor system (grain mass flow and moisture sensors). The data were recorded every second, which produced a support (footprint) of 6 × 1 m^2 depending on the forward speed of the machine.

For 2017, the last year of analysis, a yield monitoring map was unavailable. The data were measured only at the 104 sampling points.

2.3. Satellite Data

Remote sensing images acquired by the Operational Land Imager (OLI) instrument aboard the Landsat-8 satellite and by the Multi-Spectral Instrument (MSI) aboard the Sentinel-2A satellite were used in the study. Landsat-8/OLI captures images of the earth's surface in nine spectral bands at 30 m spatial resolution (15 m for panchromatic band) while Sentinel-2A/MSI captures images in 13 spectral bands at 10 m, 20 m and 60 m spatial resolution. After cloud and shadow screening, a total of 11 Landsat-8 and five Sentinel-2A (Table 1 images of the study area from 1 March 2013 to 1 June 2017 were selected.

Table 1. Acquired dates of Landsat-8 and Sentinel-2 (2014–2017).

	LANDSAT-8	SENTINEL-2
2014	19 March, 20 April	NA
2015	14 April, 30 April	NA
2016	13 March, 9 April, 18 May, 27 May	23 May
2017	2 March, 12 April, 30 May	9 March, 29 March, 8 April, 18 May

The Landsat-8/OLI images were downloaded using USGS (earthexplore.com) that provides data corrected from atmospheric effects.

Sentinel-2A/MSI images were atmospherically corrected for surface reflectance using the European Space Agency's (ESA) Sen2Cor algorithm (http://step.esa.int/main/third-party-plugins-2/sen2cor), which processes ESA's Level-1C top-of-atmosphere reflectance to atmospherically-corrected bottom-of-atmosphere (BoA) reflectance (Level-2A).

Lastly, NDVI [37] was calculated for Landsat-8 images using band 5 (NIR) and band 4 (RED), and for Sentinel-2A images using band 8 (NIR) and band 4 (RED) according to the formula:

$$\text{NDVI} = (\text{NIR} - \text{RED})/(\text{NIR} + \text{RED}) \quad (1)$$

Providing two estimates with different support: 30×30 m^2 for Landsat-8 and 10×10 m^2 for Sentinel-2A.

2.4. Data Analysis

Yield map data were firstly normalized to 13% grain moisture content (hereinafter referred to as raw data) and then processed following the Vega et al. [13] protocol. The geographic coordinates of each dataset were converted into UTM Cartesian coordinates, specifying the zone (33, north) and the ellipsoid (WGS84). The 3 years maps in shapefile format (SHP) were pre-processed following a workflow using GeoDa software [38] for Moran index calculation for outliers identification, QGIS [39] for vegetation indices (VIs) calculation, Vesper software [40] for geostatistical interpolation, Matlab [41] for data statistical analysis.

Yield monitoring data underwent a pre-processing procedure in order to automatically identify and delete incorrect values through the two following steps suggested by Vega et al. [13].

Step 1: A threshold was applied by removing the yield values of less than 0.1 t/ha and then yield data points up to 10 m from the edge were removed in order to avoid edge effects. Lastly, yield data out of mean ±3 SD were automatically detected and deleted. This filter was used to prevent changes in inflation as a result of an incorrect estimate of very low data.

Step 2: Moran's local index of spatial autocorrelation and Moran's plot were applied to detect spatial outliers [42,43].

Lastly, the yield map for each year and dataset was assessed separately (raw data, Step1 and Step1 + Step2) by means of ordinary kriging, evaluating spatial variability using a semivariogram of the variable (yield monitoring).

Vesper software was used and an exponential model was fitted to the experimental variogram and model parameters: Nugget (micro-scale variation or measurement error), sill (asymptotic value approximately corresponding to sample variance) and effective range (distance at which 95% sill is reached) were estimated.

Yield values from the prediction map based on raw yield data and Step1 + Step2 yield data were then extracted in the neighborhood (3 m radius) of 104 yield sampled grid points and compared with them.

Step1 + Step2 yield monitoring data were interpolated using block kriging over a block of 10×10 m^2 or 30×30 m^2 to assess the spatial relationship between yield and the two types of remote sensing data (Sentinel-2A and Landsat-8, respectively) and to report on the optimal timing at which spectral measurements should be taken in durum wheat to maximize the correlation with yield (2014, 2015, 2016). The same procedure was adopted for hand yield samplings (block of 10×10 m^2 or 30×30 m^2) to assess the spatial relationship between yield and remote sensing data for 2017.

The Pearson coefficient, which is an index that measures the degree of correlation between linearly related variables and ranges between −1 and +1, was used to assess the linear relationship between yield monitoring data and yield sampling data, and between the NDVI and yield monitoring data.

For each regression analysis the correlation coefficient, regression coefficients (intercept and slope) and their corresponding probability levels were estimated, to test the statistical significance (null hypothesis equal to zero).

The performance of the protocol for automating error removal from yield monitoring data was evaluated using the root-mean-square error (RMSE).

2.5. Modelling

To better understand the correlations between remote sensing data and yield and go into detail about crop development and spectral signature, we used the Delphi crop growth model to perform field-scale simulations. The Delphi model was chosen due to the recent validations of its ability to simulate crop growth, yield and product quality conducted in the same study area [1,5]. The Delphi model is based on a FORTRAN-based mechanistic model [44] calibrated for durum wheat in Mediterranean conditions. Plant transpiration and soil evaporation, water and nitrogen soil-plant cycle are incorporated in the model. Input weather data at daily time scale are: Air temperature (maximum, minimum and average), global shortwave radiation, rainfall, wind speed (average) and relative humidity (average). Input data of the main physiological parameters of the durum wheat cultivar, sowing date and number of seeds/m^2, the soil hydrological profile, soil total nitrogen content profile, agronomic data on quality and quantity of nitrogen and roots growth data are also required.

Due to the strong correlation between leaf area index (LAI) and aboveground dry biomass [45], because aboveground dry biomass of plants generally determines LAI, the Delphi model was implemented to calculate LAI for each crop season. The model also predicted the heading, anthesis, maturity dates and length of time between these phases.

These pieces of information were used for detailed analysis of interannual variability and to better understand the different relationships between yield and NDVI over the years.

The weather data input to perform the Delphi simulation were acquired by the weather station located at Foggia, 41°30′00.4″ N, 15°30′46.4″ E, 69 m a.s.l., while sowing date, number of seeds/m^2 and nitrogen fertilization application data were set according to the data reported in Section 2.1. No changes were made to the Delphi model, as it had already been calibrated, validated and tested over 11 crop seasons for this region [5].

3. Results and Discussion

3.1. Yield Map and Yield Sample

A yield map is the basis for understanding yield variability within a field, analyzing its causes and improving management to increase profit [46].

A number of errors may be associated with common yield data collection: The yield monitor may not shut off at the field end and will register 0 value until harvestable crop again moves into the combine; the combine grain-flow system may plug temporarily, especially if the crop has lodged or weeds interfere with continuous grain flow; a time lag can occur between the time the crop is cut and the time its yield is measured in the grain flow [14,15,47]. Researchers have reported that 10% to 50% of observations reveal measurement errors [13,48].

The raw yield data of all three crop seasons used in this study showed high positive skewness coefficient (Table 2 and Figure 2). However, after Step1 and Step1 + Step2, the yield probability distributions were practically symmetric and the statistics were not biased by the presence of atypical data. After removing statistical outliers, the variable distribution tended to be more symmetric, without affecting the spatial structuring.

Table 2. Statistical features of yield monitoring datasets for uncleaned and cleaned yield data. Yield monitoring vs. yield sampling based on interpolated data.

Year	Number of Yield Monitoring Data			Skewness			Yield Monitoring vs. Yield Sampling	
	RAW	Step1	Step2	RAW	Step1	Step2	RAW	Step1 + Step2
2014	6369	4703 −26.2%	4641 −1.32%	17.72	2.72	2.71	$r = 0.11$ p-value = 0.25 RMSE = 1.14 t/ha	$r = 0.40$ p-value < 0.0001 RMSE = 1.05 t/ha
2015	6351	4737 −25.4%	4648 −1.88%	21.58	0.65	0.44	$r = 0.37$ p-value < 0.0001 RMSE = 0.68 t/ha	$r = 0.50$ p-value < 0.0001 RMSE = 0.59 t/ha
2016	6699	5084 −24.1%	4967 −2.30%	23.94	0.39	0.38	$r = 0.43$ p-value < 0.0001 RMSE = 0.84 t/ha	$r = 0.49$ p-value < 0.0001 RMSE = 0.82 t/ha

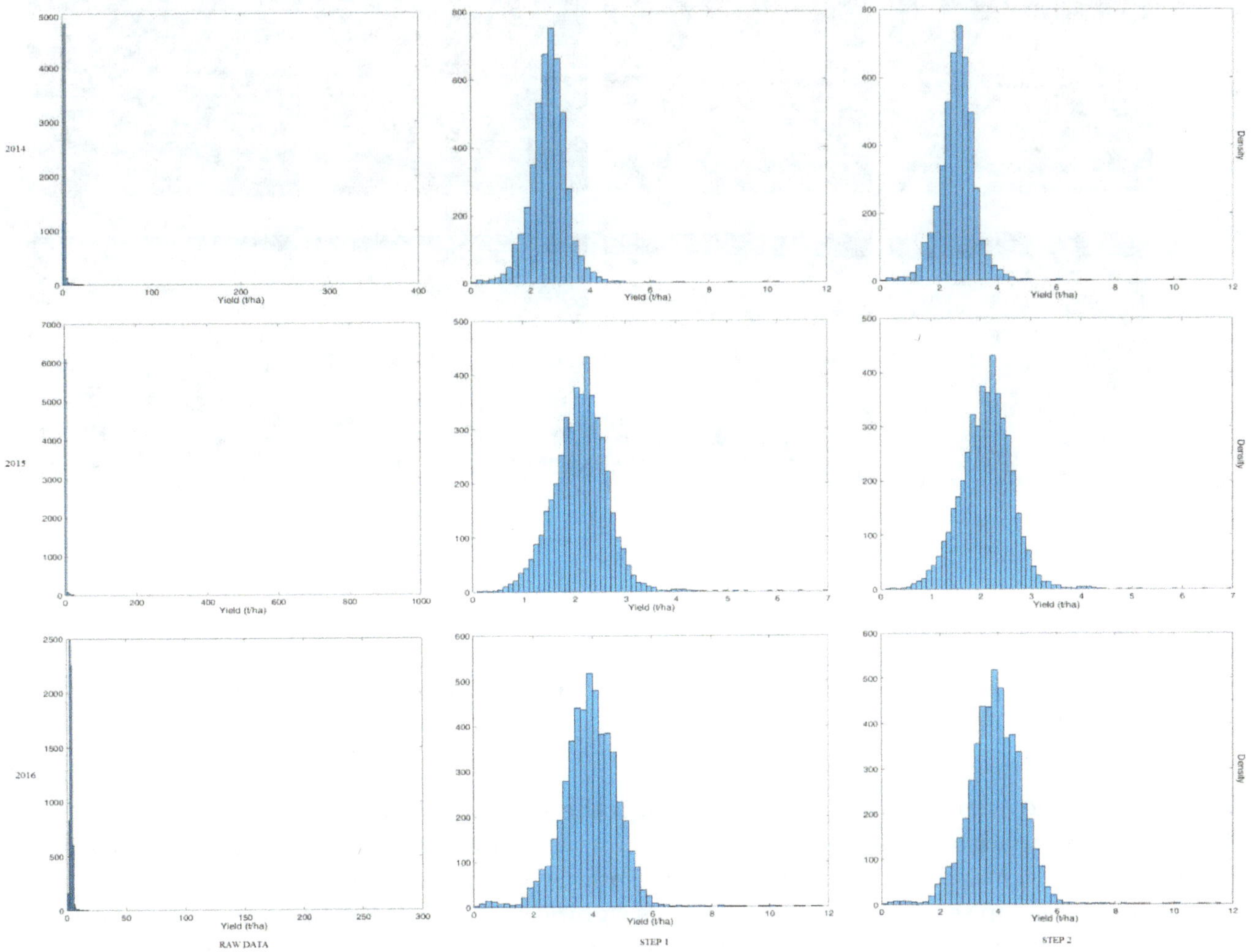

Figure 2. Distribution of yield monitoring datasets for uncleaned and cleaned yield data (data not interpolated).

In fact, after Step1 + Step2 the skewness had values close to zero for 2015 and 2016 crop seasons (Table 2), while for 2014 the final skewness of 2.71 was due to a longer tail on the right side of the data distribution. After Step1 between 24.1% and 26.2% of points were removed, whilst after Step2 a further 1.3% to 2.3% of the dataset was removed. Following the Step1 + Step2 protocol, the percentage of points removed was from 25.9% to 27.1%, close to the range previously reported in the literature [13,47–50].

Figure 3 presents a visual analysis of the location of data removed from each year after both Step1 and Step2: First of all, the highest quantity of removed points came from the filtering of edges, secondly all the data points with overlapping coordinates and lastly, data points identified as outliers through local Moran's index of spatial autocorrelation (Step2).

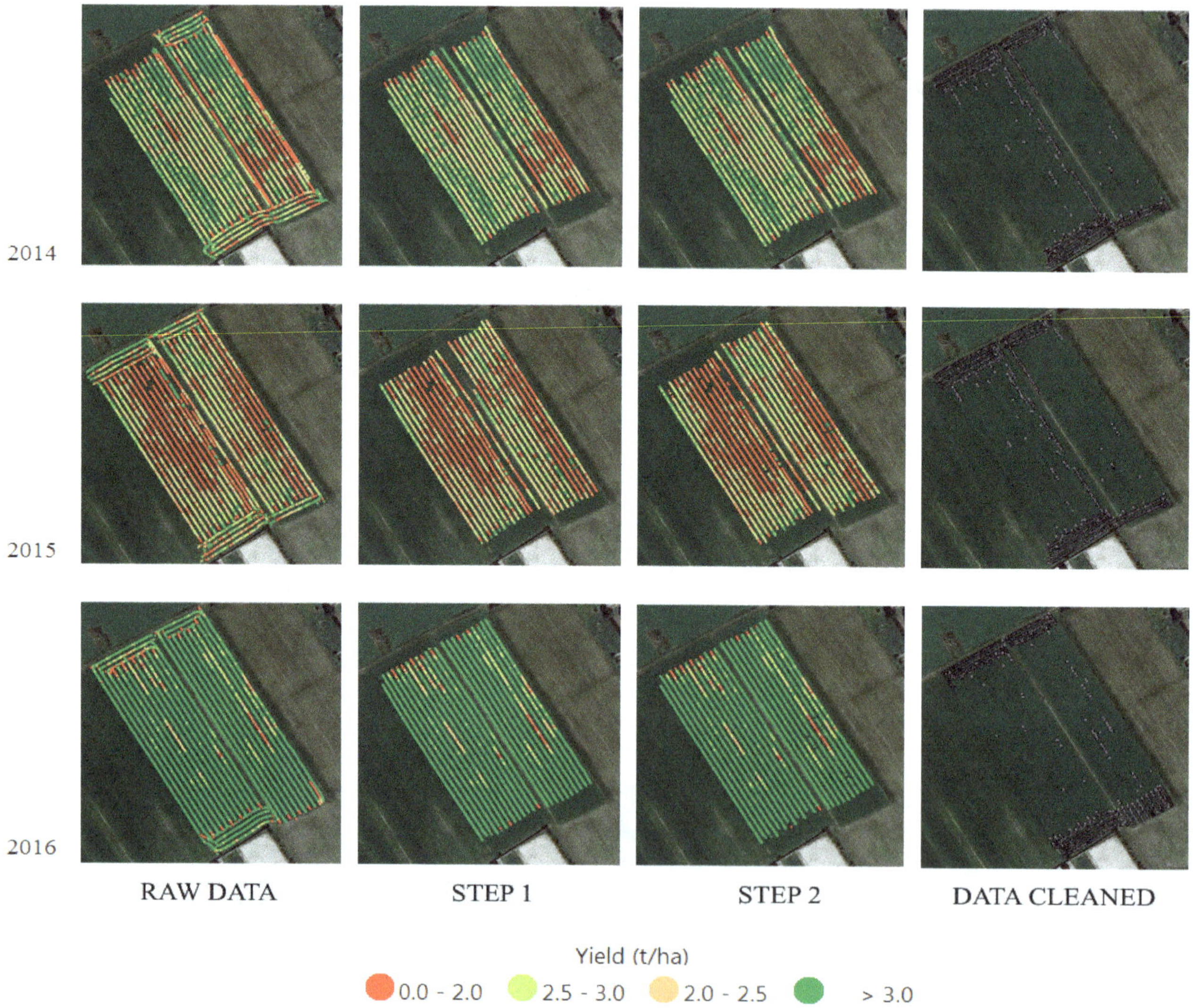

Figure 3. Maps showing yield monitoring datasets: Raw data (left), Step1 and Step2 data (center), data cleaned (right). Raw and Steps 1–2 maps are colored according to quartiles of yield distributions (data not interpolated).

Certainly, the cleaning protocol did not affect the main patterns present in the raw data and allowed both comparison with the sampled yield data and, after interpolation, comparison with the data observed by Landsat-8 (at 30 m of resolution) and Sentinel-2 (at 10 m).

For all three crop seasons, the correlation, its significance and RMSE improved in passing from raw yield monitoring vs. sampled yield to the comparison Step1 + Step2 yield monitoring vs. sampled yield (Table 2). The best improvement was achieved for 2014, where from being non-significant, we found a significant correlation and with a reduction of the RMSE which, however, remained the highest compared to other years. For both 2015 and 2016, the initial correlations between raw data and sampled data were significant and with low RMSE. In any case, the data cleaning procedures improved the performances in terms of both correlation and RMSE.

The different behavior of the 2014 season compared to the other two may be due to intense weeding during the first year of the trial, which caused a large within-field variability of yield.

Given the nature of the comparison between two sets of data that do not refer strictly to the same harvested plants, the worst results obtained for 2014 seem to be clearly linked to the greater variability of within-field yield, unlike crop seasons with uniformly low within-field yield (2015) or uniformly high yield (2016) (Table 3). This corresponds well with Arslan and Colvin [51], who reported that high accuracy cannot be achieved by spot measurements; however, the overall yield trend can be determined. The correlations between spatial data are strongly scale-dependent [52] and of course depend on the coincidence of location between the sampled and monitored data, that in our study is not possible since the hand yield sampling was destructive (before yield mapping).

Table 3. Comparison between yield (t/ha) sampled and monitoring data (mean, max, min and std) (2014–2017).

	2014		**2015**		**2016**		**2017**	
	Sample (t/ha)	**Monitor (t/ha)**	**Sample (t/ha)**	**Monitor (t/ha)**	**Sample (t/ha)**	**Monitor (t/ha)**	**Sample (t/ha)**	**Monitor (t/ha)**
Mean	3.07	2.65	2.31	2.11	3.88	3.90	5.00	N/A
Min	0.42	0.18	0.87	0.15	2.15	0.12	3.23	N/A
Max	5.69	10.48	3.64	6.90	6.01	11.62	7.43	N/A
Std	1.01	1.12	0.57	0.53	0.71	0.91	0.94	N/A

The correlation coefficients found for all three crop seasons (ranging from 0.40 to 0.50) were in general lower but similar to the findings of Ingeli et al. [16]. The latter is the only published paper to have compared two sources of yield data, the hand sampled data as independent variable and yield monitoring data as dependent variable. For five different crop seasons, the authors found correlations ranging between 0.3 to 0.9 but reporting on a small number of hand sampled data (18 + 3 replications) spread over a larger field area (16 ha) unlike our case study with 104 samples in a smaller area (5 ha).

3.2. Yield and NDVI

For the whole period, few Landsat-8 and Sentinel-2 images were used because most of those acquired were useless due to the presence of clouds. For the first two years (2014 and 2015) the images are limited to two in a fairly short time window and in any case always before crop heading stage. In 2016 there are four useful images spread over a much wider period (March–May), while three images are available in 2017 for the same period.

The scientific and operational life of Sentinel-2 started in July 2015, so the useful passages only relate to 2016 and 2017. In 2016 it is possible to use only one image and it was taken at the end of May. In 2017 there are four useful images in a time window similar to that of the Landsat-8 images for 2017 (Table 3).

A comparison of the NDVI with yield monitoring values (Step1 + Step2) (for 2014–2015–2016, Figures 4–7) reveals significant positive linear relationships (r ranging from 0.54 to 0.74) explaining most of the within-field variability in 2014 with the image acquired in April ($R^2 = 0.55$) and in 2016 with the image acquired in March ($R^2 = 0.55$). In all other cases, although the correlations are significant, R^2 are lower than 0.5.

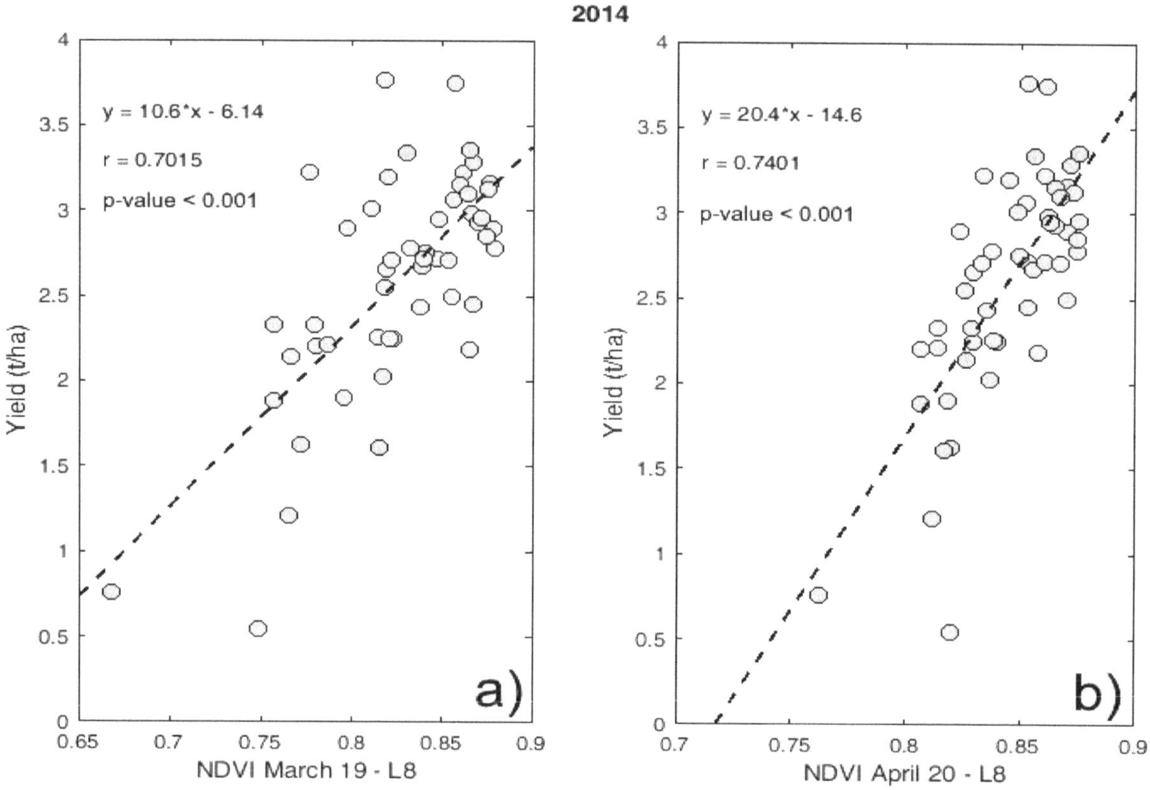

Figure 4. Relationship between observed yield (yield monitor) and NDVI (Landsat-8) for 2013–2014 crop season in (**a**) 19 March and (**b**) 20 April.

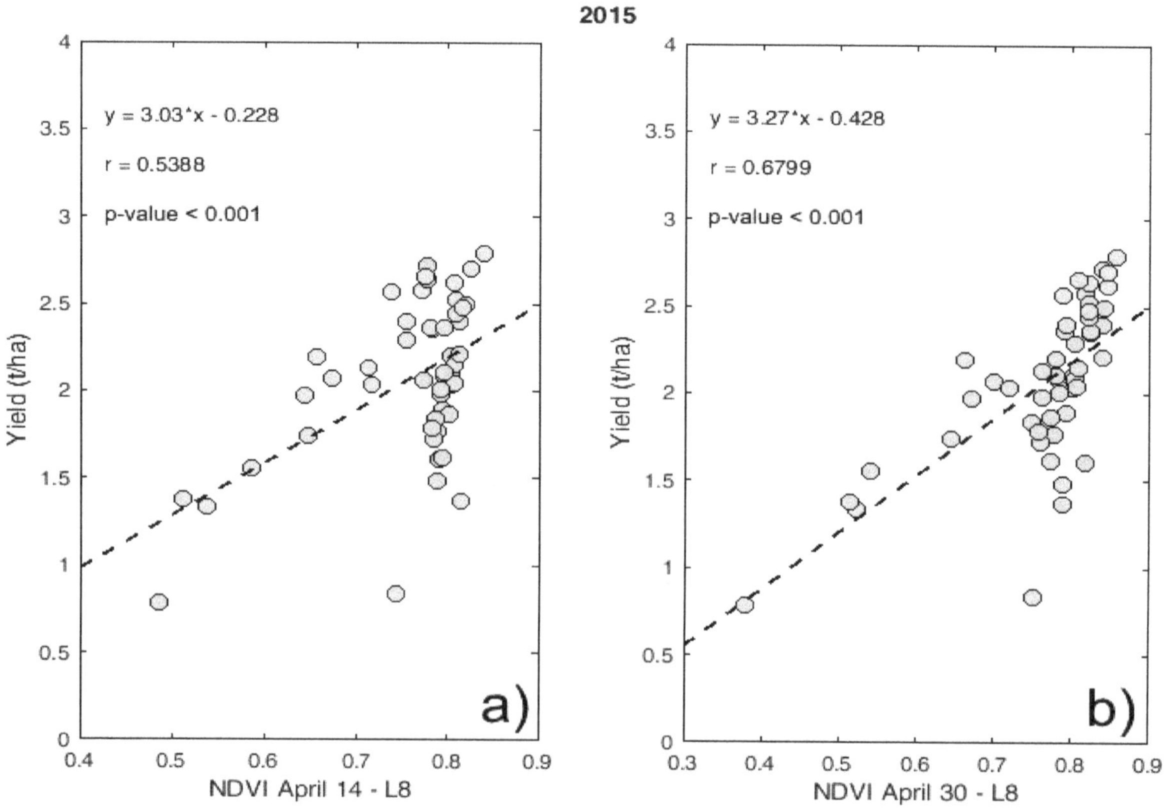

Figure 5. Relationship between observed yield (yield monitor) and NDVI (Landsat-8) for 2014–2015 crop season in (**a**) 14 April and (**b**) 30 April.

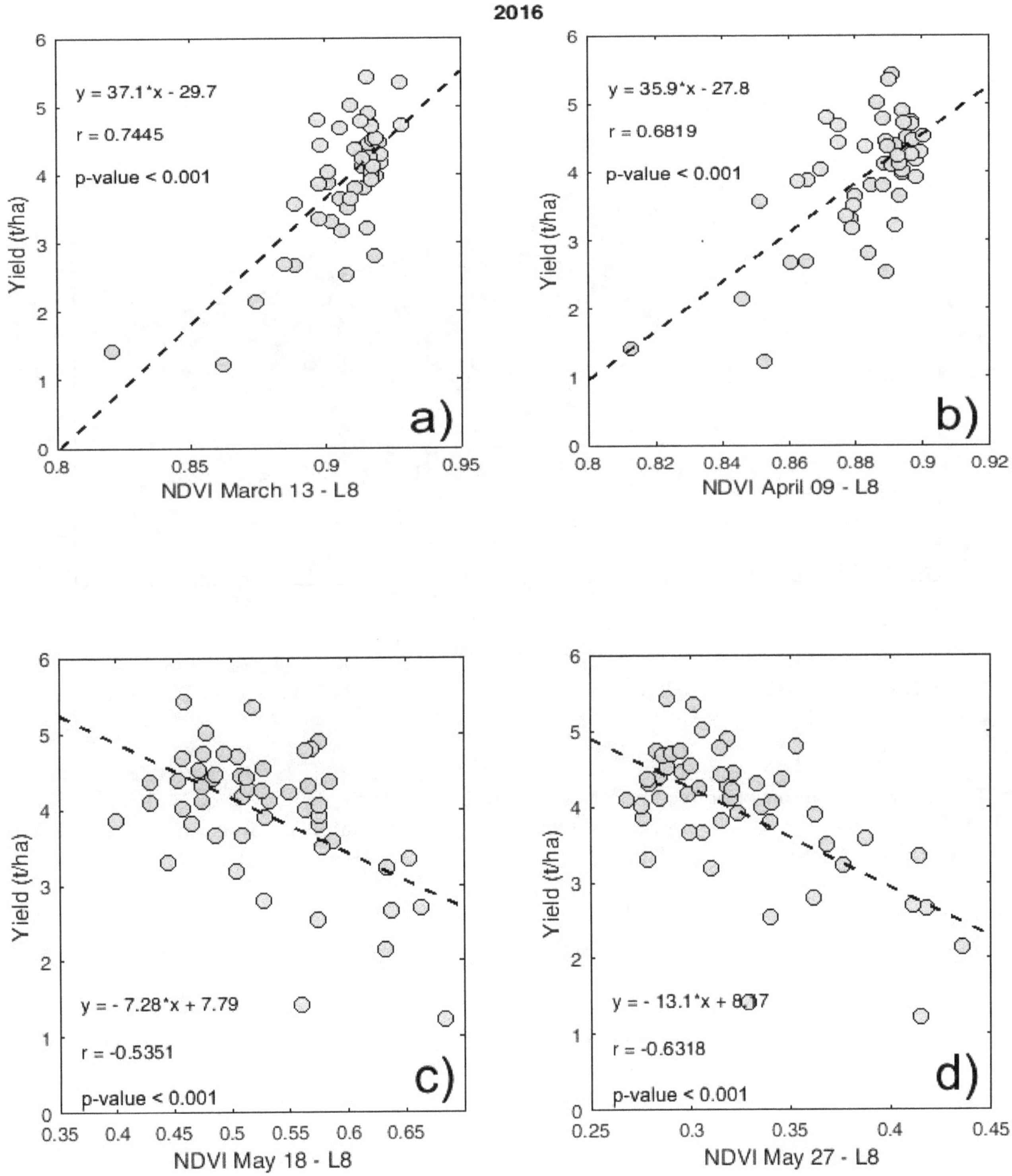

Figure 6. Relationship between observed yield (yield monitor) and NDVI (Landsat-8) for 2015–2016 crop season in (**a**) 13 March, (**b**) 9 April, (**c**) 18 May and (**d**) 27 May.

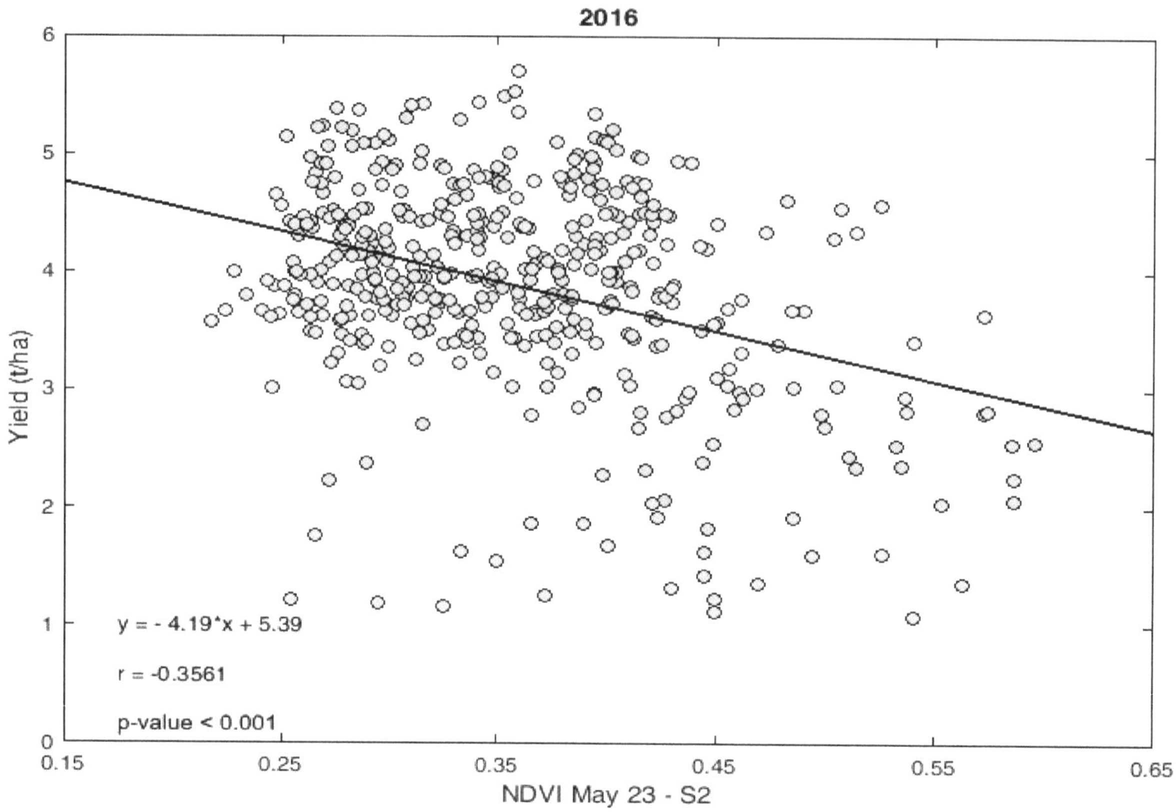

Figure 7. Relationship between observed yield (yield monitor) and NDVI (Sentinel-2) for 2015–2016 crop season.

These results, in terms of both correlations and timing, are in line with Mahey et al. [53]. Freeman et al. [54] found NDVI and wheat grain yield to be highly correlated, establishing the potential to predict yield with remotely sensed data as reported subsequently in several studies for a variety of crop types [55–58].

Freeman et al. [54] also indicated that yield estimates for wheat may be made two months prior to harvest.

Instead, only for 2016, from post-flowering to grain filling, we report weaker but significant negative correlations between NDVI and yield. This is for the Landsat-8 (18 and 27 May, Figure 6) images and the only one available for Sentinel-2 (23 May, Figure 7).

Although negative correlations between NDVI and crop yield are reported in the literature for potato late in the season [55] and for canola, after bolting and once the plants start transitioning to the reproductive stages [59], there are few similar findings for cereal crops when analyzing single or multi cultivars [60–63]. All these latter authors found a negative correlation under severe stress conditions, such as high temperature and drought, during grain filling.

Conversely, in our case study and for 2015–2016 crop season, this unique behavior of NDVI that from strongly positively correlation swings negative to more than −0.6 late in the season is mainly due to opposite climatic conditions (cool-moist) that characterized crop development and above all the period from heading to maturity (Table 4).

Table 4. Climate conditions, phenological length for heading to maturity (H to M) and anthesis to maturity (A to M), leaf area index (LAI) function derivative rate of change for the four crop seasons from anthesis to maturity.

		Number of Days	Total Rainfall (mm)	Air Temperature (°C)	LAI Rate of Change
2014	H to M	79	125.4	20.51	−0.099
	A to M	57	89.6	22.33	
2015	H to M	59	64.0	21.34	−0.142
	A to M	41	64.0	21.64	
2016	H to M	83	129.2	18.88	−0.085
	A to M	64	81.8	19.62	
2017	H to M	67	104.6	19.46	−0.131
	A to M	46	62.4	20.58	

In fact, the longest duration of heading to maturity and anthesis to maturity was observed in 2016 (Table 4) as well as the highest amount of rainfall from heading to maturity (129.2 mm) and lowest mean air temperature both from heading to maturity and anthesis to maturity (18.88 °C and 19.62 °C respectively). These conditions resulted in a general delayed leaf senescence and prolonged late grain filling as a sort of stay green effect [1,64] that is confirmed by the lowest decline rate of LAI in 2016 (Figure 8, Table 4). The derivatives of the modelled LAI function from heading to maturity has a rate of change for 2016 of −0.085, so is lower compared to other years. In the absence of water-stress, as for 2016, stay green is not always correlated with yield (in wheat [65,66] and in sorghum [67]) and can even be associated with reduced yield. For instance, in irrigated wheat and in rice in China, stay green was associated with slow export of leaf carbohydrate to the grain, increased lodging, and harvest difficulties due to delayed ripening, all of which can contribute to reduce yield [68,69]. In our case study, it is likely that this did not occur uniformly due to site-specific soil plant interactions, and the areas within the field that exhibited higher NDVI values during maturation then had translocation problems to the grains. This is confirmed by the fact that the areas with the lowest production (highest NDVI) in 2016 fall in the same low production areas as 2014 and 2015 (Figure 3). Contrary to what happened in 2014, the presence of weeds was not reported for 2016.

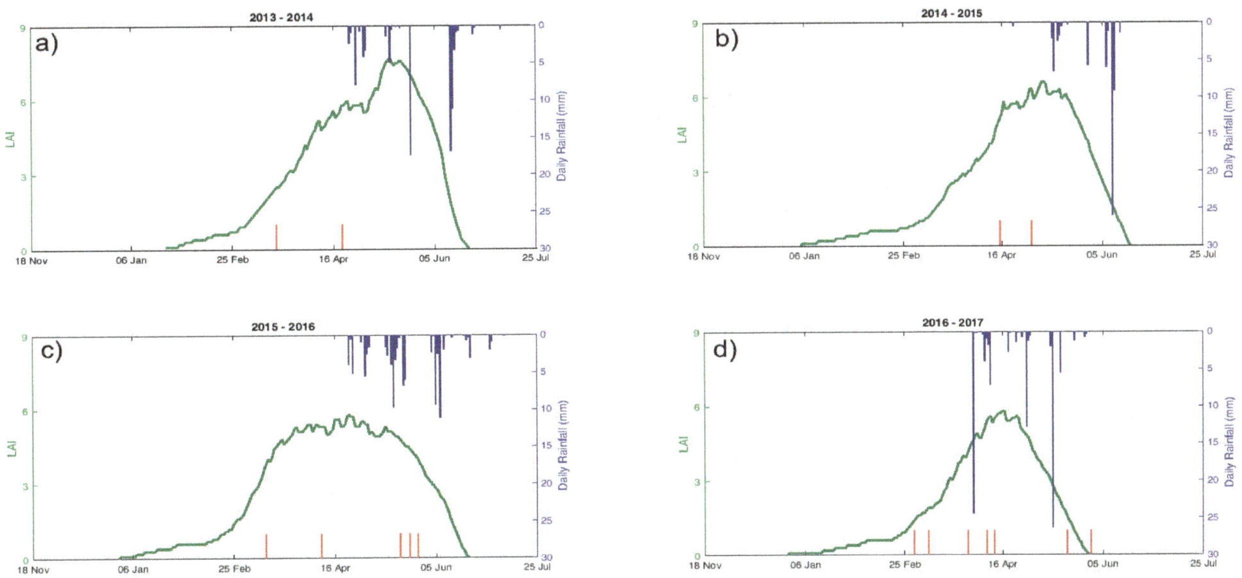

Figure 8. Modelled LAI (green), observed rainfall after heading (blue) and satellite observation (red) for (**a**) 2013–2014, (**b**) 2014–2015, (**c**) 2015–2016 and (**d**) 2016–2017 seasons.

For 2017, having no yield monitoring data, it was possible to compare NDVI data only with hand sampled data (Figures 9 and 10). Also, in this case the highest correlations are observed in the months of March and April, then tend to decrease in May (Sentinel-2, Figure 10) or become non-significant for the Landsat-8 passage in late May (Figure 9). The low and non-significant relationship in late May is probably due to the fact that the drying process had already started unevenly in some areas of the field [33].

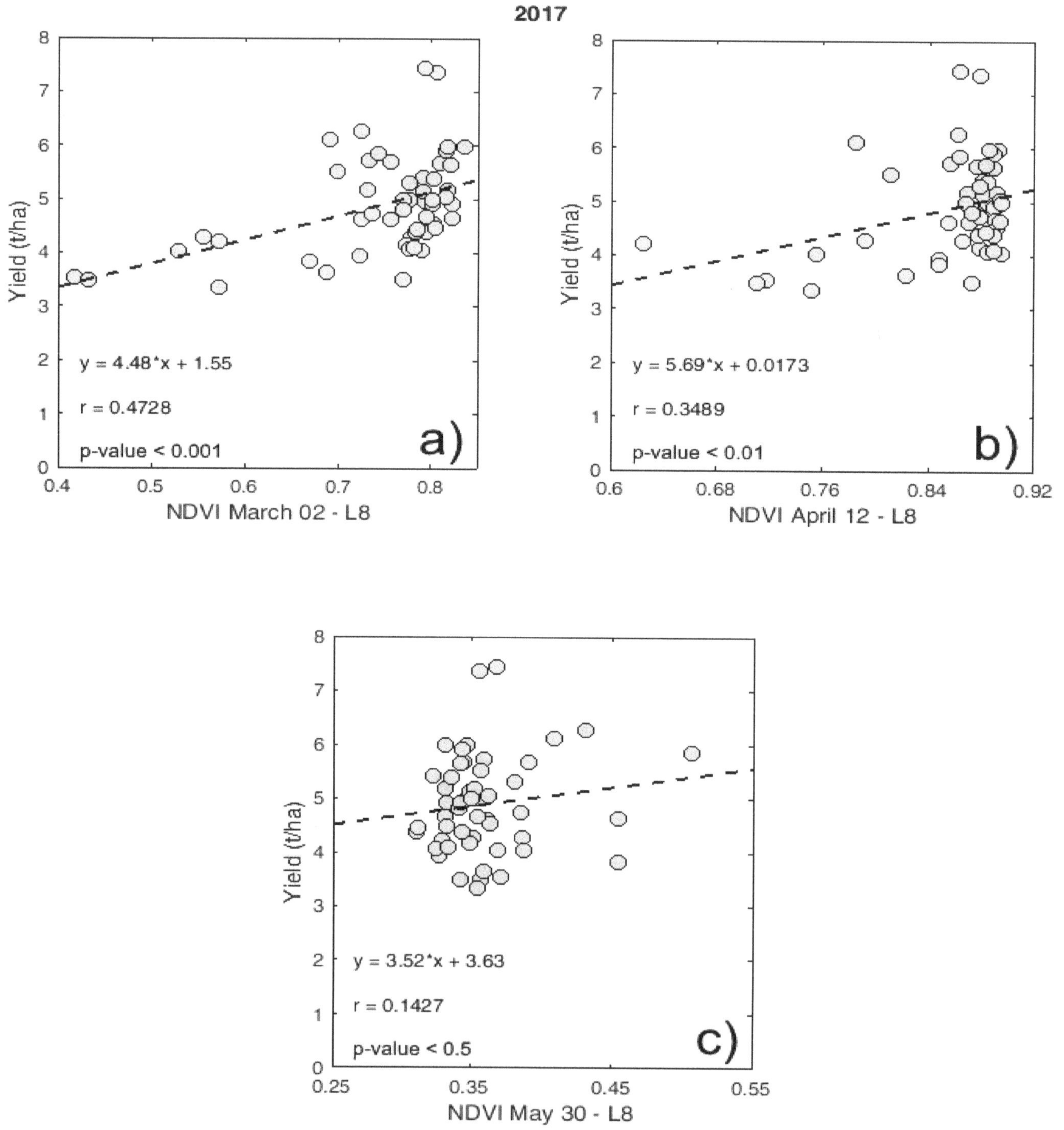

Figure 9. Relationship between observed yield (sampled) and NDVI (Landsat-8) for 2016–2017 crop season in (**a**) 2 March, (**b**) 12 April and (**c**) 30 May.

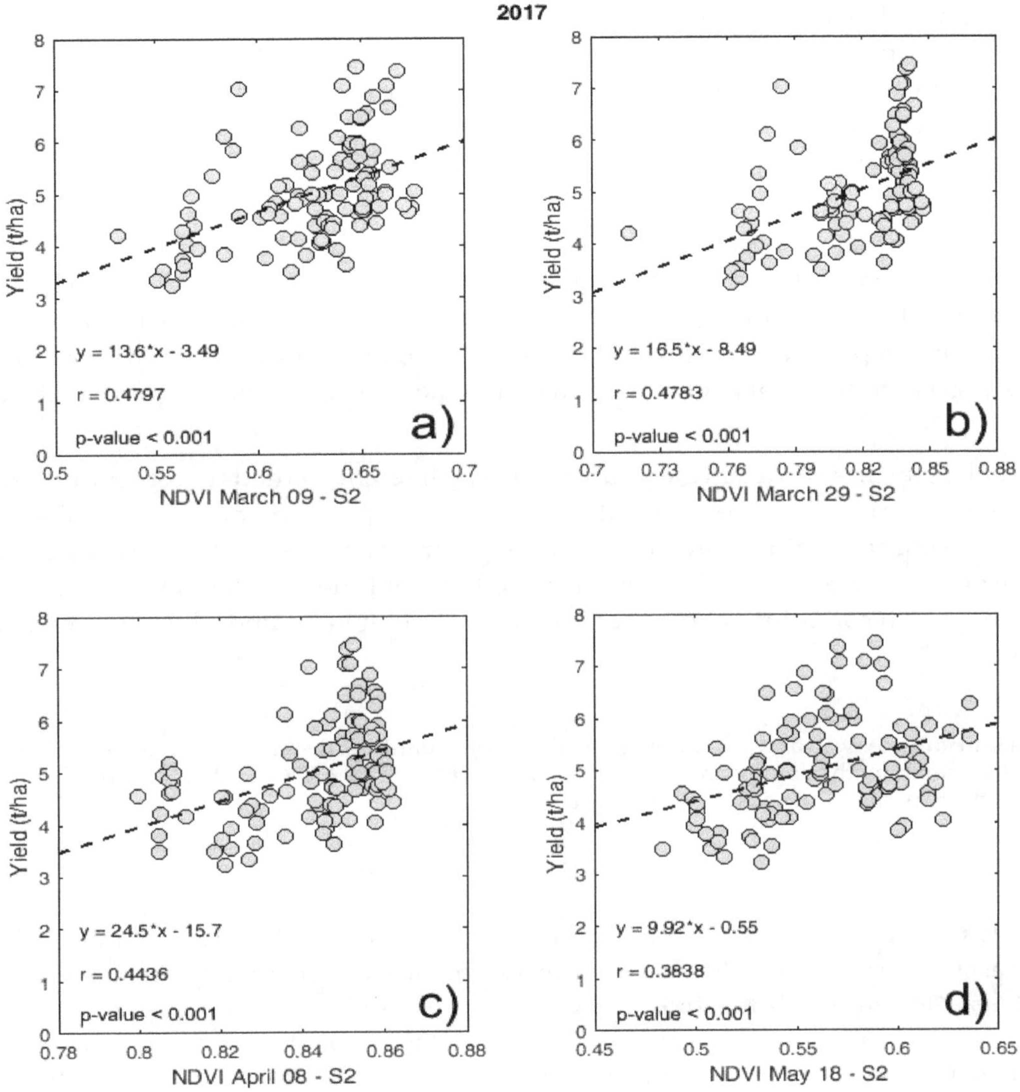

Figure 10. Relationship between observed yield (sampled) and NDVI (Sentinel-2) for 2016–2017 crop season in (**a**) 9 March, (**b**) 29 March, (**c**) 8 April and (**d**) 18 May.

Unfortunately, all the empirical relationships determined over the whole study period cannot be applied elsewhere, since a universal conversion from vegetation indices to yield values does not exist, as pointed out by Georgi et al. [9]. Many efforts have been made to determine this relationship [17,70,71], with results indicating that replicability is mostly limited by crop type and climate zone, confirming our case study findings. Our results highlight the potential use of remote sensing imagery (Sentinel-2 and Landsat-8) for within-field and interannual durum wheat yield assessment under Mediterranean conditions. Although it is not possible to retrieve absolute yield values, the results show the capacity of the NDVI to describe within-field yield levels providing objective criteria, also in terms of potential performance, on which to base nutrient management zones for soil sampling and variable-rate nutrient application, especially thanks to the availability of multiple years of data. This is also facilitated by the fact that, even if multiple surveys are done during crop development, NDVI and yield are strongly correlated at stem elongation and heading stages, which are among the most important for agronomic management to support and improve durum wheat yield and quality.

4. Conclusions

The first part of the study is mainly practice oriented, testing a state of the art protocol for error removal from yield monitoring data and comparing the cleaning map with hand field sampling data. The cleaning process improved measurement accuracy of spatial variability, which is key for adopting precision farming techniques to make daily fieldwork more efficient and increase agricultural productivity. In light of this, in the second part of the study, the usefulness of remote sensing information collected during the optimal period for characterizing within-field spatial variability of durum wheat productivity has been assessed.

Findings suggest that the best time to relate NDVI to durum wheat yields under rainfed conditions in the Mediterranean area is the period leading up to 90–60 days before harvest (March–April). At the same time the results, based on a four year yield dataset, support the conclusion that a unique NDVI-yield relationship cannot be achieved and applied to different years or environments, but year by year can suggest the best management approach while taking farmers' requirements into account.

Additional research is needed in the future to: (i) test different methods of comparing heterogeneous data (different supports, spatial resolution), (ii) address the performance of other VIs and promote them among end users. Furthermore, in case of long periods between satellite images due to cloud cover, the use of a crop simulation model has proved to be of fundamental importance to simulate the crop stage and growth conditions and better understand differences underlying correlations between yield and VIs.

Author Contributions: A.V.V. and A.C. designed the experiment. P.T., A.C., S.F.D.G., A.V.V., D.V. and A.M. formulated the research methodology and wrote the manuscript. P.T., S.F.D.G. and A.M. provided necessary data analysis. All authors reviewed and edited the draft.

References

1. Toscano, P.; Gioli, B.; Genesio, L.; Vaccari, F.P.; Miglietta, F.; Zaldei, A.; Crisci, A.; Ferrari, E.; Bertuzzi, F.; La Cava, P.; et al. Durum wheat quality prediction in Mediterranean environments: From local to regional scale. *Eur. J. Agron.* **2014**, *61*, 1–9. [CrossRef]
2. Royo, C.; Nazco, R.; Villegas, D. The climate of the zone of origin of Mediterranean durum wheat (Triticum durum Desf.) landraces affects their agronomic performance. *Genet. Resour. Crop Evol.* **2014**, *61*, 1345–1358. [CrossRef]
3. Tedone, L.; Alhajj Ali, S.; Verdini, L.; De Mastro, G. Nitrogen management strategy for optimizing agronomic and environmental performance of rainfed durum wheat under Mediterranean climate. *J. Clean. Prod.* **2018**, *172*, 2058–2074. [CrossRef]
4. Soddu, A.; Deidda, R.; Marrocu, M.; Meloni, R.; Paniconi, C.; Ludwig, R.; Sodde, M.; Mascaro, G.; Perra, E. Climate Variability and Durum Wheat Adaptation Using the AquaCrop Model in Southern Sardinia. *Procedia Environ. Sci.* **2013**, *19*, 830–835. [CrossRef]
5. Toscano, P.; Ranieri, R.; Matese, A.; Vaccari, F.P.; Gioli, B.; Zaldei, A.; Silvestri, M.; Ronchi, C.; La Cava, P.; Porter, J.R.; et al. Durum wheat modeling: The Delphi system, 11 years of observations in Italy. *Eur. J. Agron.* **2012**, *43*, 108–118. [CrossRef]
6. Giuliani, M.M.; Giuzio, L.; de Caro, A.; Flagella, Z. Relationships between nitrogen utilization and grain technological quality in durum wheat: II. Grain yield and quality. *Agron. J.* **2011**, *103*, 1668–1675. [CrossRef]
7. Bowman, M.S.; Zilberman, D. Economic factors affecting diversified farming systems. *Ecol. Soc.* **2013**, *18*, 33. [CrossRef]
8. Mulla, D.J. Twenty five years of remote sensing in precision agriculture: Key advances and remaining knowledge gaps. *Biosyst. Eng.* **2013**, *114*, 358–371. [CrossRef]
9. Georgi, C.; Spengler, D.; Itzerott, S.; Kleinschmit, B. Automatic delineation algorithm for site-specific management zones based on satellite remote sensing data. *Precis. Agric.* **2018**, *19*, 684–707. [CrossRef]
10. Thilakarathna, M.; Raizada, M. Challenges in Using Precision Agriculture to Optimize Symbiotic Nitrogen Fixation in Legumes: Progress, Limitations, and Future Improvements Needed in Diagnostic Testing. *Agronomy* **2018**, *8*, 78. [CrossRef]

11. Ping, J.L.; Dobermann, A. Processing of yield map data. *Precis. Agric.* **2005**, *6*, 193–212. [CrossRef]
12. Thöle, H.; Richter, C.; Ehlert, D. Strategy of statistical model selection for precision farming on-farm experiments. *Precis. Agric.* **2013**, *14*, 434–449. [CrossRef]
13. Vega, A.; Córdoba, M.; Castro-Franco, M.; Balzarini, M. Protocol for automating error removal from yield maps. *Precis. Agric.* 2019. [CrossRef]
14. Lyle, G.; Bryan, B.A.; Ostendorf, B. Post-processing methods to eliminate erroneous grain yield measurements: Review and directions for future development. *Precis. Agric.* **2014**, *15*, 377–402. [CrossRef]
15. Arslan, S. A grain flow model to simulate grain yield sensor response. *Sensors* **2008**, *8*, 952–962. [CrossRef]
16. Ingeli, M.; Galambošová, J.; Macák, M.; Rataj, V. Study on Correlation of Data from Yield Monitoring System and Hand Samples. *Acta Technol. Agric.* **2015**, *18*, 10–13. [CrossRef]
17. Doraiswamy, P.C.; Hatfield, J.L.; Jackson, T.J.; Akhmedov, B.; Prueger, J.; Stern, A. Crop condition and yield simulations using Landsat and MODIS. *Remote Sens. Environ.* **2004**, *92*, 548–559. [CrossRef]
18. Funk, C.; Budde, M.E. Phenologically-tuned MODIS NDVI-based production anomaly estimates for Zimbabwe. *Remote Sens. Environ.* **2009**, *113*, 115–125. [CrossRef]
19. Johnson, D.M. An assessment of pre- and within-season remotely sensed variables for forecasting corn and soybean yields in the United States. *Remote Sens. Environ.* **2014**, *141*, 116–128. [CrossRef]
20. Gao, F.; Anderson, M.; Daughtry, C.; Johnson, D. Assessing the variability of corn and soybean yields in central Iowa using high spatiotemporal resolution multi-satellite imagery. *Remote Sens.* **2018**, *10*, 1489. [CrossRef]
21. Skakun, S.; Vermote, E.; Roger, J.-C.; Franch, B. Combined Use of Landsat-8 and Sentinel-2A Images for Winter Crop Mapping and Winter Wheat Yield Assessment at Regional Scale. *AIMS Geosci.* **2017**, *3*, 163–186. [CrossRef] [PubMed]
22. Azzari, G.; Jain, M.; Lobell, D.B. Towards fine resolution global maps of crop yields: Testing multiple methods and satellites in three countries. *Remote Sens. Environ.* **2017**, *202*, 129–141. [CrossRef]
23. Jin, Z.; Azzari, G.; Lobell, D.B. Improving the accuracy of satellite-based high-resolution yield estimation: A test of multiple scalable approaches. *Agric. For. Meteorol.* **2017**, *247*, 207–220. [CrossRef]
24. Zhong, L.; Yu, L.; Li, X.; Hu, L.; Gong, P. Rapid corn and soybean mapping in US Corn Belt and neighboring areas. *Sci. Rep.* **2016**, *6*, 36240. [CrossRef] [PubMed]
25. Lambert, M.J.; Traoré, P.C.S.; Blaes, X.; Baret, P.; Defourny, P. Estimating smallholder crops production at village level from Sentinel-2 time series in Mali's cotton belt. *Remote Sens. Environ.* **2018**, *216*, 647–657. [CrossRef]
26. Newton, I.H.; Tariqul Islam, A.F.M.; Saiful Islam, A.K.M.; Tarekul Islam, G.M.; Tahsin, A.; Razzaque, S. Yield Prediction Model for Potato Using Landsat Time Series Images Driven Vegetation Indices. *Remote Sens. Earth Syst. Sci.* **2018**, *1*, 29–38. [CrossRef]
27. López-Lozano, R.; Casterad, M.A.; Herrero, J. Site-specific management units in a commercial maize plot delineated using very high resolution remote sensing and soil properties mapping. *Comput. Electron. Agric.* **2010**, *73*, 219–229. [CrossRef]
28. Nawar, S.; Corstanje, R.; Halcro, G.; Mulla, D.; Mouazen, A.M. Delineation of Soil Management Zones for Variable-Rate Fertilization: A Review. In *Advances in Agronomy*; Academic Press: Cambridge, MA, USA, 2017; pp. 175–245.
29. Han, J.; Wei, C.; Chen, Y.; Liu, W.; Song, P.; Zhang, D.; Wang, A.; Song, X.; Wang, X.; Huang, J. Mapping above-ground biomass ofwinter oilseed rape using high spatial resolution satellite data at parcel scale under waterlogging conditions. *Remote Sens.* **2017**, *9*, 3. [CrossRef]
30. Dong, T.; Shang, J.; Liu, J.; Qian, B.; Jing, Q.; Ma, B.; Huffman, T.; Geng, X.; Sow, A.; Shi, Y.; et al. Using RapidEye imagery to identify within-field variability of crop growth and yield in Ontario, Canada. *Precis. Agric.* **2019**, 1–20. [CrossRef]
31. Maes, W.H.; Steppe, K. Perspectives for Remote Sensing with Unmanned Aerial Vehicles in Precision Agriculture. *Trends Plant Sci.* **2019**, *24*, 152–164. [CrossRef]
32. Al-Gaadi, K.A.; Hassaballa, A.A.; Tola, E.; Kayad, A.G.; Madugundu, R.; Alblewi, B.; Assiri, F. Prediction of potato crop yield using precision agriculture techniques. *PLoS ONE* **2016**, *11*, e0162219. [CrossRef] [PubMed]
33. Escolà, A.; Badia, N.; Arnó, J.; Martínez-Casasnovas, J.A. Using Sentinel-2 images to implement Precision Agriculture techniques in large arable fields: First results of a case study. *Adv. Anim. Biosci.* **2017**, *8*, 377–382. [CrossRef]

34. Jeppesen, J.H.; Jacobsen, R.H.; Jørgensen, R.N.; Halberg, A.; Toftegaard, T.S. Identification of High-Variation Fields based on Open Satellite Imagery. *Adv. Anim. Biosci.* **2017**, *8*, 388–393. [CrossRef]
35. Soil Survey Staff. *Soil Taxonomy: A Basic System of Soil Classification for Making and Interpreting Soil Surveys*, 2nd ed.; USDA-Soil Survey Staff, Ed.; United States Department of Agriculture: Washington, DC, USA, 1999.
36. Ventrella, D.; Stellacci, A.M.; Castrignanò, A.; Charfeddine, M.; Castellini, M. Effects of crop residue management on winter durum wheat productivity in a long term experiment in Southern Italy. *Eur. J. Agron.* **2016**, *77*, 188–198. [CrossRef]
37. Deering, D.W. Rangeland reflectance characteristics measured by aircraft and spacecraft sensors. *Diss. Abstr. Int. B* **1979**, *39*, 3081–3082.
38. Anselin, L.; Syabri, I.; Kho, Y. GeoDa: An introduction to spatial data analysis. *Geogr. Anal.* **2006**, *38*, 5–22. [CrossRef]
39. QGIS. QGIS Geographic Information System. Open Source Geospatial Foundation Project. 2017. Available online: https://qgis.org/en/site/(accessed on 8 August 2019).
40. Minasny, B.; McBratney, A.B.; Whelan, B.M. *VESPER Version 1.62*, Australian Centre for Precision Agriculture, The University of Sydney: Sydney, Australia, 2005.
41. Mathworks Inc. *MATLAB and Statistics Toolbox Release*, Mathworks Inc.: Natick, MA, USA, 2016.
42. Amelin, L. Local Indicators of Spatial Association-LISA. *Geogr. Anal.* **1995**, *27*, 93–115.
43. Anselin, L. The Moran Scatterplot as an ESDA Tool to Assess Local Instability in Spatial Association. In *Spatial Analytical Perspectives on GIS*; Fischer, M., Scholten, H., Unwin, D., Eds.; Taylor and Francis: London, UK, 1996; pp. 111–125.
44. Porter, J.R.; Porter, J.R. AFRCWHEAT2 A model of the growth and development of wheat incorporating responses to water and nitrogen. *Eur. J. Agron.* **1993**, *2*, 69–82. [CrossRef]
45. Su, L.; Wang, Q.; Wang, C.; Shan, Y. Simulation models of leaf area index and yield for cotton grown with different soil conditioners. *PLoS ONE* **2015**, *10*, e0141835. [CrossRef] [PubMed]
46. Wang, M.H. Field Information Collection and Process Technology. *Agric. Mech.* **1999**, *7*, 22–24.
47. Leroux, C.; Jones, H.; Clenet, A.; Dreux, B.; Becu, M.; Tisseyre, B. A general method to filter out defective spatial observations from yield mapping datasets. *Precis. Agric.* **2018**, *19*, 789–808. [CrossRef]
48. Sudduth, K.A.; Drummond, S.T. Yield editor: Software for removing errors from crop yield maps. *Agron. J.* **2007**, *99*, 1471–1482. [CrossRef]
49. Simbahan, G.C.; Dobermann, A.; Ping, J.L. Site-specific management—Screening yield monitor data improves grain yield maps. *Agron. J.* **2004**, *96*, 1091–1102. [CrossRef]
50. Sun, W.; Whelan, B.; McBratney, A.B.; Minasny, B. An integrated framework for software to provide yield data cleaning and estimation of an opportunity index for site-specific crop management. *Precis. Agric.* **2013**, *14*, 376–391. [CrossRef]
51. Arslan, S.; Colvin, T.S. Grain yield mapping: Yield sensing, yield reconstruction, and errors. *Precis. Agric.* **2002**, *3*, 135–154. [CrossRef]
52. Young, L.J.; Gotway, C.A. Linking spatial data from different sources: The effects of change of support. *Stoch. Environ. Res. Risk Assess.* **2007**, *21*, 589–600. [CrossRef]
53. Mahey, R.K.; Singh, R.; Sidhu, S.S.; Narang, R.S. The use of remote sensing to assess the effects of water stress on wheat. *Exp. Agric.* **1991**, *27*, 423–429. [CrossRef]
54. Freeman, K.W.; Raun, W.R.; Johnson, G.V.; Mullen, R.W.; Stone, M.L.; Solie, J.B. Late-season prediction of wheat grain yield and grain protein. *Commun. Soil Sci. Plant Anal.* **2003**, *34*, 1837–1852. [CrossRef]
55. Johnson, D.M. A comprehensive assessment of the correlations between field crop yields and commonly used MODIS products. *Int. J. Appl. Earth Obs. Geoinf.* **2016**, *52*, 65–81. [CrossRef]
56. Kogan, F.; Kussul, N.; Adamenko, T.; Skakun, S.; Kravchenko, O.; Kryvobok, O.; Shelestov, A.; Kolotii, A.; Kussul, O.; Lavrenyuk, A. Winter wheat yield forecasting in Ukraine based on Earth observation, meteorologicaldata and biophysical models. *Int. J. Appl. Earth Obs. Geoinf.* **2013**, *23*, 192–203. [CrossRef]
57. Mkhabela, M.S.; Bullock, P.; Raj, S.; Wang, S.; Yang, Y. Crop yield forecasting on the Canadian Prairies using MODIS NDVI data. *Agric. For. Meteorol.* **2011**, *151*, 385–393. [CrossRef]
58. Becker-Reshef, I.; Vermote, E.; Lindeman, M.; Justice, C. A generalized regression-based model for forecasting winter wheat yields in Kansas and Ukraine using MODIS data. *Remote Sens. Environ.* **2010**, *114*, 1312–1323. [CrossRef]

59. Weber, J.; Gazali, I. 2017 Annual Report for the Agriculture Demonstration of Practices and Technologies (ADOPT). 2017. Available online: https://sa241e23dba898335.jimcontent.com/download/version/1526570262/module/10558155383/name/20150375%20ADOPT%20Report%20Nitrogen%20Benefits%20of%20adapted%20grain%20legumes%20to%20succeeding%20crops%20in%20NW%20SK.pdf (accessed on 8 August 2019).
60. Kyratzis, A.C.; Skarlatos, D.P.; Menexes, G.C. Assessment of Vegetation Indices Derived by UAV Imagery for Durum Wheat Phenotyping under a Water Limited and Heat Stressed Mediterranean Environment. *Front. Plant Sci.* **2017**, *8*, 1–14. [CrossRef] [PubMed]
61. Lopes, M.S.; Saglam, D.; Ozdogan, M.; Reynolds, M. Traits associated with winter wheat grain yield in Central and West Asia. *J. Integr. Plant Biol.* **2014**, *56*, 673–683. [CrossRef]
62. Corresponding, J.B.; Casadesus, J.; Nachit, M.M.; Araus, J.L. Factors affecting the grain yield predicting attributes of spectral reflectance indices in durum wheat: growing conditions, genotype variability and date of measurement. *Int. J. Remote Sens.* **2005**, *26*, 2337–2358.
63. Rutkoski, J.; Poland, J.; Mondal, S.; Autrique, E.; Pérez, L.G.; Crossa, J.; Reynolds, M.; Singh, R. Canopy Temperature and Vegetation Indices from High-Throughput Phenotyping Improve Accuracy of Pedigree and Genomic Selection for Grain Yield in Wheat. *Genes Genomes Genet.* **2016**, *6*, 2799–2808. [CrossRef]
64. Spano, G.; Di Fonzo, N.; Perrotta, C.; Platani, C.; Ronga, G.; Lawlor, D.W.; Napier, J.A. Physiological characterization of 'stay green' mutants in durum wheat. *J. Exp. Bot.* **2003**, *54*, 1415–1420. [CrossRef]
65. Lopes, M.S.; Reynolds, M.P. Stay-green in spring wheat can be determined by spectral reflectance measurements (normalized difference vegetation index) independently from phenology. *J. Exp. Bot.* **2012**, *63*, 3789–3798. [CrossRef]
66. Gregersen, P.L.; Culetic, A.; Boschian, L.; Krupinska, K. Plant senescence and crop productivity. *Plant Mol. Biol.* **2013**, *82*, 603–622. [CrossRef] [PubMed]
67. Jordan, D.R.; Hunt, C.H.; Cruickshank, A.W.; Borrell, A.K.; Henzell, R.G. The relationship between the stay-green trait and grain yield in elite sorghum hybrids grown in a range of environments. *Crop Sci.* **2012**, *52*, 1153–1161. [CrossRef]
68. Gong, Y.-H.; Zhang, J.; Gao, J.-F.; Lu, J.-Y.; Wang, J.-R. Slow Export of Photoassimilate from Stay-Green Leaves during Late Grain-Filling Stage in Hybrid Winter Wheat (Triticum aestivum L.). *J. Agron. Crop Sci.* **2005**, *191*, 292–299. [CrossRef]
69. Yang, J.; Zhang, J. Grain filling of cereals under soil drying. *New Phytol.* **2006**, *169*, 223–236. [CrossRef] [PubMed]
70. Shanahan, J.F.; Schepers, J.S.; Francis, D.D.; Varvel, G.E.; Wilhelm, W.W.; Tringe, J.M.; Schlemmer, M.R.; Major, D.J. Use of remote-sensing imagery to estimate corn grain yield. *Agron. J.* **2001**, *93*, 583–589. [CrossRef]
71. Dalla Marta, A.; Grifoni, D.; Mancini, M.; Orlando, F.; Guasconi, F.; Orlandini, S. Durum wheat in-field monitoring and early-yield prediction: Assessment of potential use of high resolution satellite imagery in a hilly area of Tuscany, Central Italy. *J. Agric. Sci.* **2015**, *153*, 68–77. [CrossRef]

6

Detection of Spatial and Temporal Variability of Wheat Cultivars by High-Resolution Vegetation Indices

Stefano Marino * and Arturo Alvino

Department of Agricultural, Environmental and Food Sciences (DAEFS), University of Molise, Via De Sanctis, 86100 Campobasso, Italy; alvino@unimol.it

* Correspondence: stefanomarino@unimol.it

Abstract: An on-farm research study was carried out on two small-plots cultivated with two cultivars of durum wheat (Odisseo and Ariosto). The paper presents a theoretical approach for investigating frequency vegetation indices (VIs) in different areas of the experimental plot for early detection of agronomic spatial variability. Four flights were carried out with an unmanned aerial vehicle (UAV) to calculate high-resolution normalized difference vegetation index (NDVI) and optimized soil-adjusted vegetation index (OSAVI) images. Ground agronomic data (biomass, leaf area index (LAI), spikes, plant height, and yield) have been linked to the vegetation indices (VIs) at different growth stages. Regression coefficients of all samplings data were highly significant for both the cultivars and VIs at anthesis and tillering stage. At harvest, the whole plot (W) data were analyzed and compared with two sub-areas characterized by high agronomic performance (H) yield 20% higher than the whole plot, and low performances (L), about 20% lower of yield related to the whole plot). The whole plot and two sub-areas were analyzed backward in time comparing the VIs frequency curves. At anthesis, more than 75% of the surface of H sub-areas showed a VIs value higher than the L sub-plot. The differences were evident also at the tillering and seedling stages, when the 75% (third percentile) of VIs H data was over the 50% (second percentile) of the W curve and over the 25% (first percentile) of L sub-plot. The use of high-resolution images for analyzing the frequency value of VIs in different areas can be a useful approach for the detection of agronomic constraints for precision agriculture purposes.

Keywords: UAV; vegetation indices; relative frequencies; yield; precision agriculture; cultivars

---

1. Introduction

Monitoring the spatial and temporal variability of wheat within a season is crucial to decision-making in precision farming. Precision agriculture is a modern farming management concept using digital techniques to monitor and optimize agricultural production processes. Precision agriculture can play an important role in enhancing crop yield and ensuring sustainability [1]. Among the tools used to acquire information, unmanned aerial vehicles (UAVs) equipped with visible and near-infrared cameras, provide, in a fast and easy way, field data for precision agriculture applications [2,3]. The resolution of information from satellite data typically ranges from 5 to 30 m pixels and is unsuitable in agronomy trials given the limitations of real-time monitoring and accuracy [4]. In contrast to satellite imagery and aircraft-based remote sensing, UAVs can be used frequently during the entire growth period [5]. Furthermore, vegetation indices (VIs) of UAV imagery have the same ability as ground-based recordings to quantify crop responses to experimental treatments [6]. UAVs are a useful technology for crop monitoring at different scales and can be used for agronomic experiments where space, resource, and time constraints limit manual sampling [7]. Furthermore, UAVs provide images at small pixel sizes (3–5 cm^2 pixels), with a higher resolution than an aerial or satellite platform [8–10]. Moreover, a field can be frequently surveyed to study ongoing different phenological development phenomena [3]. Unmanned aerial vehicles equipped with near-infrared (NIR) and multispectral sensors have been useful in the research environment for determining principal spectral patterns and wavebands that relate to plant stress, through the estimation of vegetation indices (VIs), which are based on formulations fitted with the canopy light reflected at different wavelengths [11,12]. Starting from wavebands and spectral patterns, different VIs have been developed and were related to vegetation canopies including plant nutrient status, plant growth rate, physiological conditions, and crop yields [13–16]. Among VIs, the normalized difference vegetation index (NDVI) and optimized soil-adjusted vegetation index (OSAVI) are generally employed as the typical quantitative data for the estimation indicator of crop growth. Crop phenotyping involves the measurement and evaluation of physical characteristics such as biomass, leaf area index (LAI), height [17,18] and that represents a bottleneck for the fine-tuning management of field crops [19]. Crop phenology is characterized by a set of growth stages; it is well known that a single crop trait in a single growth stage plays an important role in establishing the final grain yield [20,21]. These phenological events affect the vegetation index (VI) value and provide essential information for the detection of the agronomic practices and the identification of sub-field areas having the same yield-limiting factors or similar attributes that significantly affect crop yield [22].

Indeed, there are still many aspects to be analyzed in order to discriminate homogeneous agronomic areas, starting from spectroradiometric data, because many agronomic factors affect spatial and temporal crop variability. For precision agriculture purposes it is important to define, within each field, sub-areas having similar biotic or abiotic factors that are expected to affect yield significantly [23] and point out an automatic technique to detect homogeneous sub-areas. The most appropriate time to start crop monitoring for defining sub-plot areas remains unsolved [22]. Several cluster methods [24], among vegetation indices, and crop and soil parameters for delineating management zones maps have been used. Candiago et al., [3] proposed a new approach for analyzing the high-resolution contents of the VI images based on relative frequency distributions of VIs on vineyard and tomato crop, identifying best and worst vegetative areas, stating however that the main limitation of the study was the lack of ground measurements.

In the present paper, a new approach is proposed based on the combination of cluster analysis, relative frequency distribution of VIs and ground agronomic data of winter wheat cultivars.

Since each VI has its own peculiar suitability for specific purposes and practical applications, ground agronomic data of Odisseo (O) and Ariosto (A) at seedling, tillering, and anthesis have been linked to NDVI and OSAVI indices to verify the ability of VIs to detect yield and yield components.

Once the ability to detect crop parameters has been verified, within each cultivar, the relative frequency value of high-resolution VI data of the whole plot and of two homogeneous sub-areas identified by cluster analysis [22], one with high yield (HO and HA) and the second one with low yield (LO and LA) were investigated. The whole plot and the sub-areas were analyzed backward in time to assess the power of high-resolution VI data to detect, explain, and quantify the agronomic spatial and temporal variability.

2. Materials and Methods

2.1. Study Area and Field Measurements

An on-farm research study was carried out in Central Italy (411081.29 E, 4730618.48 N; UTM-WGS84 zone 33N Italy), in a flat area at 75 m above sea level, in the 2015 crop season. The experimental field (2 hectares) was cultivated with different plots of durum and winter wheat. Two cultivars of durum wheat, Odisseo and Ariosto, were cultivated on small plots 24 m × 30 m (720 m^2), previously cultivated with soybean. Sowing was performed in December 2014 (200 kg of seeds ha^{-1}) and the crop was harvested in July 2015. The soil was prepared with the help of minimum-tillage equipment and basic fertilization was performed with 200 kg ha^{-1} of P_2O_5 (27%). Nitrogen was applied at the end of March (tillering) with a slow release nitrogen fertilizer, 35% N; 23% SO_3, and at April (booting/stem elongation) with 150 kg ha^{-1} of fertilizer with 30% N; 23% SO_3 and with two foliar nitrogen application. The weed and pest control scouting was carried out with chemical pesticides during the crop cycle.

A standard agro-meteorological station was placed in the experimental fields and temperatures, rainfall, wind, humidity, and radiation were recorded. The minimum temperature (−1 °C) was recorded in December 2014, while the maximum (36.2 °C) was recorded in July 2015. Through the whole cropping season (December 2014–July 2015), precipitation amounted to 680.2 mm; a good half of the total precipitation was recorded in 6 weeks, from the final third of January to the first third of March. From tillering to anthesis, rainfall was scarce, while from anthesis to ripening, rainfall was considerable. Air temperatures were mild from emergence to anthesis and afterwards, maximum air temperatures were around 30 °C, as expected.

The phenological stages of the wheat crop were periodically recorded according to the Zadoks Scale (ZS), which is a standardized reference scale used to evaluate and measure plant growth stage in cereals [25]. During the crop growth, sampling was carried out for each cultivar: 10 at seedling growth 13—ZS (February 23), 12 at tillering 25—ZS (March 30), 11 at anthesis 65—ZS (May 14), and 11 at harvest—99 ZS (July 7). Plants from 0.5 m^2 were georeferenced and hand cut for the calculation of yield-related traits (biomass, green leaf area index—LAI, number of spikes, plant height, yield, and thousand kernels weight). For each cultivar, 44 georeferenced samplings during crop growth were collected.

Whole plant dry mass was determined after oven drying the fresh plant material at 75 °C until constant weight for 48 h. The sampling points based on visual-spatial crop variability and identified by low-resolution orthoimages of flight were processed in real time in the field. Ground truth coordinates of target locations were recorded with GPS Leica Viva GS15 (Leica Geosystems AG, Heerbrugg, Switzerland) at each sampling and flight.

Furthermore, at harvest, within each cultivar, yield data and yield components of the whole plot (720 m^2) were measured, as well as for four sub-areas, 10 m^2 each was characterized by the different yield levels in Figure 1 (H = high yield and L = low yield). The sub-areas were selected within the homogeneous areas detected using a hierarchical clustering method, Ward's minimum variance approach, as reported in a previous paper for Odisseo [25] and calculated for Ariosto. NDVI and OSAVI high-resolution data of the four areas were analyzed backward in time to deliver relative frequency histograms of VI values.

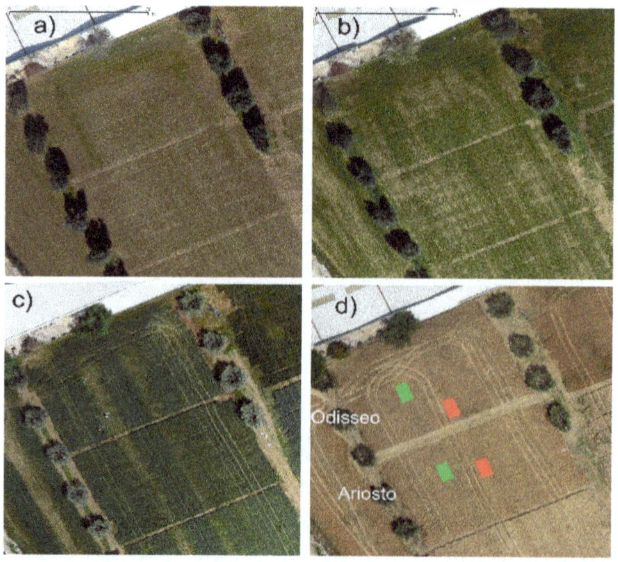

Figure 1. RGB images and samplings H (red) and L (green) areas of Odisseo (above) and Ariosto (below) at seedling (**a**), tillering (**b**), anthesis (**c**), and harvest (**d**) stages.

2.2. UAV System and Flight Missions

Four flight missions were carried out in the same dates to measure the destructive plant samples: for seedling (February 23), tillering (March 30), anthesis (May 14), harvest (July 7), with eBee UAV (senseFly, Cheseaux-sur-Lausanne, Switzerland). The eBee, designed as a fixed-wing UAV for application in precision agriculture (payload of 150 g). All flights were carried out in stable ambient light conditions from 11:30 a.m. to 1:00 p.m., with excellent visibility and wind below 5 m s^{-1}, at a flight altitude of 97 m (A.G.L.). The imaged area of the experimental field, including the surroundings, is about 15 hectares so that 96 camera stations (single flight) and 15 min flight were needed. The autopilot analyzes (continuously) the inertial measurement unit (IMU) and onboard GPS data to control every aspect of the eBee's flight. The integration of the UAV and sensors with GPS and IMU enables obtaining direct georeferencing imaging after image processing. At any acquisition date, two flights were carried out, the first one by a Canon Powershot S110 photo camera (visible spectrum, RGB-red/green/blue) for visible RGB image (orthophoto) to run a rapid analysis for visual crop variability. The second flight used a Canon Powershot S110 NIR camera (near infra-red, NIRGB—near infra-red/green/blue) which provides maximum absorption peaks at 550 nm (green), 625 nm (red), and 850 nm (NIR) wavelengths, respectively, allowing the computation of VIs.

The S110 RGB and S110 NIR camera characteristics were: weight 0.7 kg, resolution 12 million pixels, sensor size 5.58 × 7.44 mm^2, pixel pitch 1.33 μm, and format images RAW JPEG. The S110 RGB camera acquired the true-color, while the S110 NIR acquired the false-color image data at 0.55 μm (green), 0.625 μm (red), and 0.85 μm (NIR) bands.

To avoid geometric distortion due to low altitude, 96 overlapping pictures from each camera and flight were used for mosaicking to produce an ortho-image. An 80% frontal overlap and an 80% side overlap were used.

To relate and orient UAV imagery to the ground, ten ground control points (GCPs) were distributed across the field at the beginning of the season, to obtain photogrammetric imagery with uniform vertical and horizontal accuracy. The GCPs were 25 cm × 25 cm square, with a specific albedo for camera calibration (atmospheric corrections). The GCP coordinates were ensured with a Leica Viva GS15 (Leica Geosystems AG, Heerbrugg, Switzerland) GPS (horizontal accuracy of 0.025 m-vertical accuracy of 0.035 m). Fixed targets were used for a more accurate geo-referencing of UAV aerial imagery and for overlaying the measurements from multiple dates.

2.3. Data Processing

The acquired images were processed by the eMotion software (senseFly, Cheseaux-sur-Lausanne, Switzerland) to generate low-resolution visible (RGB) images of the crops. The eBee's supplied software to build a project using the drone's geo-tagged images. The project is loaded on a laptop in Pix4Dmapper Ag (senseFly, Cheseaux-sur-Lausanne, Switzerland) to run a rapid analysis for visual crop variability.

The multiple overlapped images of the whole plot were stitched and ortho-rectified to create the geo-referenced ortho-mosaicked image. In the laboratory, data processing (orthomosaicking) of acquired images were performed with Postflight Terra 3D software package, a customized version of the Pix4D digital photogrammetric solution specifically optimized for the eBee, to generate ortho-images. Postflight Terra 3D incorporates a scale-invariant feature transform (SIFT) algorithm to match key points across multiple images [26] and process data in three key steps: (1) initial processing, (2) point cloud densification, and (3) orthomosaic generation.

Ortho-rectification by aero-triangulation and mosaicking were elaborated for processing, starting from the exterior position and orientation parameters provided by the UAV internal system (roll, pitch, and yaw angles). Orthoimages were produced from the flights, with a pixel resolution of 5 cm. Ten GCPs (Leica Viva GS15 (Leica Geosystems AG, Heerbrugg, Switzerland) were used to complete the external camera orientation. The orthomosaic was georeferenced to UTM-WGS84 zone 33N Italy. The final outputs were an RGB (Visible) GeoTIFF with a resolution of 5 cm^2 pixels.

The normalized difference vegetation index (NDVI) and the soil-adjusted vegetation index (SAVI) layers were generated in a raster calculator from extracted red (R) and near infra-red (NIR) channels. The index calculator function of Postflight Terra 3D was used for generating VI maps. A certificate of calibration of Canon S110 NIR was uploaded in the software to optimize internal camera parameters, such as focal length, principal points, and lens distortions. The ten GCPs with a known albedo for Red, Green, and NIR channel (reflectance panel) were used to calibrate the camera to achieve uniform quality of image (exposure and brightness) and for atmospheric correction in the software section "processing options," point 3 DSM, orthomosaic, index and for creating VIs map. The resolution of the reflectance map has been set at 5 cm^2 pixels, the whole map was elaborated for each VI and date, and the NDVI and OSAVI formula were selected (index map function). The final outputs were an NDVI GeoTIFF and an OSAVI GeoTIFF, both with a resolution of 5 cm^2 pixels. GeoTIFF images and georeferenced sampling data were processed for agronomic purpose with QGIS 2.8.1.

2.4. Vegetation Indices

The normalized difference vegetation index (NDVI) was calculated according to Equation (1):

$$\text{NDVI} = (\text{NIR} - \text{RED})/(\text{NIR} + \text{RED}) \tag{1}$$

The NDVI ranges from −1.0 to 1.0, where positive values indicate increasing greenness and negative values indicate non-vegetated features. It has some disadvantages though such as saturation in later growth stages [27,28].

The optimized soil-adjusted vegetation index (OSAVI) was calculated according to Equation (2). The OSAVI index was proposed by Rondeaux et al. [29] using reflectance in the red and near-infrared bands; through the following formulation:

$$\text{OSAVI} = (\text{NIR} - \text{RED})/(\text{NIR} + \text{RED} + 0.16) \tag{2}$$

where 0.16 is the soil adjustment coefficient, selected according to the optimal value to minimize soil background variations.

2.5. Statistical Analysis

Regression analysis, coefficients of determination, significance levels, and RMSE were computed on two sets of geo-referred data: ground crop samplings (LAI, Biomass, and spikes for square meter) and remote UAV data using the statistical package Origin PRO 8 (Origin Lab Corporation, Northampton, MA, USA). H and L sub-area were selected starting from homogeneous areas identified in a previous paper [22] as follow: the yield-related traits and VIs data were analyzed using the hierarchical clustering Ward's minimum variance approach [30] to classify observations into groups, in which the group members have common properties. Statistical procedures were computed using OriginPRO 8. The difference between H and L sub-areas were recorded by the analysis of variance.

3. Results and Discussion

Ground agronomic data (biomass, green leaf area index, spikes, and plant height) of each cultivar have been correlated to high-resolution multispectral images (NDVI and OSAVI). The correlation of the whole crop cycle data was performed in accordance with van Ittersum et al. [31], who states that it is essential to study the relationship among VIs and crop traits, to evaluate and estimate the yield potential and the yield gap (Table 1). In the present study, polynomial regressions were found to be the simplest adjustment to report the time-course variability along the growth stage, starting from seedling to harvest. Moreover, significant relationships with different R^2 values were found for both the indices and crop growth traits. The spectro-radiometric response of the wheat canopy was not linear during the growing period in accordance with Aparicio et al. [32] and Dang et al. [33], since the high sensitivity of the VIs when LAI is lower (early season) and less sensitive after the canopy closes.

The relationships among VIs and crop parameters were highly significant: the R^2 for NDVI vs. biomass reached 0.953 for Odisseo and 0.859 for Ariosto, while the R^2 of OSAVI vs. biomass was 0.943 for Odisseo and 0.857 for Ariosto. The relationships among each VI and LAI showed an R^2 of 0.910 for NDVI Odisseo and R^2 of 0.777 for Ariosto, and it was 0.879 for OSAVI Odisseo and 0.763 for Ariosto. The relationships among each VI and height showed an R^2 of 0.828 for NDVI Odisseo and 0.817 for OSAVI Odisseo and R^2 of 0.518 for both NDVI and OSAVI for Ariosto. Ground truth georeferenced data of the number of spikes per square meter at harvest stages were also related to NDVI and OSAVI data; significant relationships were found with an R^2 of 0.788 for Odisseo and 0.769 for Ariosto recorded for NDVI index and an R^2 of 0.790 for Odisseo and of 0.597 for Ariosto recorded for OSAVI index.

Briefly, all R^2 of yield components were highly correlated (p-value < 0.001) with VIs, and are higher for Odisseo than the Ariosto.

Table 1. Polynomial regression analysis of vegetation index (VI) data and crop parameters each based on 44 data points. LAI = leaf area index, RMSE = root mean square error. Significance level: p-value < 0.001.

	Odisseo				**Ariosto**			
	NDVI		**OSAVI**		**NDVI**		**OSAVI**	
	R^2	**RMSE**	R^2	**RMSE**	R^2	**RMSE**	R^2	**RMSE**
Biomass	0.953	0.06	0.943	0.062	0.859	0.075	0.857	0.082
LAI	0.910	0.059	0.879	0.076	0.777	0.095	0.763	0.106
Spikes	0.788	0.067	0.790	0.075	0.769	0.042	0.597	0.065
Height	0.828	0.083	0.817	0.102	0.518	0.105	0.519	0.127

3.1. Agronomic Data and VIs at Different Crop Growth Stages

In the whole plot, the biomass at seedling stage ranged from 68 to 156 g m^{-2} for both cultivars and that of LAI (0.12 to 0.47) (Table 2), while VIs ranged from 0.26 to 0.57 for NDVI and from 0.38 to 0.65 for OSAVI, for both cultivars. The light differences recorded within yield components as well as VIs value gave no significant relationships among VIs and yield components (Table 3). This was due to a low amount of accumulated biomass and LAI values, as well as the difficulties of simple indices to discriminate between soil and vegetation as found also by many authors, such as Aparicio et al. [34] and Chlingaryan et al. [24].

At the tillering stage, the spatial variability of crop traits was highly evident; thus, the most productive data was shown by the biomass and LAI which was four times higher than the less favored zones (Table 2).

In the most productive zones, VI values ranged from 0.85 to 0.99 (for both cultivars), which revealed 2.5 times more than the lower value (Figure 2).

The relationships between VIs and crop parameters were highly significant for both cultivars (Table 3). For biomass, the R^2 reached 0.88 for Odisseo and 0.66 for Ariosto for LAI, the R^2 reached 0.85 for Odisseo and 0.69 for Ariosto. The plant height and VI regression were not significant for Ariosto. These results are in accordance with findings of several authors at tillering, for example Dalla Marta et al. [35] found a highly significant (positive) relationship with biomass weight, while Reyniers and Vrindts [36] and Magney et al. [37] found respectively an R^2 of 0.76 and $R^2 = 0.62$ between NDVI and biomass.

At anthesis, the high spatial variability of the crop traits value and VI values were still evident for both cultivars (Table 2, Figure 2). VIs were highly correlated to crop traits as reported in Table 3; the values of the R^2 for NDVI vs. biomass reached 0.89 for Odisseo and 0.82 for Ariosto, while the R^2 of OSAVI vs. biomass was 0.80 for Odisseo and 0.78 for Ariosto ($p < 0.001$).

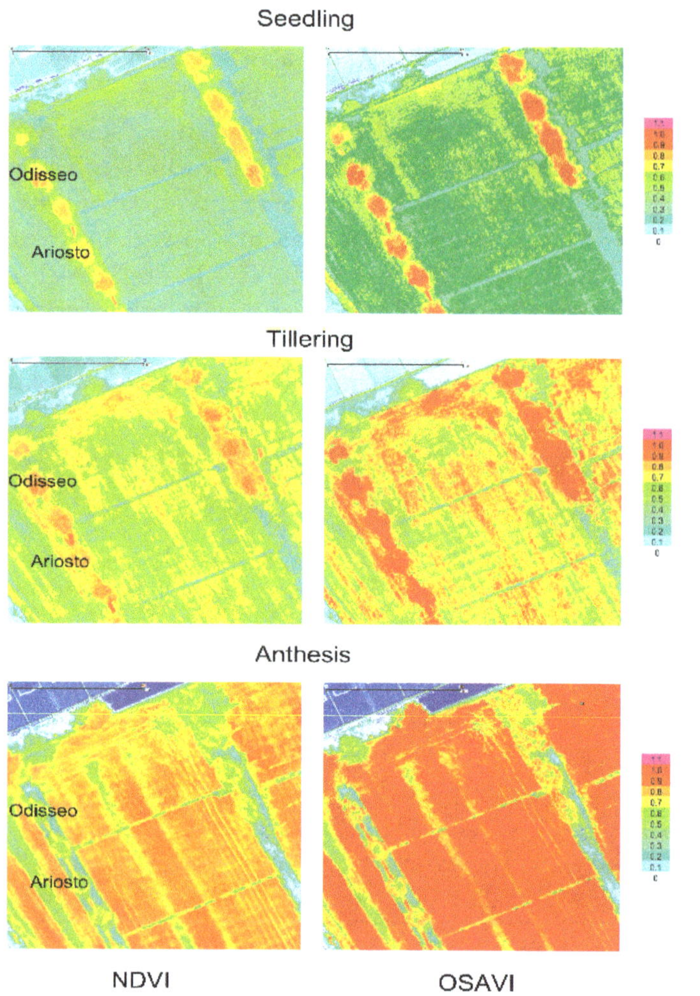

Figure 2. Normalized difference vegetation index (NDVI) and optimized soil-adjusted vegetation index (OSAVI) images for Odisseo (upper plot) and Ariosto (lower plot), at seedling, tillering, and anthesis growth stages.

Table 2. Range value (min and max), mean and standard deviation (SD) of yield components (biomass, LAI, plant height, and spike) and vegetation indices (VIs) (NDVI and OSAVI) measured at seedling, tillering, and anthesis stages.

Growth Stages	Yield Components and VIs	Odisseo				Ariosto			
		Min	Max	Mean	SD	Min	Max	Mean	SD
Seedling	Biomass (g m^{-2})	72.2	156.1	102.1	30.0	67.5	145.0	97.8	26.6
	LAI	0.167	0.472	0.267	0.086	0.123	0.424	0.269	0.089
	NDVI	0.266	0.467	0.383	0.083	0.263	0.569	0.414	0.098
	OSAVI	0.386	0.596	0.499	0.087	0.381	0.650	0.532	0.088
Tillering	Biomass (g m^{-2})	116	558	332	149	185.2	740.8	363.05	185.9
	LAI	0.328	1.610	0.922	0.409	0.500	2.259	1.106	0.566
	Plant height (cm)	34.5	54.5	45.8	5.43	28.3	45.6	39.1	5.2
	NDVI	0.338	0.828	0.639	0.165	0.434	0.857	0.637	0.145
	OSAVI	0.400	0.978	0.755	0.195	0.500	0.991	0.743	0.166
Anthesis	Biomass (g m^{-2})	282	1116	761	258	587.5	1453.9	911.5	251.6
	LAI	0.732	2.56	1.722	0.546	1.32	2.69	1.94	0.459
	Spikes (n° m^{-2})	217	527	411	91	258	651	415	119
	Plant height (cm)	39.0	73.0	63.2	10.7	60.0	73.0	69.5	3.9
	NDVI	0.527	0.838	0.753	0.092	0.654	0.876	0.814	0.073
	OSAVI	0.631	1.004	0.892	0.112	0.725	1.067	0.955	0.105

Table 3. Polynomial regression analysis of VIs data and crop parameters (whole plots) during crop cycle (seedling, tillering, and anthesis). LAI = leaf area index, RMSE = root mean square error, n.s. = not significant.

Growth Stages	VIs	Yield Components	Odisseo			Ariosto		
			R^2	RMSE	*p*-Value	R^2	RMSE	*p*-Value
Seedling	NDVI	Biomass	n.s.	0.0588	0.039	n.s.	0.082	0.012
		LAI	n.s.	0.0577	0.034	n.s.	0.079	0.097
	OSAVI	Biomass	n.s.	0.5175	0.032	n.s.	0.071	0.092
		LAI	n.s.	0.0684	0.079	n.s.	0.076	0.144
Tillering	NDVI	Biomass	0.882	0.063	0.001	0.665	0.077	0.007
		LAI	0.852	0.070	0.001	0.689	0.089	0.005
		Plant height	0.866	0.091	0.002	n.s.	0.141	0.320
	OSAVI	Biomass	0.937	0.851	0.001	0.672	0.105	0.007
		LAI	0.844	0.085	0.001	0.687	0.103	0.005
		Plant height	0.859	0.110	0.002	n.s.	0.16	0.250
Anthesis	NDVI	Biomass	0.893	0.033	0.001	0.818	0.007	0.001
		LAI	0.843	0.040	0.001	0.694	0.073	0.422
		Spikes	0.725	0.060	0.001	0.768	0.044	0.001
		Plant height	0.736	0.070	0.030	0.820	0.046	0.010
	OSAVI	Biomass	0.804	0.756	0.001	0.777	0.055	0.002
		LAI	0.745	0.063	0.002	0.672	0.106	0.470
		Spikes	0.680	0.074	0.008	0.697	0.060	0.001
		Plant height	0.680	0.091	0.061	0.803	0.070	0.016

These results are in accordance with the Marti et al. [38], who found the highest correlation values between growth and NDVI measurements when performed around anthesis, such as with Villegas et al. [39] that reported the highest values of R^2 for the relationship between NDVI and crop dry weight at anthesis. Cabrera-Bosquet et al. [40] have demonstrated a strong linear regression between durum wheat and VIs. On the contrary, Dalla Marta et al. [35] found completely different results, with no correlations observed between the crop parameters and the indices, due to VIs saturation.

3.2. Backward Analysis

Table 4 reports yield and yield components at harvest for the complete plot (W), as well as data of the two sub-areas, H_O and H_A with high yield and L_O and L_A with low yield.

Table 4. Yield (t ha^{-1}), biomass (g m^{-2}), spike number (n° m^{-2}), and plant height (cm) of Odisseo and Ariosto at harvest stage. Mean data are related to the whole plot (W) and to the sub-areas; H_o (high yield of Odisseo) and H_A (high yield of Ariosto), L_O (low yield Odisseo), and L_A (low yield Ariosto). The *p*-level values are related to H and L significant differences within each cultivar.

	Odisseo				Ariosto			
	W	H_O (SD)	L_O (SD)	*p*-level	W	H_A (SD)	L_A (SD)	*p*-level
Yield (t ha^{-1})	5.73	6.77 (0.83)	4.66 (0.61)	<0.0064	5.05	6.17 (0.44)	3.94 (0.25)	1.26×10^{-4}
Biomass (g m^{-2})	760	1029 (106)	500 (208)	<0.0398	911	1171 (200)	682 (77)	<0.0039
Spikes (n° m^{-2})	412	509 (74.3)	320 (27.3)	<0.0034	360	465 (25)	230 (31)	2.39×10^{-5}
Plant height (cm)	63.2	72.2 (2.95)	50.5 (8.5)	<0.0236	69.5	72.5 (5.57)	66.25 (4.78)	n.s.

The yield data of both cultivars were in accordance with those reported by Visioli et al. [41] in Italy. The crop yield of the whole plots was 5.73 t ha^{-1} for Odisseo (W_O) and 5.05 t ha^{-1} for Ariosto (W_A). The yield of H_O and H_A were about 20% higher than the whole plot yield, which in turn were 20% higher than L_O and L_A. Yield components were ranked according to the yield levels: biomass and spikes per square meter of H were about 25–35% higher than the W and 37–50% higher than L sub-plot areas. Differences in plant heights were less marked.

The subplot detected at harvest was analyzed backward in time from anthesis to seedling.

At anthesis, the analysis of VI value frequencies showed the maximum evidence of different crop status of H and L sub-areas and W plots. More than the 75% of the surface of the sub-areas showed a value higher (H sub-plot) or lower (L sub-plot) than W plot (Figure 3). The H and L areas curves showed a low overlapping. The H curve appeared narrow and tall. Data were confirmed by all the yield components that were significantly different from each other in the two sub-areas and W plot.

At tillering, the histogram of VIs frequency (Figure 4) shows differences among the whole plot data, both H and L areas. These differences were lower than those detected at the anthesis stage.

For both the cultivars and indices, the H zones showed values of the frequency which were higher than the 75% of H_O and H_A, and L_A, L_O VI values were higher and lower respectively than the median value of the whole plot.

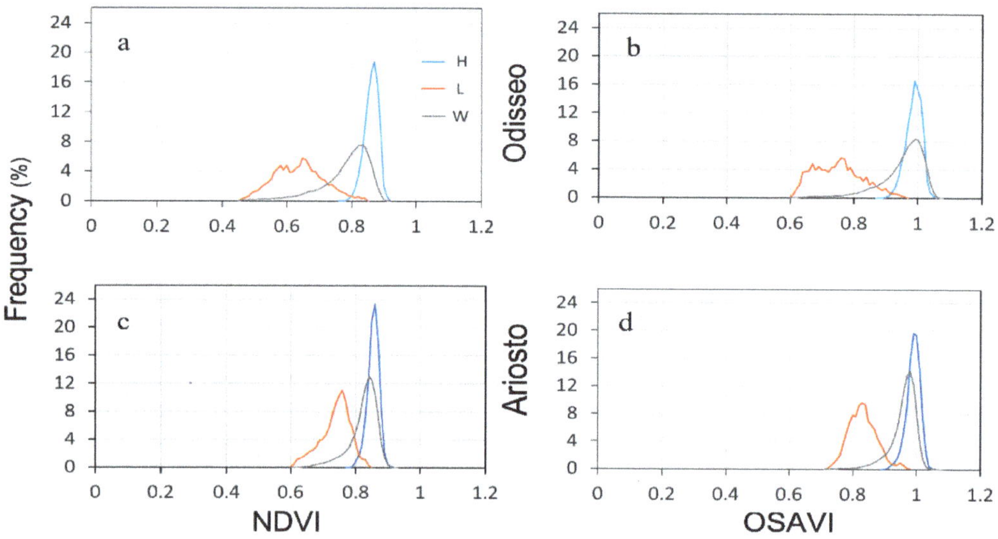

Figure 3. Relative frequency curves of the two VIs, NDVI (**a–c**) and OSAVI (**b–d**) within each cultivar, Odisseo (**a–b**) and Ariosto (**c–d**). In grey, the whole plot (W) curves have been shown, in red the low-yield sub-areas (L), and in blue the high-yield sub-areas (H) at the anthesis stage.

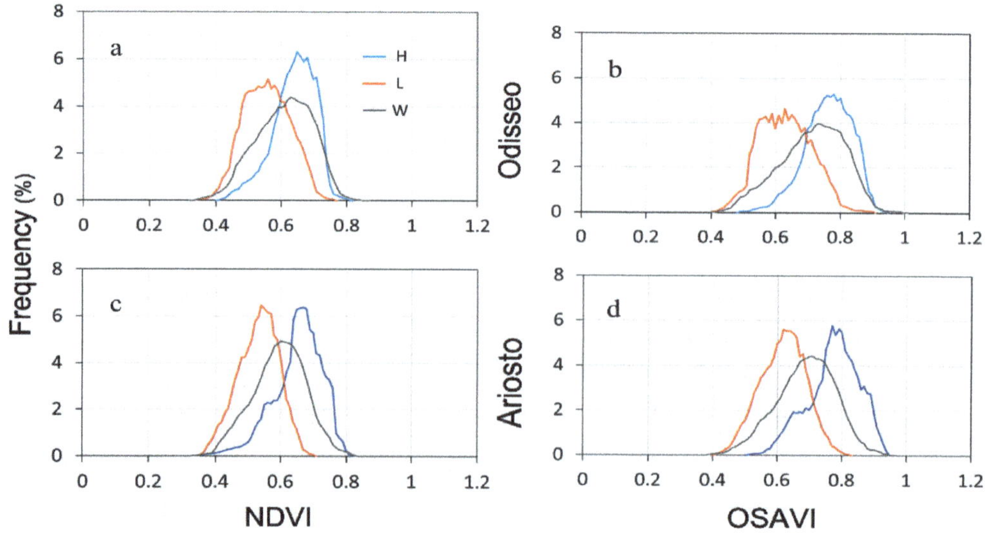

Figure 4. Relative frequency curves of the two VIs, NDVI (**a–c**) and OSAVI (**b–d**) within each cultivar, Odisseo (**a–b**) and Ariosto (**c–d**). In grey, the whole plot (W) curves have been shown, in red the low-yield sub-areas (L), and in blue the high-yield sub-areas (H) at the tillering stage.

At the seedling stage, the relative frequency curves of the two VIs are presented in Figure 5. In all the graphs, the curves appeared to be very close, except for the OSAVI of Ariosto (d) at the seedling stage.

In all graphs, about 75% of the H data (third percentile) were higher than the second percentile of the W curve as well as the first percentile of L data. The highest differences between H and L curves were detected for the OSAVI index, Ariosto (d), because the index that considers soil brightness are perform slightly better than the NDVI, as reported in different studies [42,43].

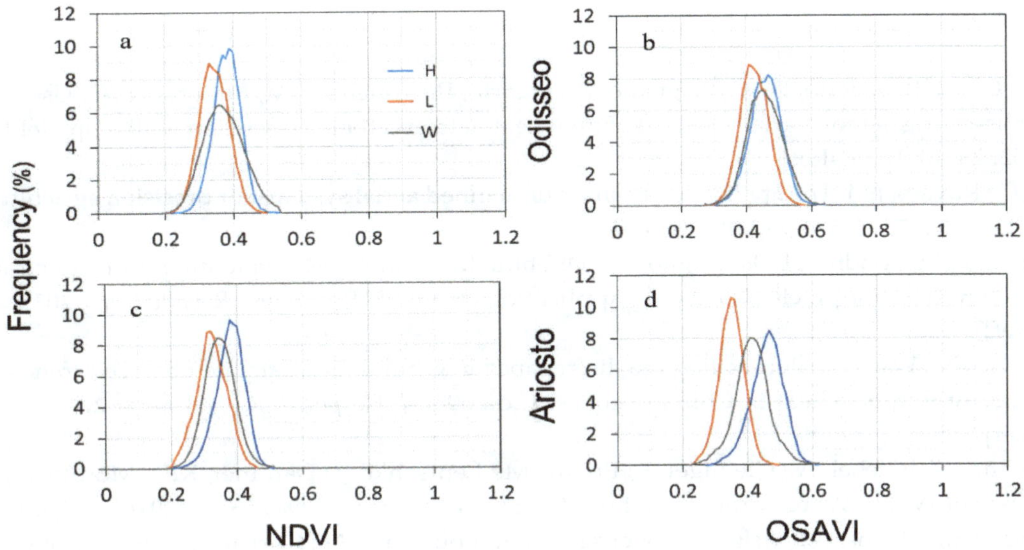

Figure 5. Relative frequency curves of the two VIs, NDVI (**a–c**) and OSAVI (**b–d**) within each cultivar, Odisseo (**a–b**) and Ariosto (**c–d**). In grey, the whole plot (W) curves have been shown, in red the low-yield sub-areas (L), and in blue the high-yield sub-areas (H) at the seedling stage.

4. Conclusions

In the last few years, UAVs have been used as tools that provide a high number of information per square meter for precision agriculture management and its development. The potentials of the information provided by the drone images have not yet been fully explored. In this paper, we have analyzed the frequency distribution curves of high-resolution VIs images of the two cultivars of durum wheat.

Within each cultivar, two yield sub-areas were detected by cluster analysis at the harvest stage and analyzed backward at different growth stages, in comparison with the whole plot data. The crop yield of the whole plots revealed 5.73 t ha^{-1} for Odisseo (W_O) and 5.05 t ha^{-1} for Ariosto (W_A). The yield for H_O and H_A recorded 20% higher than the whole plot yield, which in turn were 20% higher than L_O and L_A, moreover the biomass and the spikes per square meter of H was noted 25–35% higher than the W and 37–50% higher than L sub-plot areas, confirming the ability of cluster analysis to detect groups with the same variability.

The analysis of frequencies of high-resolution VI images showed that the 75% of H data (third percentile) were recorded higher than the second percentile of the W curve as well as for the first percentile of L data. The maximum evidence of different crop status of H and L sub-areas and W plots were detected at anthesis by both indices and cultivars.

The analysis of frequency curves provided significant differences at the seedling stage where agronomic parameters showed only slight differences which were revealed as not significant following the traditional approach (regression between agronomic traits and VIs). For Ariosto, the seedling OSAVI index showed an interesting differentiation among areas, confirming the ability of OSAVI to consider the soil brightness.

The analysis of frequencies of high-resolution VI images provided detailed information on crop spatial and temporal variability and can allow developing a new approach for automatic detection of suitable vegetation indices, and spatial and temporal crop variability for precision agriculture practices.

Author Contributions: The authors, S.M. and A.A. contributed equally to this work.

Acknowledgments: The authors are grateful to Giancarmine Montalto and Emiliano Camarda (Molise Geodetica) who were in charge of the UAV flights.

References

1. Cavallo, G.; De Benedetto, D.; Castrignanò, A.; Quarto, R.; Vonella, A.V.; Buttafuoco, G. Use of geophysical data for assessing 3D soil variation in a durum wheat field and their association with crop yield. *Biosyst. Eng.* **2016**, *152*, 28–40. [CrossRef]
2. Zhang, C.; Kovacs, J.M. The application of small unmanned aerial systems for precision agriculture: A review. *Precis. Agric.* **2012**, *13*, 693–712. [CrossRef]
3. Candiago, S.; Remondino, F.; De Giglio, M.; Dubbini, M.; Gattelli, M. Evaluating multispectral images and vegetation indices for precision farming applications from UAV images. *Remote Sens.* **2015**, *7*, 4026–4047. [CrossRef]
4. Guo, L.; An, N.; Wang, K. Reconciling the discrepancy in ground- and satellite-observed trends in the spring phenology of winter wheat in China from 1993 to 2008. *J. Geophys. Res. Atmos.* **2016**, *121*, 1027–1042. [CrossRef]
5. Schirrmann, M.; Giebel, A.; Gleiniger, F.; Pflanz, M.; Lentschke, J.; Dammer, K.H. Monitoring Agronomic Parameters of Winter Wheat Crops with Low-Cost UAV Imagery. *Remote Sens.* **2016**, *8*, 706. [CrossRef]
6. Rasmussen, J.; Ntakos, G.; Nielsen, J.; Svensgaard, J.; Poulsen, R.N.; Christensen, S. Are vegetation indices derived from consumer-grade cameras mounted on UAVs sufficiently reliable for assessing experimental plots? *Eur. J. Agron.* **2016**, *74*, 75–92. [CrossRef]
7. Duan, T.; Chapman, S.C.; Guo, Y.; Zheng, B. Dynamic monitoring of NDVI in wheat agronomy and breeding trials using an unmanned aerial vehicle. *F. Crop. Res.* **2017**, *210*, 71–80. [CrossRef]
8. Abdollahnejad, A.; Panagiotidis, D.; Surový, P.; Ulbrichová, I. UAV capability to detect and interpret solar radiation as a potential replacement method to hemispherical photography. *Remote Sens.* **2018**, *10*, 423. [CrossRef]
9. Alvino, A.; Marino, S.; Alvino, A.; Marino, S. Remote Sensing for Irrigation of Horticultural Crops. *Horticulturae* **2017**, *3*, 40. [CrossRef]
10. Mulla, D.J. Twenty five years of remote sensing in precision agriculture: Key advances and remaining knowledge gaps. *Biosyst. Eng.* **2013**, *114*, 358–371. [CrossRef]
11. Kyratzis, A.C.; Skarlatos, D.P.; Menexes, G.C.; Vamvakousis, V.F.; Katsiotis, A. Assessment of Vegetation Indices Derived by UAV Imagery for Durum Wheat Phenotyping under a Water Limited and Heat Stressed Mediterranean Environment. *Front. Plant Sci.* **2017**, *8*, 1–14. [CrossRef] [PubMed]
12. Liebisch, F.; Kirchgessner, N.; Schneider, D.; Walter, A.; Hund, A. Remote, aerial phenotyping of maize traits with a mobile multi-sensor approach. *Plant Methods* **2015**, *11*. [CrossRef] [PubMed]
13. Basso, B.; Cammarano, D.; Cafiero, G.; Marino, S.; Alvino, A. Cultivar discrimination at different site elevations with remotely sensed vegetation indices. *Ital. J. Agron.* **2011**, *6*, 1–5. [CrossRef]
14. Marino, S.; Alvino, A. Proximal sensing and vegetation indices for site-specific evaluation on an irrigated crop tomato. *Eur. J. Remote Sens.* **2014**, *47*, 271–283. [CrossRef]
15. Marino, S.; Aria, M.; Basso, B.; Leone, A.P.; Alvino, A. Use of soil and vegetation spectroradiometry to investigate crop water use efficiency of a drip irrigated tomato. *Eur. J. Agron.* **2014**, *59*, 67–77. [CrossRef]
16. Marino, S.; Cocozza, C.; Tognetti, R.; Alvino, A. Use of proximal sensing and vegetation indexes to detect the inefficient spatial allocation of drip irrigation in a spot area of tomato field crop. *Precision Agric.* **2015**, *16*, 613–629. [CrossRef]
17. Yue, J.; Feng, H.; Jin, X.; Yuan, H.; Li, Z.; Zhou, C.; Yang, G.; Tian, Q. A comparison of crop parameters estimation using images from UAV-Mounted Snapshot Hyperspectral Sensor and High-Definition Digital Camera. *Remote Sens.* **2018**, *10*, 1138. [CrossRef]

18. Holman, F.H.; Riche, A.B.; Michalski, A.; Castle, M.; Wooster, M.J.; Hawkesford, M.J. High throughput field phenotyping of wheat plant height and growth rate in field plot trials using UAV based remote sensing. *Remote Sens.* **2016**, *8*. [CrossRef]
19. Araus, J.L.; Cairns, J.E. Field high-throughput phenotyping: The new crop breeding frontier. *Trends Plant Sci.* **2014**, *19*, 52–61. [CrossRef] [PubMed]
20. Erdle, K.; Mistele, B.; Schmidhalter, U. Spectral high-throughput assessments of phenotypic differences in biomass and nitrogen partitioning during grain filling of wheat under high yielding Western European conditions. *F. Crop. Res.* **2013**, *141*, 16–26. [CrossRef]
21. Fischer, R.A. Wheat physiology: A review of recent developments. *Crop Pasture Sci.* **2011**, *62*, 95–114. [CrossRef]
22. Marino, S.; Alvino, A. Detection of homogeneous wheat areas using multi-temporal UAS images and ground truth data analyzed by cluster analysis. *Eur. J. Remote Sens.* **2018**, *51*, 266–275. [CrossRef]
23. Damian, J.M.; De Castro Pias, O.H.; Santi, A.L.; Di Virgilio, N.; Berghetti, J.; Barbanti, L.; Martelli, R. Delineating management zones for precision agriculture applications: A case study on wheat in sub-tropical Brazil. *Ital. J. Agron.* **2016**, *11*, 171–179. [CrossRef]
24. Chlingaryan, A.; Sukkarieh, S.; Whelan, B. Machine learning approaches for crop yield prediction and nitrogen status estimation in precision agriculture: A review. *Comput. Electron. Agric.* **2018**, *151*, 61–69. [CrossRef]
25. Zadoks, J.C.; Chang, T.T.; Konzak, C.F. A decimal code for the growth stages of cereals. *Weed Res.* **1974**, *14*, 415–421. [CrossRef]
26. Küng, O.; Strecha, C.; Beyeler, A.; Zufferey, J.-C.; Floreano, D.; Fua, P.; Gervaix, F. The accuracy of automatic photogrammetric technique on ultra-light UAV imagery. *Int. Arch. Photogramm. Remote Sens. Spat. Inf. Sci.* **2012**, *XXXVIII-1/C22*, 125–130. [CrossRef]
27. Lopresti, M.F.; Di Bella, C.M.; Degioanni, A.J. Relationship between MODIS-NDVI data and wheatyield: A case study in Northern Buenos Airesprovince, Argentina. *Inf. Process. Agric.* **2015**, *2*, 73–84. [CrossRef]
28. Colomina, I.; Molina, P. Unmanned aerial systems for photogrammetry and remote sensing: A review. *ISPRS J. Photogramm. Remote Sens.* **2014**, *92*, 79–97. [CrossRef]
29. Rondeaux, G.; Steven, M.; Baret, F. Optimization of soil-adjusted vegetation indices. *Remote Sens. Environ.* **1996**, *55*, 95–107. [CrossRef]
30. Ward, J.H. Hierarchical Grouping to Optimize an Objective Function. *J. Am. Stat. Assoc.* **1963**, *58*, 236–244. [CrossRef]
31. Van Ittersum, M.K.; Cassman, K.G.; Grassini, P.; Wolf, J.; Tittonell, P.; Hochman, Z. Yield gap analysis with local to global relevance—A review. *Field Crop. Res.* **2013**, *143*, 4–17. [CrossRef]
32. Aparicio, N.; Villegas, D.; Casadesus, J.; Araus, J.L.; Royo, C. Spectral Vegetation Indices as Nondestructive Tools for Determining Durum Wheat Yield. *Agron. J.* **2000**, *92*, 83–91. [CrossRef]
33. Dang, Y.P.; Pringle, M.J.; Schmidt, M.; Dalal, R.C.; Apan, A. Identifying the spatial variability of soil constraints using multi-year remote sensing. *F. Crop. Res.* **2011**, *123*, 248–258. [CrossRef]
34. Aparicio, N.; Villegas, D.; Royo, C.; Casadesus, J.; Araus, J.L. Effect of sensor view angle on the assessment of agronomic traits by ground level hyper-spectral reflectance measurements in durum wheat under contrasting Mediterranean conditions. *Int. J. Remote Sens.* **2004**, *25*, 1131–1152. [CrossRef]
35. Dalla Marta, A.; Grifoni, D.; Mancini, M.; Orlando, F.; Guasconi, F.; Orlandini, S. Durum wheat in-field monitoring and early-yield prediction: Assessment of potential use of high resolution satellite imagery in a hilly area of Tuscany, Central Italy. *J. Agric. Sci.* **2015**, *153*, 68–77. [CrossRef]
36. Reyniers, M.; Vrindts, E. Measuring wheat nitrogen status from space and ground-based platform. *Int. J. Remote Sens.* **2006**, *27*, 549–567. [CrossRef]
37. Magney, T.S.; Eitel, J.U.H.; Huggins, D.R.; Vierling, L.A. Proximal NDVI derived phenology improves in-season predictions of wheat quantity and quality. *Agric. For. Meteorol.* **2016**, *217*, 46–60. [CrossRef]
38. Marti, J.; Bort, J.; Slafer, G.A.; Araus, J.L. Can wheat yield be assessed by early measurements of Normalized Difference Vegetation Index? *Ann. Appl. Biol.* **2007**, *150*, 253–257. [CrossRef]
39. Villegas, D.; Aparicio, N.; Blanco, R.; Royo, C. Biomass accumulation and main stem elongation of Durum wheat grown under Mediterranean conditions. *Ann. Bot.* **2001**, *88*, 617–627. [CrossRef]

40. Cabrera-Bosquet, L.; Molero, G.; Stellacci, A.; Bort, J.; Nogués, S.; Araus, J. NDVI as a potential tool for predicting biomass, plant nitrogen content and growth in wheat genotypes subjected to different water and nitrogen conditions. *Cereal Res. Commun.* **2011**, *39*, 147–159. [CrossRef]
41. Visioli, G.; Bonas, U.; Dal Cortivo, C.; Pasini, G.; Marmiroli, N.; Mosca, G.; Vamerali, T. Variations in yield and gluten proteins in durum wheat varieties under late-season foliar versus soil application of nitrogen fertilizer in a northern Mediterranean environment. *J. Sci. Food Agric.* **2018**, *98*, 2360–2369. [CrossRef] [PubMed]
42. Marino, S.; Alvino, A. Hyperspectral vegetation indices for predicting onion (*Allium cepa* L.) yield spatial variability. *Comput. Electron. Agric.* **2015**, *116*. [CrossRef]
43. Marino, S.; Basso, B.; Leone, A.P.; Alvino, A. Agronomic traits and vegetation indices of two onion hybrids. *Sci. Hortic.* **2013**, *155*, 56–64. [CrossRef]

7

Crop Row Detection through UAV Surveys to Optimize On-Farm Irrigation Management

Giulia Ronchetti [1,*], Alice Mayer [2], Arianna Facchi [2], Bianca Ortuani [2] and Giovanna Sona [1]

1 Department of Civil and Environmental Engineering, Politecnico di Milano, 20133 Milan, Italy; giovanna.sona@polimi.it
2 Department of Agricultural and Environmental Sciences – Production, Landscape, Agroenergy, Università degli Studi di Milano, 20133 Milan, Italy; alice.mayer@unimi.it (A.M.); arianna.facchi@unimi.it (A.F.); bianca.ortuani@unimi.it (B.O.)
* Correspondence: giulia.ronchetti@polimi.it

Abstract: Climate change and competition among water users are increasingly leading to a reduction of water availability for irrigation; at the same time, traditionally non-irrigated crops require irrigation to achieve high quality standards. In the context of precision agriculture, particular attention is given to the optimization of on-farm irrigation management, based on the knowledge of within-field variability of crop and soil properties, to increase crop yield quality and ensure an efficient water use. Unmanned Aerial Vehicle (UAV) imagery is used in precision agriculture to monitor crop variability, but in the case of row-crops, image post-processing is required to separate crop rows from soil background and weeds. This study focuses on the crop row detection and extraction from images acquired through a UAV during the cropping season of 2018. Thresholding algorithms, classification algorithms, and Bayesian segmentation are tested and compared on three different crop types, namely grapevine, pear, and tomato, for analyzing the suitability of these methods with respect to the characteristics of each crop. The obtained results are promising, with overall accuracy greater than 90% and producer's accuracy over 85% for the class "crop canopy". The methods' performances vary according to the crop types, input data, and parameters used. Some important outcomes can be pointed out from our study: NIR information does not give any particular added value, and RGB sensors should be preferred to identify crop rows; the presence of shadows in the inter-row distances may affect crop detection on vineyards. Finally, the best methodologies to be adopted for practical applications are discussed.

Keywords: UAV; crop row; precision agriculture; DEM; RGB; vegetation indices

1. Introduction

According to the most recent projections presented by the Intergovernmental Panel on Climate Change (IPCC), the variation in precipitation is altering hydrological systems in many agricultural areas of the planet, affecting water resources in terms of both quantity and quality [1]. In particular, climate change is leading to more frequent extreme events (droughts and heat waves, occurring more often and lasting longer), to the alteration of spatial and temporal precipitation (less precipitation concentrated in a few heavy rainfall events), as well as to an increase in air temperatures and crop water needs in many geographical areas.

In Mediterranean countries, freshwater resources are currently highly exploited due to the rapid population growth and intensive water use in agriculture, industry, and tourism activities. In many areas of Europe, including Italy, the effects of climate change are already detectable and have led to the need to irrigate crops and areas traditionally not irrigated [2].

In the context of Precision Agriculture (PAg), particular attention is given to the optimization of on-farm irrigation management, since water resources for agricultural use have become scarcer due to the combined effect of climate change and the increased competition among different water uses [1,3]. Moreover, due to the intensification of extreme weather events, irrigation is becoming an important tool to guarantee adequate quality standards to agricultural products [4]. The optimization of on-farm irrigation management through variable rate irrigation systems is based on the detection of the spatial variability of soil and crop properties. Variable rate irrigation is aimed at managing water inputs to match the spatial variability of the water requirements found in the field, by providing irrigation in different amounts depending on real crop requirements. This approach represents a valid solution to increase water use efficiency and water savings, and for certain crops, variable water application might lead also to an increase of yield and product quality [5–7].

Currently, the detection of within-field spatial variability of crops can be easily carried out through many different technologies, including proximal and remote sensing techniques, sensors mounted on agricultural machinery, georeferencing systems, and geographical information systems [8]. Among them, the noteworthy development of Unmanned Aerial Vehicles (UAVs) registered in the last few years has allowed filling the gap between remote sensing and terrestrial techniques in agricultural applications [9]. Using UAVs is a good compromise between the large coverage obtainable with remote platforms (mainly satellite and aircraft) and the accuracy of the terrestrial data, with advantages in terms of time consumption and the costs of the surveys.

By allowing collecting information closely from above the crop canopy, UAVs have introduced a new point of view for agricultural surveys, giving rise to several applications. As early as 2008, Nebiker et al. [10] proposed the prototype of a multispectral sensor suitable to be mounted on a UAV and conducted experiments with it to assess vegetation health with promising results. Starting from this experience, a variety of studies can be found in the literature about the effectiveness of UAV surveys conducted for PAg purposes. The main applications involve in-field weed mapping, vegetation growth monitoring, crop water stress analysis, and optimization of irrigation management [9,11].

The major strength of UAV surveys is to provide information in a rapid, non-destructive way with a high spatial resolution that can detect within-field variation in detail. Nevertheless, when UAV imagery is used to monitor crop variability in row-crops, a post-processing procedure must be set up to identify and extract crop rows from soil background and weeds, thus generating a crop mask. This issue is crucial to manage precision irrigation properly, by adjusting water supplies according to crop water requirements, considering the actual crop water status and vigor and their spatial variability. Moreover, the crop row detection is crucial to reliably assess the effectiveness of variable rate irrigation in optimizing the crop growth by reducing crop variability. In both of these cases, a non-accurate recognition of crop rows could induce misleading assessment of crop status, then erroneous evaluation of water amounts to be applied through irrigation, as well as erroneous validation of irrigation management. The same highlights are also valid for fertilization practices based on prescription maps, obtained from NDVI maps acquired through UAV surveys. Fertilization is more effective the more accurate the crop row detection is. As a matter of fact, what has been reported is valid in fruit-viticulture and horticulture, where nutrients are applied by fertigation through drip irrigation systems. On the other hand, crop row detection can be useful even for herbaceous crops sown in rows, such as maize. In the case of maize, procedures to generate crop masks are fundamental to manage precision fertilization or to control weeds in the early growth stages, before the canopy closure when the crop rows are still recognizable [12,13].

Different studies can be found in the literature, proposing (semi)automatic methods, using image-processing techniques on single-band images, maps of Vegetation Indices (VIs), or Digital Elevation Models (DEMs), to detect crop canopy [14]. Poblete-Echeverría et al. [15] compared the performance of four classification methods, including standard and well known methods (i.e., K-means and VIs' thresholding) and machine learning methods (i.e., artificial neural networks and random forest), to detect vine canopy using ultra-high-resolution RGB imagery acquired with a conventional camera mounted on a low-cost UAV. Marques et al. [16] presented a UAV-based automatic method to detect chestnut trees, by using RGB and CIR (Color Infrared) orthomosaics combined with the canopy height model. In [17], potato plant objects were extracted from bare soil using the excess green index and Otsu thresholding methods.

This study focuses on the crop row detection and extraction by analyzing and post-processing images acquired through a UAV. Different methodologies are tested and compared, as well as several segmentation methods, such as supervised classifications, Bayesian segmentation, and thresholding algorithms, developed ad hoc for this purpose. Extraction algorithms are applied both on geometric products (i.e., digital elevation model) and vegetation indices' maps. As already presented by other studies [15,16,18], the proposed methods exploit existing indicators, including NDVI and RGB derived indices. In this way, this study aims at demonstrating the importance and effectiveness of UAV systems in precision agriculture and providing to less experienced users the opportunity to use them. As an added value, using simple tools and proving their usefulness would allow the diffusion of these methodologies to a wide audience, even outside the academic environment. All the described methods are semi-automatic and require little human intervention for parameter choosing and fine-tuning. The assessment of the methods was performed on three different crop types, grapevine, pear, and tomato, to analyze the suitability of the proposed methods with respect to the characteristics of each crop. Considering the application of the proposed algorithms to different agro-ecosystems, each with its own peculiarities, represents the real novelty of this study with respect to the existing literature.

2. Materials

2.1. Study Sites

This study is part of the two year project NUTRIPRECISO (RDP-EU, Measure 1.2.01, Lombardy Region), which involved the Department of Agricultural and Environmental Sciences (DiSAA), University of Milan, the Department of Civil and Environmental Engineering (DICA), Politecnico di Milano, and CREA (Consiglio per la ricerca in agricoltura e l'analisi dell'economia agraria). The project was aimed at developing and disseminating practices of PAg in fruit viticulture and horticulture. Consequently, in this study, simple and reliable methodologies, moreover using commonly available data, are presented to be proposed to the farmers and potential users.

Three different types of crops were investigated in the NUTRIPRECISO project: grapevine, pear, and tomato; therefore, three different areas in the Lombardy Region were chosen as study sites, to be representative of each crop. The location of the vineyard is shown in Figure 1a, while the pear orchard and tomato sites are reported in Figure 1b. In the following, the main characteristics of the three study sites are described.

2.1.1. Vineyard

The vineyard, with an extent of 1 ha, is located in Olfino di Monzambano, in the province of Mantova, the heart of the Morainic Hills region (Northern Italy).
The vineyard is nearly flat and located at an altitude of about 88 m a.s.l. A soil survey with an Electro-Magnetic Induction (EMI) sensor followed by a traditional pedological survey showed that the field, despite its small dimension, was characterized by four different soil types with coarse soils in the northwestern part and fine heavy soils in the eastern part [19].

According to the ARPA (Regional Environmental Protection Agency) agro-meteorological station located at Ponti sul Mincio (about 6.5 km from the vineyard), the climatic conditions are warm and mild with average maximum and minimum temperatures during summer of about 30 and 20 °C, respectively, and rainfall concentrated in spring and autumn, with a mean annual value of 765 mm (values calculated as the average over the period 1993–2017), which provide favorable conditions for the grapevine to grow.

The vineyard variety was Chardonnay, cultivated in rows oriented along the east-west axis, with a distance of plants on the row of 0.8 m and a distance between the rows of 2.4 m, while the row height was about 1.3 m. The plant cover fraction in the phase of maximum development of the canopy was estimated to be about 25% (equivalent to a row width of 0.6 m). The soil, both under the rows and between the rows, was grass-covered with periodic mowing to regulate the excessive development of vegetation [19].

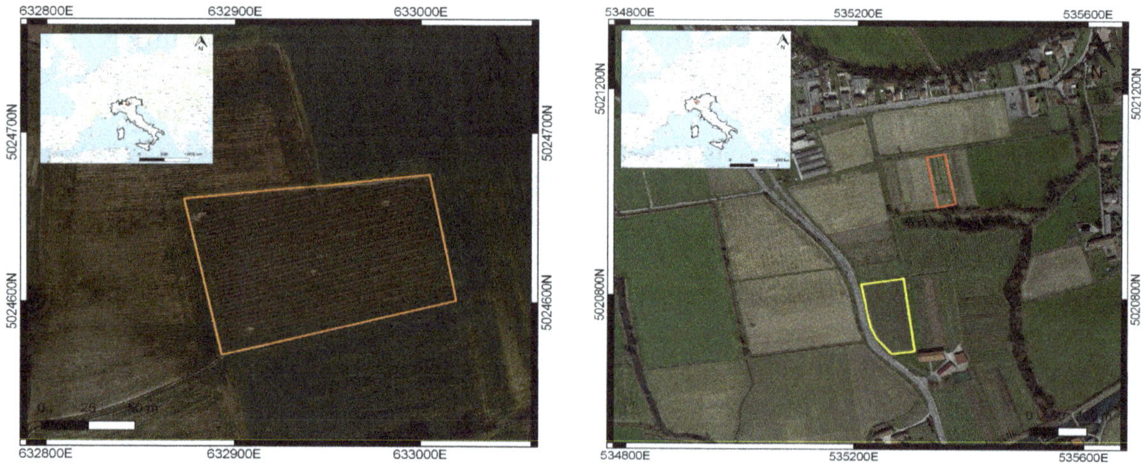

(a) (b)

Figure 1. The experimental sites: (**a**) the vineyard; (**b**) in yellow the pear orchard and in red the tomato field. Coordinate Reference System (CRS): WGS84/UTM Zone 32N. Map data: ©OpenStreetMap contributors.

2.1.2. Pear Orchard

The pear orchard was situated in the "Dotti" farm, a research facility of the University of Milan, located in Arcagna locality, Montanaso Lombardo (province of Lodi (LO)).

The orchard covers an area of about 1 ha and is flat, positioned at about 80 m a.s.l. According to the data recorded in the period 1993–2017 at the nearest ARPA agro-meteorological station (located at about 12 km southeast from the experimental site in Cavenago d'Adda), the climate is characterized by two rainy periods, respectively in April and September, while the highest temperatures occur in July, when rain is minimum.

The soil is loam with more clay in deeper horizons. In the orchard, four different pear varieties were cultivated, namely Williams, Abate, Kaiser, and Conference varieties, distributed in 17 rows with an inter-row distance of about 4 m, while the distance between plants on the row was about 1.5 m, depending on the variety; the soil, under the rows and between the rows, was grass-covered with periodic cutting.

2.1.3. Tomato Field

The third site included in the project is an area of about one hectare cultivated with industrial tomato "Pietra Rossa F1", inside the CREA research center in Montanaso Lombardo (LO). The site is characterized by loamy soils with an increase of clay with the depth, while climatic conditions are the same as the ones above described for the pear orchard site. Tomatoes were 0.3 m high at images' acquisition and cultivated in parallel rows of 0.5 m width with a distance between rows of about 1.5 m.

2.2. UAV Surveys and Photogrammetric Processing

Two UAV surveys were conducted on each study site, during the agricultural season of 2018. For each site, in the first survey, the Parrot Bebop 2 (Parrot S.A., Paris, France) was used to acquire

RGB images, and the Parrot Sequoia camera (Parrot S.A., Paris, France) was mounted on the Parrot Disco (Parrot S.A., Paris, France) to collect multispectral imagery during the second survey. The photogrammetric processing of all the surveys was performed with Pix4Dmapper Pro software (Pix4D S.A., Prilly, Switzerland), Version 4.1.1. The details of the surveys and their processing are described in the following sections.

2.2.1. Vineyard

The RGB survey of the vineyard took place on 28 June 2018, while the multispectral survey, six days later on 4 July. The flight height of the Parrot Bebop 2 used in the RGB survey was set to 40 m Above Ground Level (AGL), while the Parrot Disco flew at 60 m (AGL) during the multispectral survey. The same overlaps among images were fixed for the two surveys: longitudinal overlapping equal to 80% and transversal equal to 70%. An amount of 130 and 560 images were collected during the RGB and multispectral survey, respectively. According to the sensors' characteristics (i.e., Parrot Bebop 2 fisheye camera and Sequoia camera), the final Ground Sample Distance (GSD) of the acquired images was about 0.1 m for both cases. As suggested in [20], the coordinates of nine Ground Control Points (GCPs) were measured with the Global Navigation Satellite System (GNSS) receiver Leica Viva GS14 (Leica Geosystems, Heerbrugg, Switzerland) in Network Real-Time Kinematic (NRTK) mode, to ensure the georeferencing of the photogrammetric products with high accuracy. According to the results of [20], eight GCPs were distributed all around the perimeter of the vineyard, while the ninth target was placed in the middle of the field (Figure 2). In all the study sites, some of the GCPs were adopted as Check Points (CPs) during the photogrammetric processing, to assess process accuracy.

Figure 2. Ground Control Points' (GCPs) distribution for the surveys on the vineyard. Map data: ©Google Satellite.

At the end of two independent photogrammetric processing, performed with Pix4Dmapper Pro software, a Digital Surface Model (DSM), an RGB orthophoto, and a four band multispectral orthophoto were produced, with GSD of around 0.1 m, to exploit for the crop row detection. The DSM generated from the multispectral survey had lower quality than the RGB one; therefore, it was not considered in further analysis. Table 1 summarizes the residuals on GCPs after photogrammetric processing, while in Figure 3, DSM and orthophotos of the vineyard are shown.

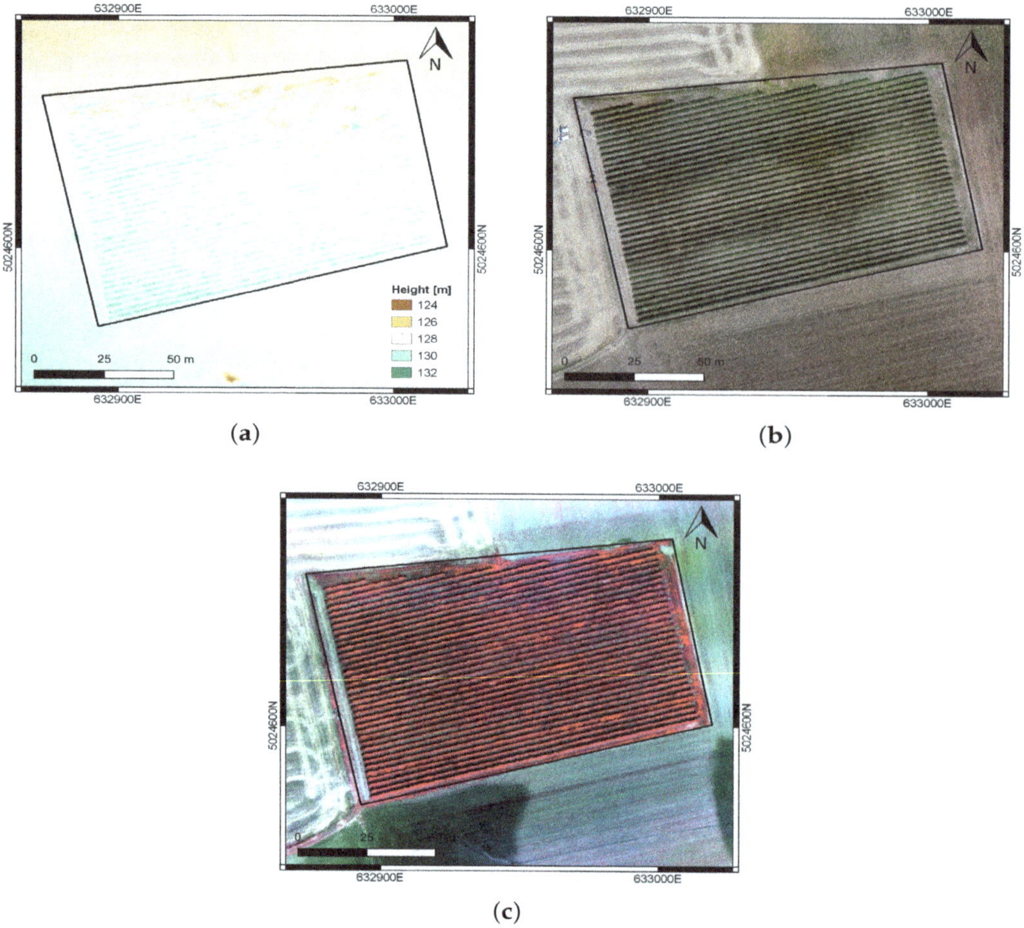

Figure 3. Vineyard site: DSM (**a**) and orthophoto (**b**) produced through photogrammetric processing of the RGB dataset; false color orthophoto (**c**), generated from the multispectral dataset.

Table 1. Vineyard site: residuals on the GCPs after bundle block adjustment.

Label	Easting (m)	Northing (m)	Height (m)
v1	−0.029	−0.008	0.088
v2	−0.007	−0.012	−0.012
v3	−0.025	−0.030	−0.063
v4	0.008	0.025	0.062
v5	0.015	0.005	0.000
v6	0.022	−0.013	0.024
v7	0.030	−0.013	0.027
v8	0.020	−0.015	−0.086
v9	0.012	0.008	0.012
RMSE	**0.021**	**0.016**	**0.052**

2.2.2. Pear Orchard

In the pear orchard site, the RGB and multispectral surveys were conducted on 26 June and 2 July, respectively. The characteristics of the flights were the same as the surveys performed on the vineyard site: longitudinal overlap among images equal to 80%, transversal overlap equal to 70%, flight height set at 40 m and 60 m for the multirotor UAV and the fixed-wings UAV, respectively, thus ensuring a GSD of the images of about 0.1 m. 140 images were acquired during the RGB survey, while 540 during the multispectral one. During the UAV flights, seven targets were placed on the terrain to be used as GCPs, well distributed all around the orchard, as shown in Figure 4.

Figure 4. Ground Control Points' (GCPs) distribution for the surveys on the pear orchard. Map data: ©Google Satellite.

As the case of the vineyard, the DSM and the orthophotos were produced in Pix4Dmapper Pro with a spatial resolution of 0.1 m (Figure 5). The residuals on the GCPs computed after the photogrammetric workflow are reported in Table 2.

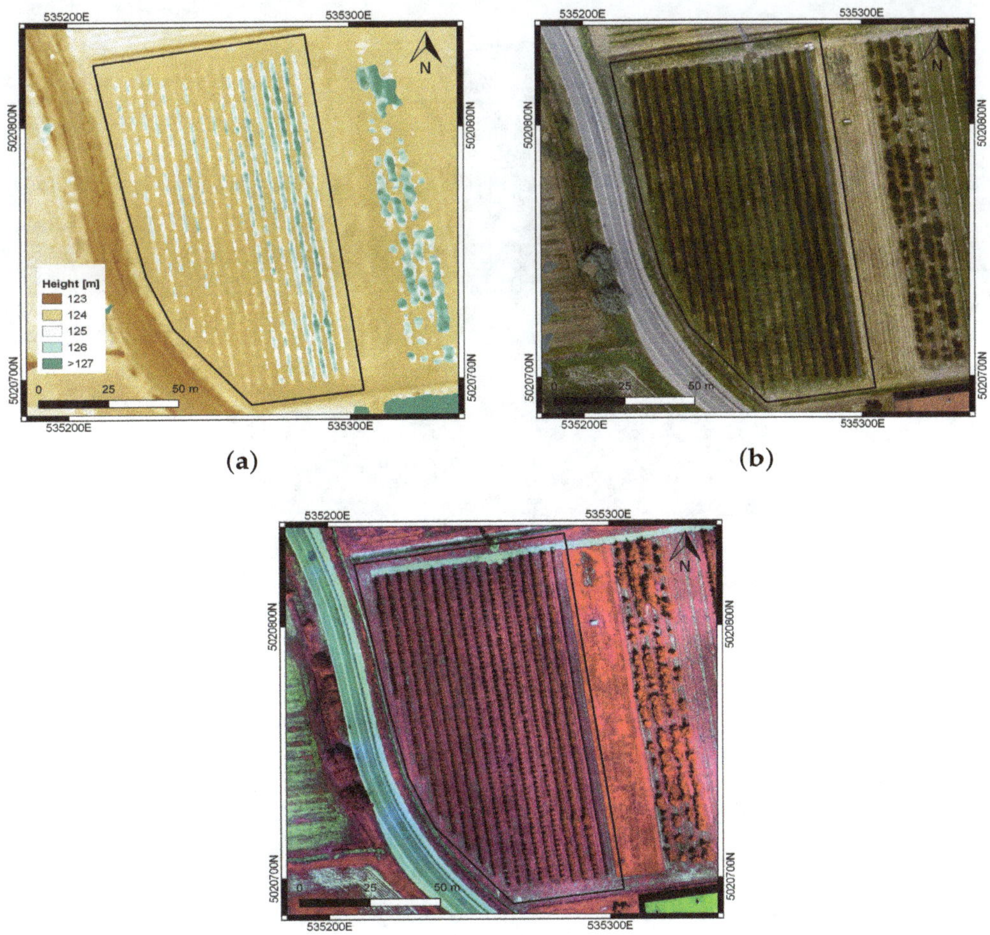

(a) (b) (c)

Figure 5. Pear orchard site: DSM (**a**) and orthophoto (**b**) produced through photogrammetric processing of the RGB dataset; false color orthophoto (**c**), generated from the multispectral dataset.

Table 2. Pear orchard site: residuals on the GCPs after bundle block adjustment.

Label	Easting (m)	Northing (m)	Height (m)
p1	−0.050	−0.017	0.007
p2	−0.032	−0.004	−0.017
p3	0.018	−0.060	0.021
p4	0.028	−0.089	0.113
p5	0.003	0.023	−0.290
p6	0.100	0.053	0.018
p7	−0.085	0.003	0.054
RMSE	**0.056**	**0.047**	**0.119**

2.2.3. Tomato Field

Given the proximity of the two sites, the tomato field was surveyed on the same days as the pear orchard, with the same equipment and flight characteristics. During the RGB survey, 96 images were acquired, as well as 388 images came from the multispectral flight. From these datasets, one DSM and two orthophotos, having a GSD equal to 0.1 m, were generated after bundle block adjustment in Pix4Dmapper Pro. The various photogrammetric products were georeferenced by means of six GCPs, whose center coordinates were measured with GNSS-NRTK on the dates of the surveys. The distribution of the GCPs on the tomato site and their computed residuals are shown in Figure 6 and Table 3, respectively. The photogrammetric products are reported in Figure 7.

Figure 6. Ground Control Points' (GCPs) distribution for the surveys on the tomato site. Map data: ©Google Satellite.

Table 3. Tomato site: residuals on the GCPs after bundle block adjustment.

Label	Easting (m)	Northing (m)	Height (m)
t1	−0.098	0.071	−0.101
t2	0.095	0.068	0.046
t3	0.028	0.71	0.077
t4	−0.026	−0.127	−0.054
t5	−0.019	−0.154	0.149
t6	−0.053	−0.038	0.114
RMSE	**0.062**	**0.096**	**0.097**

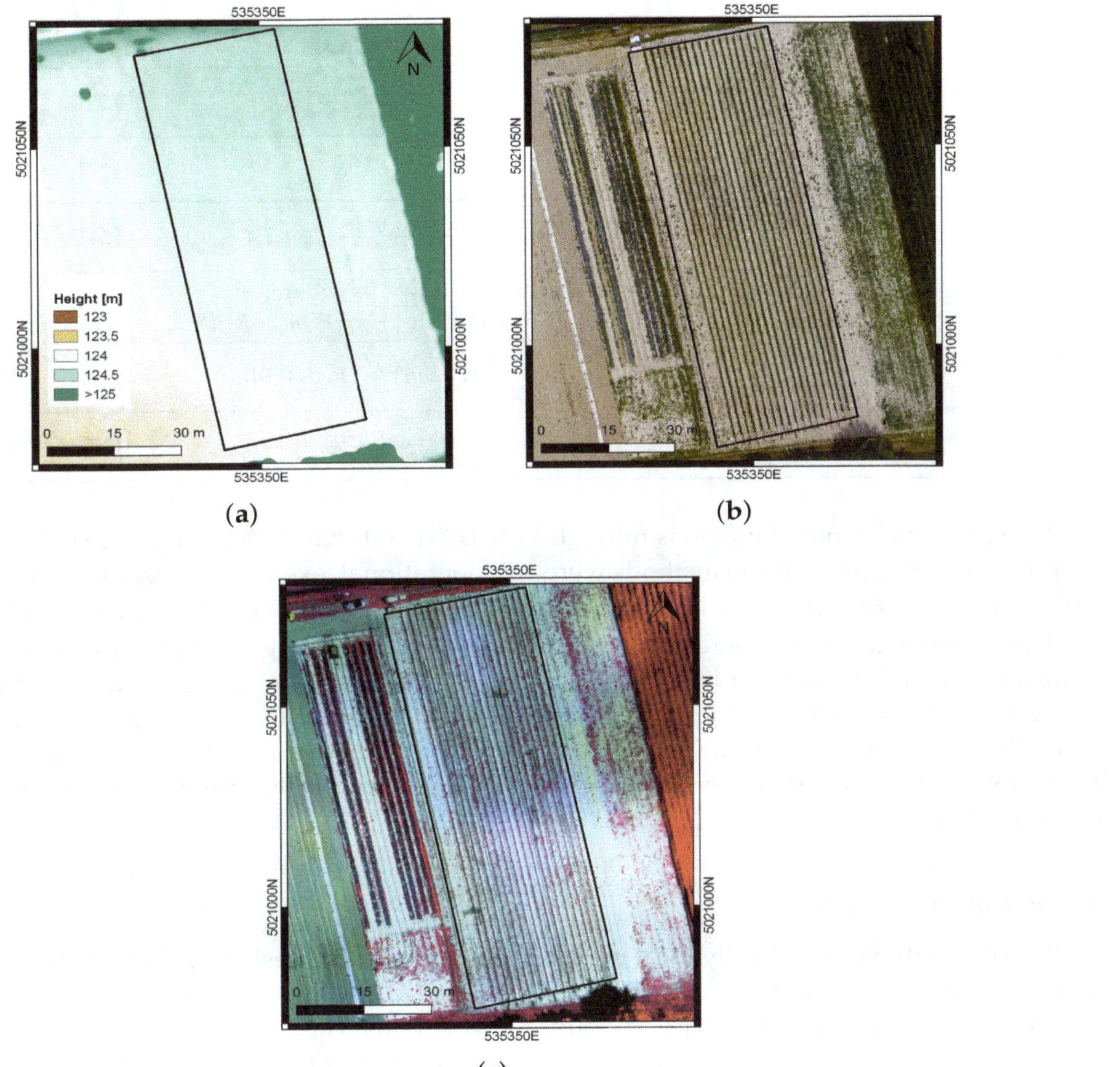

Figure 7. Tomato site: DSM (**a**) and orthophoto (**b**) produced through photogrammetric processing of RGB dataset; false color orthophoto (**c**), generated from the multispectral dataset.

3. Crop Row Detection Methods

Differentiating between crop canopy and background can be very challenging. Moreover, the different crop types considered in this study led to the need for many methods to extract crop rows, each one more suitable for a specific crop type. In the following sections, five different detection methods were proposed; some of them were taken from the existing literature (i.e., classification algorithms and Bayesian segmentation), while two methods, labeled as thresholding algorithms, were developed ad hoc for the purposes of the project.

In order to achieve the best possible crop mask to be used to identify crop rows on orthophotos, Vegetation Indices (VIs) were chosen as the inputs of the detection methods, as already proposed by many authors [15,16,21]. Considering vegetation, most of those indices take into account Red (R) and NIR reflectance bands (ρ_λ): the greater is the difference between ρ_R and ρ_{NIR}, the greater is the amount of green and healthy vegetation in that particular pixel. Among all the possible VIs, only those composed of spectral bands that sensors involved in the surveys could provide were used in this study (Table 4).

Table 4. Vegetation Indices (VIs) used in this study.

Index	Name	Formula	References
NDVI	Normalized Difference Vegetation Index	$\frac{NIR-Red}{NIR+Red}$	[22]
SR	Simple Ratio	$\frac{NIR}{Red}$	[23]
SAVI	Soil-Adjusted Vegetation Index	$\frac{NIR-Red}{NIR+Red+L}(1+L)$	[24]
ARVI	Atmospherically Resistant Vegetation Index	$\frac{NIR-RB}{NIR+RB}$ where: $RB = Red - \gamma(Blue - Red)$	[25]
ExG	Excess Green	$2(Green) - (Red + Blue)$	[26]
G%	Normalized Green Channel Brightness	$\frac{Green}{Red+Green+Blue}$	[27]

To exploit the proposed methods fully, also the DSM and RGB orthophoto were individually used as inputs. This ensured that the methods were still operational, even in the cases where only imagery resulting from UAVs supporting only RGB sensors was available, as for the Parrot Bebop 2.

The assessment of the final results was performed by computing error matrices and classification accuracies (Overall Accuracy (OA), User's Accuracy (UA), and Producer's Accuracy (PA)), on some validation samples manually identified on orthophotos, through visual inspection. In particular, the quality of the crop detection was defined according to the value of PA of the class crop canopy: the greater is the PA, the lower is the probability to omit crop pixels; therefore, to underestimate the detected crop rows.

3.1. Thresholding Algorithms

Two algorithms (specifically local maxima extraction and threshold selection) were developed ad hoc for the purpose of the project, by starting from [28], who proposed a method for crop rows' extraction by using as input the 3D point cloud. Both methods were generated in MATLAB R2017b [29] and are based on the concept that high pixel values generally correspond to crop rows. They are illustrated in the following sections.

3.1.1. Local Maxima Extraction

This method aims at generating a binary raster, where non-null values refer to the presence of the crop canopy. First, the input raster (e.g., VI map or DSM) is divided into square cells (macro-cells), then inside each macro-cell, the crop pixels are identified as those corresponding to a percentage of pixels with the highest values. It is a semi-automatic algorithm, where the user has to define the dimensions of the cell and the percentage value.

This method is sensitive to the user's choices: in particular, the macro-cell dimension should be selected in order to include both crop and ground pixels, thus being close to the distance between rows or slightly larger. Macro-cell size should be neither too small nor too big with respect to the distance between rows. If it is too small, a wrong selection of pixels is performed whatever the chosen percentage is. When the cell includes only crop pixels, whatever percentage not equal to 100% causes an underestimation of crop pixels; on the other side, when the cell overlaps only ground pixels, they would be selected as crop pixels, thus producing an over-estimation in the crop mask. The overestimation arises also when the dimension of the macro-cell is too big, because the probability of selecting pixels belonging to the ground increased with the chosen cell size.

3.1.2. Threshold Selection

This method produces a binary crop mask, by selecting as crop pixels all pixels with values higher than a reference value. The challenging part of this algorithm is the definition of this reference value. When both DSM and Digital Terrain Model (DTM) of the area are available, the Canopy Height Model (CHM) could be derived as difference between DSM and DTM, and zero could be considered as the reference value, while in other cases (i.e., VIs as input, no availability of an accurate DTM), it should be determined as described below.

Starting from the input raster, create a smoothed raster with a moving window average filter. Subtract the smoothed raster to the input raster, and define on the differences a threshold to be considered as the reference value, by visual checking of the results with an empirical trial and error approach. Pixels with values greater than the threshold are retained as crop. Even in this algorithm, the user intervention is twofold, choosing the dimensions of the moving window and the value of the threshold, and could cause problems of under-/over-estimation of crop canopy pixels.

3.2. Classification Algorithms

Two well-known classification algorithms were exploited in this study, K-means clustering and the Minimum Distance to the Mean (MDM) classifier, to be representative of both unsupervised and supervised classification algorithms. Both methods were applied in QGIS (Version 3.4) [30], to allow users not familiar with programming languages to run the algorithms thanks to a dedicated user-friendly GUI.

3.2.1. K-Means Clustering

This is a well-known algorithm for hard unsupervised thematic classification [31]. The clustering made by K-means is based on the minimization of the objective function $f(\Omega)$, defined as the Euclidean distance of samples of a cluster from the respective centroid.

The number of classes (K) are known a priori. Once K is defined, the method consists of three iterative steps. In the first step, for each class K_i, a centroid is automatically chosen. The rest of the data are assigned to K clusters based on the minimum distance criterion. The Euclidean distances of each sample from the centroids are computed, and in the second step, the sample is assigned to the cluster for which the computed distance is minimum. In the last step, centroids are re-calculated, and all the samples are re-assigned. This step is iterated until the clustering converges to a stable solution, namely when centroids of clusters do not change meaningfully.

The final configuration is stable and does not depend on the initial position of centroids arbitrarily selected. The initial configuration only influences the number of iterations necessary to reach the convergence.

3.2.2. Minimum Distance to Mean Classifier

This method finds the mean values of all the training sets and classifies all the image pixels according to the class mean they are closest. The process is performed for all image pixels, one at a time. Bounds are determined using statistics derived from the training sets, and the distance used is the Euclidean one.

As for all supervised algorithms, a crucial phase of the MDM classifier is the selection of the training samples. They are used to compute class spectral signatures, therefore must be representative of all the classes. In this study, training samples were defined by visual inspections of the UAV images and grouped in two macro-classes: crop canopy and background, which included weeds, soil, and shadow pixels.

3.3. Bayesian Segmentation

This method relies on the Bayesian approach, where any uncertainty can be considered random variables that are fully described by probability distributions [32]. Given the vector of data y and the vector of parameter x, the conditional distribution of parameters is described by the Bayes theorem [33]:

$$P(x|y) = \frac{P(x,y)}{P(y)} = \frac{P(y|x)P(x)}{P(y)} \tag{1}$$

where:
$P(x|y)$ is called the posterior probability and describes the new level of knowledge of the unknown parameters x given the observed data y.
$P(y)$ is a normalization constant used to impose that the sum of $P(y|x)$ for all possible x is equal to one.
$P(x)$, instead, represents the prior probability distribution. It describes the knowledge of the unknown parameters x without the contribution of the observed data.
$P(y|x)$ is defined as the likelihood and is a function of x. It describes the way in which the a priori knowledge is modified by data and depends on the noise distribution.

The terms in Equation (1) can be adapted to match the purpose of this study, the detection of crop rows: the posterior probability is the probability of a pixel to be part of the class crop canopy or background, and the prior probability is defined starting from the mean and standard deviation values, a priori assigned to each class, while the likelihood is described by a Gaussian distribution, in which the parameters are the mean and standard deviation of the two classes:

$$P(y_i|x_i) = \frac{1}{\sigma_{x_i}\sqrt{2\pi}} exp\left(-\frac{(y_i-\mu_{x_i})^2}{2\sigma_{x_i}^2}\right) \tag{2}$$

The final goal of Bayesian approach can be identified in finding the optimal parameters x that maximize the posterior probability distribution $P(x|y)$. This is called the Maximum A Posteriori (MAP) estimate [34], and it is defined as:

$$x^{MAP} = \arg\max_x P(x|y) \tag{3}$$

In crop row detection, it consists of assigning a unique class to each pixel of the image, depending on the posterior probabilities estimated for each pixel. In order to obtain outputs less affected by pixel noise, smoothing filters or image adjustment can be applied on the input raster.

4. Results

Considering all the detection methods, their parameters, and all the possible input rasters, the number of crop masks potentially available is very high. For the vineyard site, 191 outputs were tested and 166 and 104 for the pear orchard and tomato site, respectively. For the sake of brevity, only the best masks, representing each detection method, are reported and compared. An exhaustive analysis of all the tests performed was described in [35].

4.1. Vineyard

For the vineyard site, the parameters of each detection method that resulted in the best results are reported in Table 5.

Table 5. Vineyard site: parameters for the best results of each detection method. MDM, Minimum Distance to the Mean.

Method	Input	User's Choices
Local Maxima Extraction	G%	cell size: 5 m percentage: 30%
Threshold Selection	DSM	cell size: 3 m threshold: 0.3
K-means Clustering	RGB orthophoto	classes: 6
MDM Classifier	RGB orthophoto	classes: 2
Bayesian Segmentation	ExG, Gaussian filter ($\sigma = 3$)	Background: $\mu = 0.2, \sigma = 0.2$ Crop canopy: $\mu = 0.7, \sigma = 0.25$

For the computation of the error matrices, 104 polygons (N pixels = 42,732) were defined for the class crop canopy and 97 polygons (N pixels = 64,309) for the class background. In Table 6, the accuracies of the five selected best results are summarized, and the detail of each crop mask is shown in Figure 8.

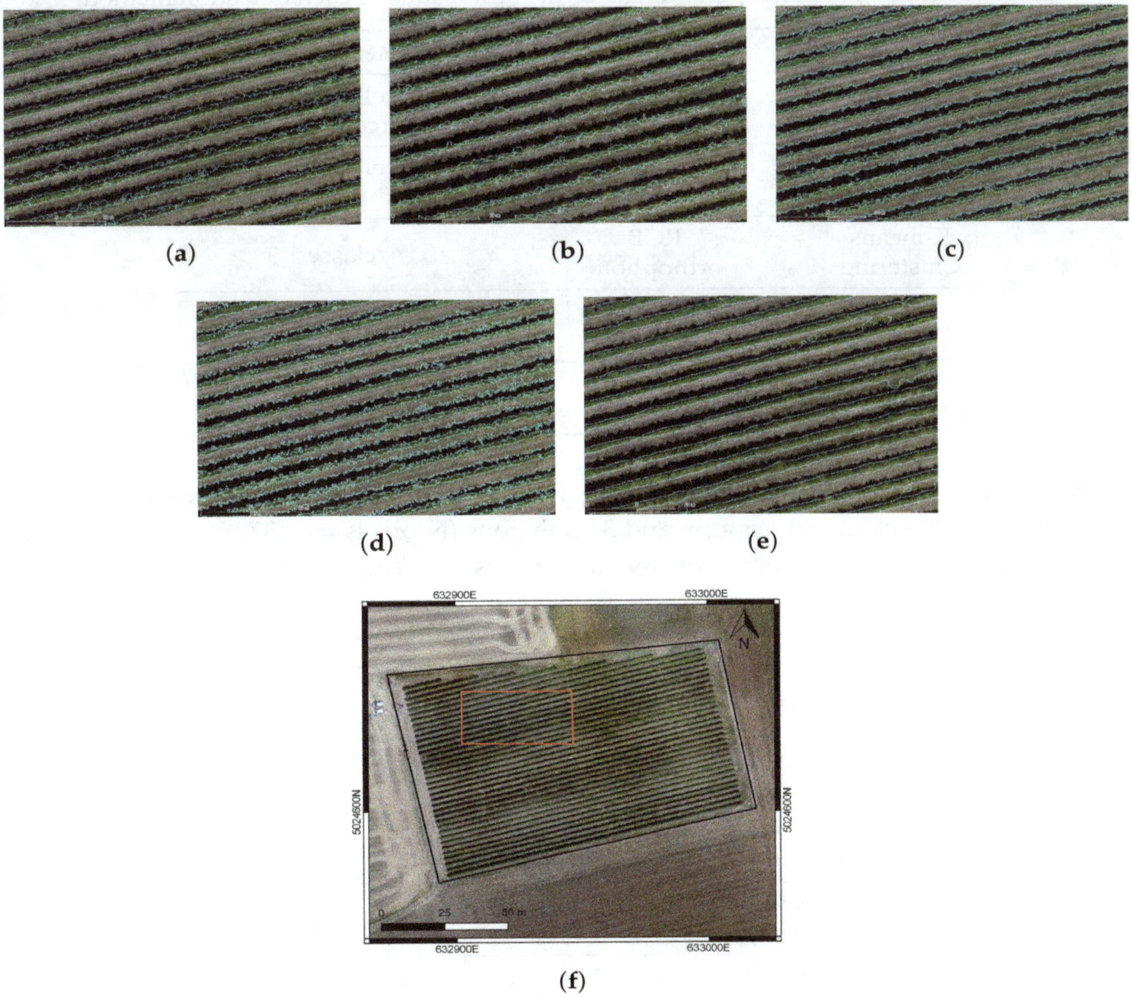

Figure 8. Vineyard site: crop row detection results for (**a**) local maxima extraction, (**b**) threshold selection, K-means clustering (**c**), MDM classifier (**d**), and the Bayesian segmentation (**e**). Figures refer to the area included in the red box in (**f**).

Table 6. Vineyard site: assessment of the best results of each detection method.

Method	OA	PA Crop Canopy	UA Crop Canopy
Local Maxima Extraction	0.94	0.95	0.91
Threshold Selection	0.76	0.41	0.99
K-means Clustering	0.82	0.73	0.80
MDM Classifier	0.87	0.84	0.83
Bayesian Segmentation	0.96	0.97	0.94

4.2. Pear Orchard

The parameters for the best results of each detection method for the pear orchard are summarized in Table 7.

Table 7. Pear orchard site: parameters for the best results of each detection method.

Method	Input	User's Choices
Local Maxima Extraction	DSM	cell size: 4 m percentage: 40%
Threshold Selection	DSM	cell size: 4 m threshold: 0
K-means Clustering	RGB orthophoto	classes: 5
MDM Classifier	RGB orthophoto	classes: 2
Bayesian Segmentation	NDVI, Gaussian filter (σ = 3)	Background: μ=0.8, σ=0.04 Crop canopy: μ=0.93, σ=0.04

The error matrices were computed starting from a validation set composed by 37 polygons (N pixels = 44,889) for the class crop canopy and 34 polygons (N pixels = 62,378) for the class background. The five selected best results and their respective accuracies are presented in Figure 9 and in Table 8.

Table 8. Pear orchard site: assessment of the best results of each detection method.

Method	OA	PA Crop Canopy	UA Crop Canopy
Local Maxima Extraction	0.92	0.88	0.93
Threshold Selection	0.95	0.97	0.92
K-means Clustering	0.95	0.90	0.99
MDM Classifier	0.87	0.68	0.99
Bayesian Segmentation	0.94	0.91	0.95

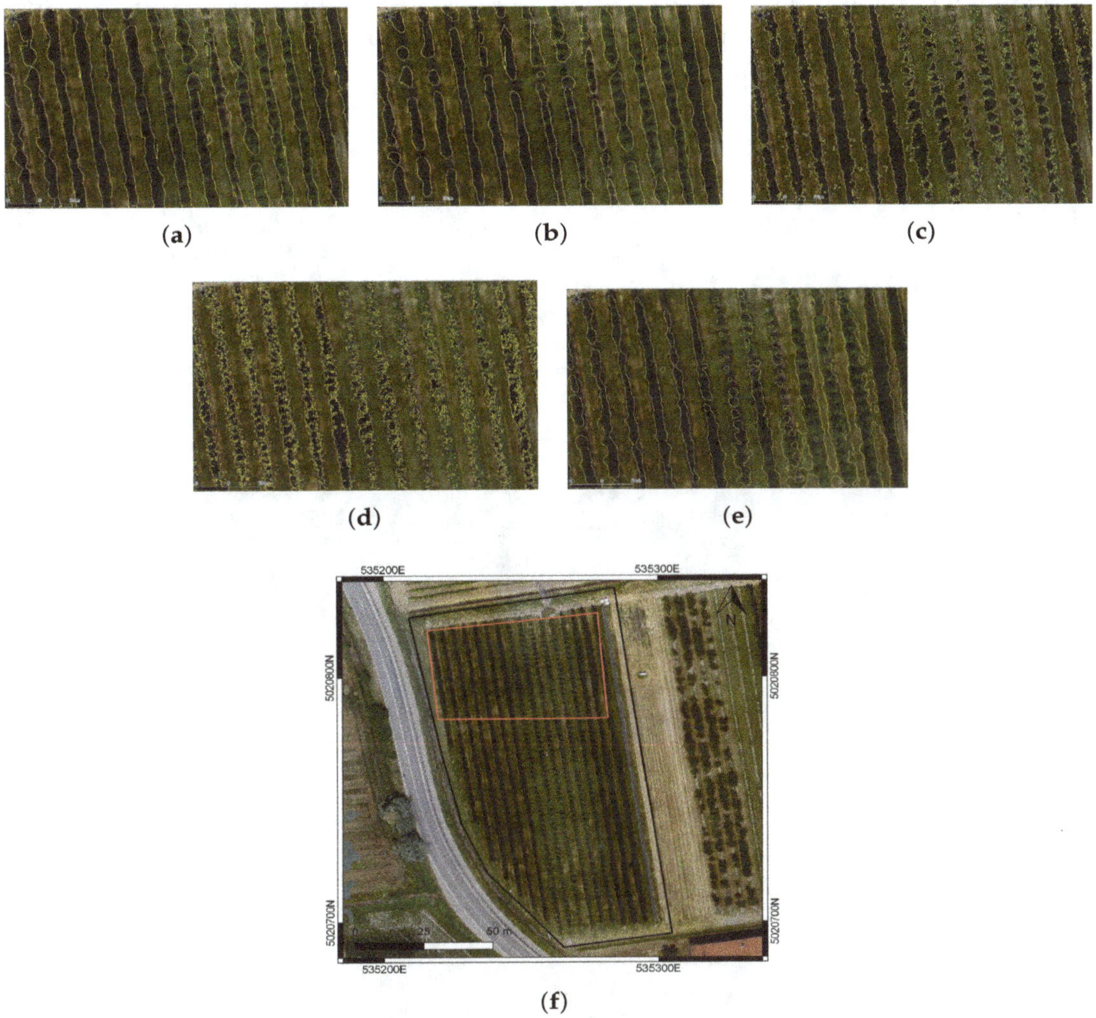

Figure 9. Pear orchard site: crop row detection results for (**a**) local maxima extraction, (**b**) threshold selection, K-means clustering (**c**), MDM classifier (**d**), and Bayesian segmentation (**e**). Figures refer to the area included in the red box in (**f**).

4.3. Tomato Field

Table 9 reports the parameters chosen for each detection method, which gave the best crop mask outputs.

Table 9. Tomato field site: parameters for the best results of each detection method.

Method	Input	User's Choices
Local Maxima Extraction	G%	cell size: 3 m percentage: 30%
Threshold Selection	DSM	cell size: 4 m threshold: 0
K-means Clustering	SAVI + NDVI	classes: 5
MDM Classifier	SAVI + NDVI	classes: 2
Bayesian Segmentation	ExG, Histogram adjustment	Background: $\mu = 0.05$, $\sigma = 0.15$ Crop canopy: $\mu = 0.65$, $\sigma = 0.35$

The assessment of the results for the tomato site was performed on 52 polygons (N pixels = 81,290) for the class crop canopy and 42 polygons (N pixels = 155,296) for the class background. The accuracy values are presented in Table 10, while Figure 10 shows the crop detection for the five selected methods.

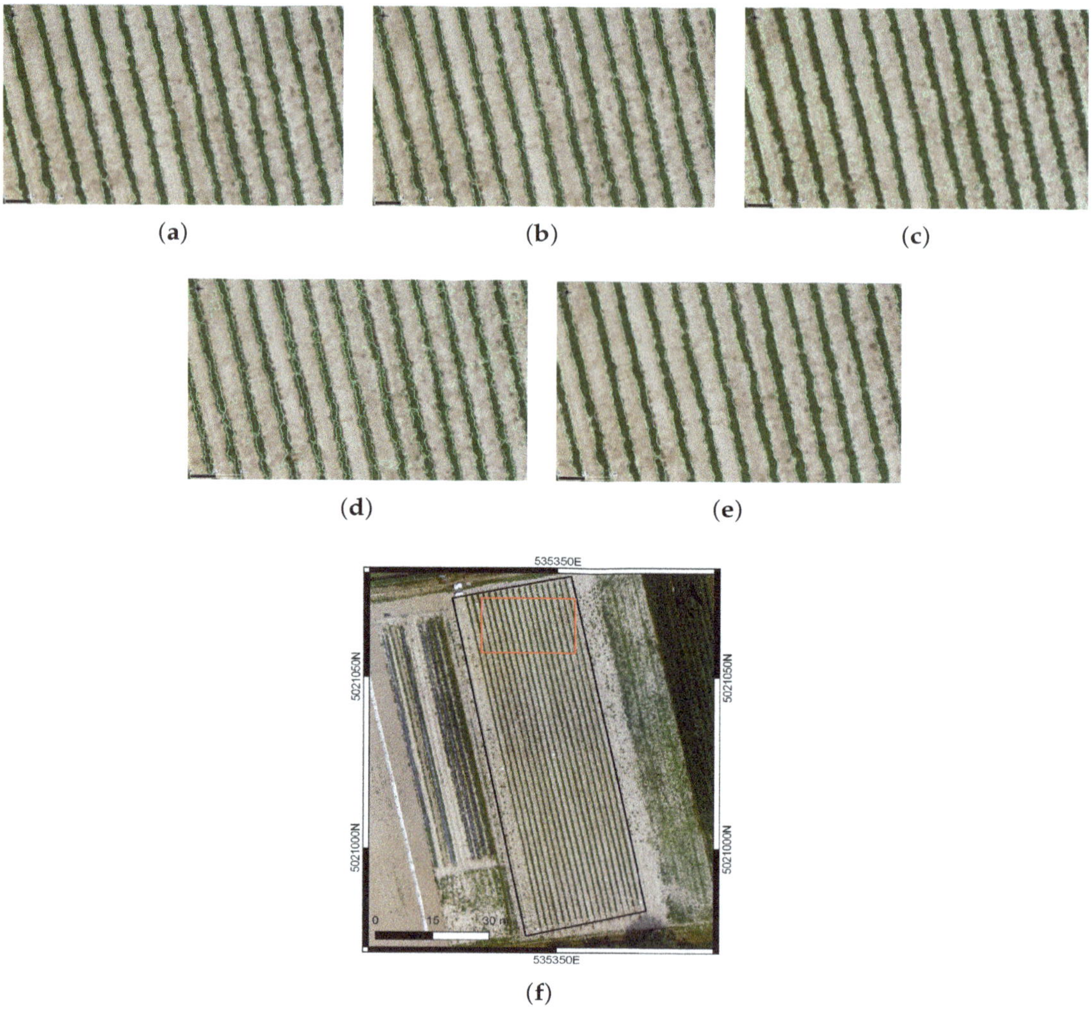

Figure 10. Tomato field site: crop row detection results for (**a**) local maxima extraction, (**b**) threshold selection, K-means clustering (**c**), MDM classifier (**d**), and Bayesian segmentation (**e**). Figures refer to the area included in the red box in (**f**).

Table 10. Tomato field site: assessment of the best results of each detection method.

Method	OA	PA Crop Canopy	UA Crop Canopy
Local Maxima Extraction	0.98	0.94	0.98
Threshold Selection	0.97	0.87	0.99
K-means Clustering	0.93	0.93	0.79
MDM Classifier	0.90	0.60	0.92
Bayesian Segmentation	0.98	0.91	0.98

5. Discussion

As general findings, it can be stated that all the methods tested in this study performed well for crop row detection, with OA close or even greater than 0.9. The vineyard site seemed to be the most challenging (OA values lower than 0.9 for some methods), due to the concurrent presence of weed, bare soil, and shadow in the inter-row distance, while the high contrast between bare soil and crop canopy facilitated the crop detection in the tomato site (OA values always higher than 0.9).

Despite the high accuracy values achieved, the proposed algorithms did not require any particular computational resources, and the calculation time is reasonable with mass-market hardware, even if it varied according to the method. Local Maxima Extraction (LME) and Bayesian Segmentation (BS) overall returned the best outputs in terms of accuracies values, but with different performances in terms of time cost and parameter setting. The first method was faster, and choosing a cell size comparable with the rows' distance, or slightly larger, and a percentage of maxima between 30% and 40% could produce high quality results in all cases. The latter required a high level of a priori knowledge, and parameters had to be ad hoc fine-tuned with a time-consuming trial and error approach. The Threshold Selection (TS) algorithm needs to be run with an accurate DSM, and the definition of the reference value is crucial and sometimes can fail, as demonstrated by the low accuracy values registered in the vineyard site (OA = 0.76), especially on fields characterized by a relevant slope. On hilly fields and non-flat areas, the use of a real CHM is necessary and cannot be bypassed by the creation of a smoothed raster, on which the reference value of terrain height is identified. Classification algorithms are easy to run, especially in the QGIS implementation, and widely used in remote sensing, but cannot reach, in all cases, the same level of accuracy as the other methods. In addition, these algorithms require considerable human intervention, either in the labeling phase, as the case of K-means clustering, or in the delineation of the training samples, as for starting the MDM Classifier.

Regarding input rasters, DSM, RGB orthophotos, and VIs obtained as a combination of RGB bands are the most adopted in the selected methods. Only the cases of Bayesian segmentation on the pear orchard site and classification algorithms for tomato site require NDVI and NDVI jointly with SAVI to obtain the best results. Therefore, NIR information does not give any particular additional value in crop row detection, and RGB sensors can perform accurate canopy extraction, as already demonstrated by other authors [15,28], saving the time and cost of the UAV surveys and processing.

According to crop characteristics, specific considerations can be stressed for each single crop. In the case of a vineyard, it is important to maintain the continuity of the crop row, when detecting the crop canopy. This characteristics was enhanced in the Bayesian segmentation, as shown in Figure 8e, also thanks to the Gaussian filter applied to the input raster before launching the algorithm. The continuity of the vine rows was also guaranteed by using the local maxima extraction algorithm as the detection method (Figure 8a), apart from some rare and sparse pixels. Indeed, the two aforementioned methods registered the highest accuracy values, in particular PA values: 0.97 and 0.95 for BS and LME, respectively, considerably greater than the PA values of the other detection methods. The major issue of detecting vine rows is the presence of shadows, weeds, and bare soil in the inter-row distance. Our results demonstrated that the shadows made the classification methods practically unusable on the vineyard. In Figure 8c,d, it is clearly visible how the pixels at the edges of the shadow areas were detected as crop pixels. Classification algorithms are unable to separate the vine canopy from its shadow on the terrain, resulting in an overestimation of the crop rows (UA values around 0.8).

In the orchard, pear trees are planted quite distant from one another (in our study site, around 1.5 m); therefore, a good detection has to identify single plants rather than rows. In these terms, classification algorithms return the best results, as visible in Figure 9c,d and confirmed by the highest values of UA practically equal to one (Table 8). On the other hand, these detection methods also returned the most noisy outputs and underestimated the presence of pear trees in the orchard, in particular the MDM classifier with a PA equal to 0.68. The height of the trees favored their extraction from the background, also in presence of weeds in the inter-row distance. To exploit these characteristics fully, it is advisable to use the DSM as the input of whatever detection method; in particular, the threshold selection algorithm gave the best outcomes in the pear orchard site, with PA value for the class crop canopy of 0.97.

As already mentioned, the tomato field site had the highest accuracy values for the crop detection, thanks to the regular alternation of bare soil and vegetation canopy. The OA values for all the tested methods were higher than 0.9; the LME algorithm, BS, and K-means also returned PA values above 0.9, while the PA of the TS method was slightly lower than 0.9 due to the use of the DSM as the input for starting the algorithm. At the time of the survey, the plants had a height of 0.3 m, equivalent to a few image GSDs; therefore, the errors in the photogrammetric processing (column height in Table 3) in this specific case affected the results of the canopy detection.

Precision viticulture is already widespread in the world, and recent articles have demonstrated the added value that remote sensing from UAV platforms can give to this sector [36]. Hence, numerous studies can be found in the literature dealing with vine canopy extraction [15,28,37,38]. The results presented in this work had accuracy values similar to those available in the literature. To the best of authors' knowledge, very few studies have already been published related to the detection of pear trees in orchards or tomato canopy, thus hampering the availability of reference values to compare the outputs. The assessment of the reliability of the illustrated results was based on similar case studies present in the literature. The detection of pear plants was performed with accuracy values slightly lower than the results obtained by [39], in two orchards in China, but close to the outcomes of the chestnut tree extraction described in [16]. The high values found for the tomato site were in agreement with the results presented in [17], about the estimation of crop emergence in potatoes.

The potential utility of this study in precision agriculture is high. The methods herein described allowed deriving from UAV imagery vegetation properties specifically related to the characteristics of the crop under investigation. In particular, in the irrigation management context, it must be taken into account that usually, irrigation for row-crops is provided through localized pressurized systems, wetting only the areas near the crop. The soil between rows is usually grass-covered, so analyzing crop maps (for example NDVI or thermal maps) of the field without masking the soil between rows could lead to errors when evaluating the water status or vigor of the crop under study. In Figures 11–13, NDVI maps for the three analyzed study sites are shown: on the left, the original maps, while on the right, the canopy maps generated after the extraction of crop rows. This information could be used in precision agriculture applications for mapping vegetation stress status or to optimize on-farm irrigation management. As an example, in [40], the use of a crop row detection method to delineate Site-Specific Management Zones (SSMZs) maps on a vineyard was described.

Moreover, it is crucial to stress that in the case of row-crops, the importance of using UAV imagery with respect to satellite imagery becomes fundamental. In satellite images, in fact, pixels have larger dimensions, and the operation of extracting crop rows is not possible. This leads to the presence of mixed pixels, for which Vegetation Indices' maps give information about both row and inter-row vegetation, therefore not very usable for agronomic inputs' management [41,42].

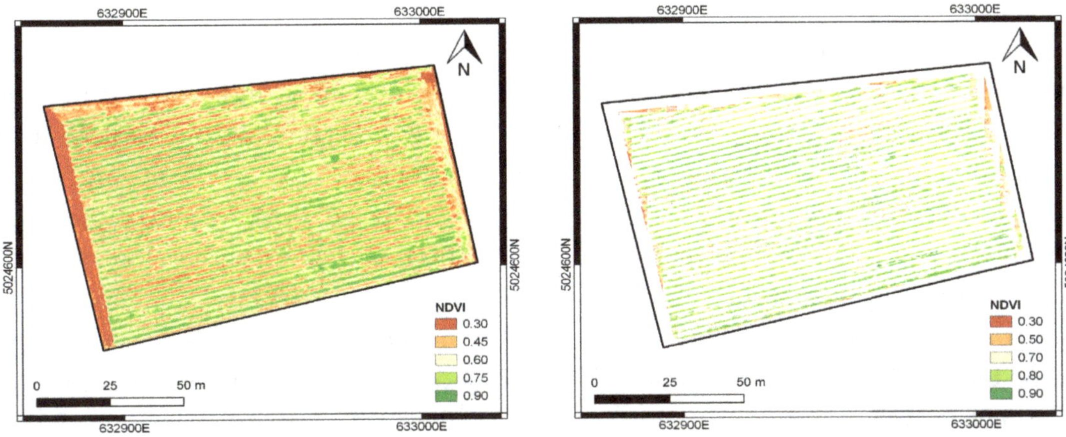

(a) (b)

Figure 11. Vineyard site: NDVI map before (**a**) and after (**b**) the crop rows' extraction.

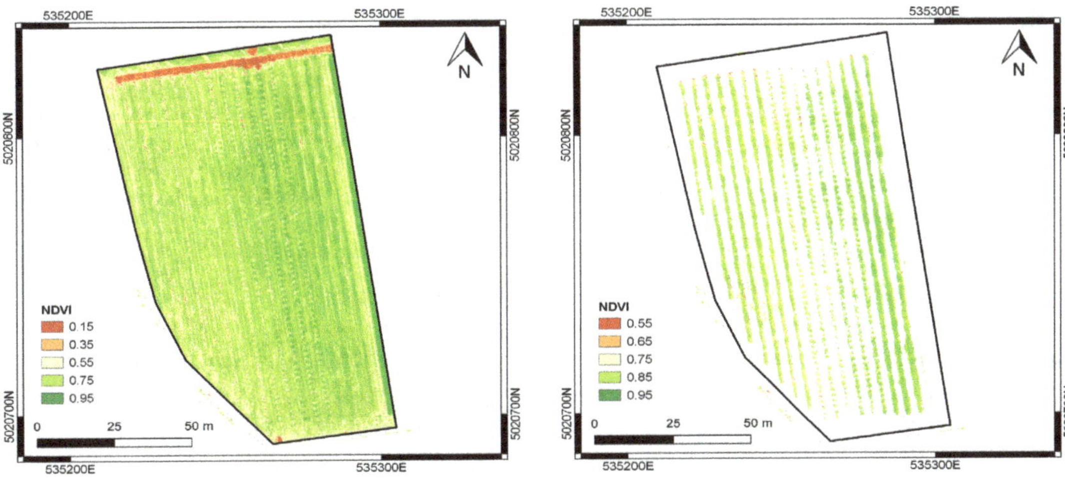

(a) (b)

Figure 12. Pear orchard site: NDVI map before (**a**) and after (**b**) the crop rows' extraction.

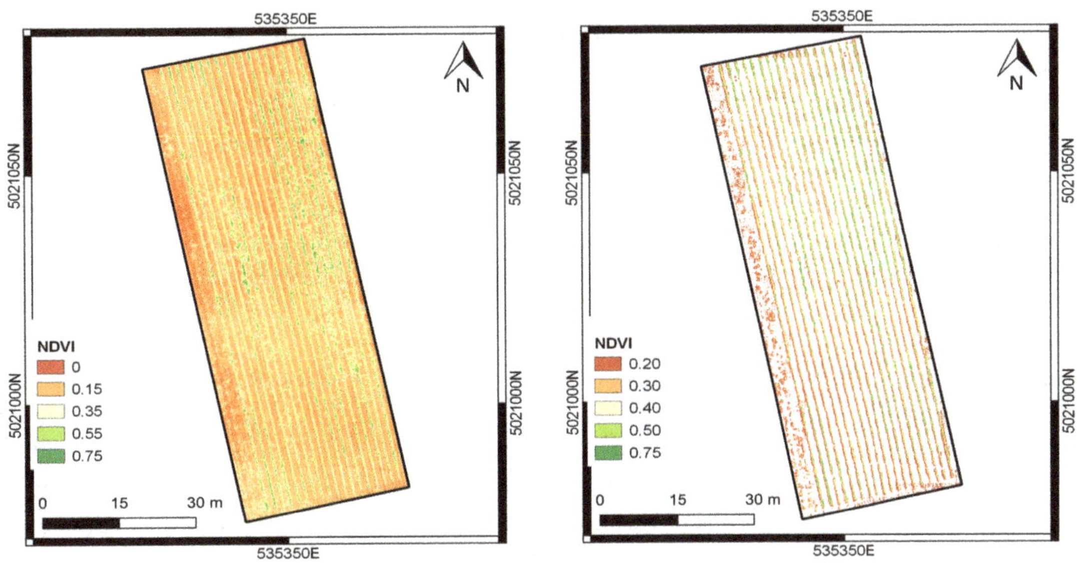

(a) (b)

Figure 13. Tomato field site: NDVI map before (**a**) and after (**b**) the crop rows' extraction.

6. Conclusions

This study demonstrates the feasibility to perform crop row detection from high-resolution UAV imagery, for different crop types, including vineyards, orchards, and horticultural crops. DSM, RGB, or multispectral orthophotos can be used as input for the detection methods; in particular, the DSM performs better with crop characterized by high heights (i.e., grapevine and pear), even in the presence of inter-row weed, but it should be avoided to detect horticultural crops (i.e., tomato). Commercial RGB sensors give high accuracy values for crop row detection; therefore, for this purpose, it is not necessary to perform surveys mounting more expensive multispectral cameras, if no additional infrared information is required. Furthermore, in the presence of shadows produced by the crop canopy on the terrain, indices based on the NIR band and classification algorithms could lead to an overestimation of the crop rows.

Although all applied methods need some level of human intervention, among all, the local maxima extraction algorithm, developed ad hoc within this study, allows reaching the best compromise in terms of time-cost, automation, and quality of results. Bayesian segmentation applied on VIs performs better than the other methods in the presence of bare soils, but it depends on a priori information.

Author Contributions: Conceptualization, G.R., A.F., B.O., and G.S.; methodology, G.R. and G.S.; software, G.R. and A.M.; validation, G.R.; data curation, G.R., A.M.; writing, original draft preparation, G.R. and A.M.; writing, review and editing, G.R., A.M., B.O., A.F., and G.S.; visualization, G.R. and A.M.; supervision, B.O., A.F., and G.S.; project administration, B.O., A.F., and G.S.; funding acquisition, B.O., A.F., and G.S. All authors read and agreed to the published version of the manuscript.

References

1. Pachauri, R.K.; Allen, M.R.; Barros, V.R.; Broome, J.; Cramer, W.; Christ, R.; Church, J.A.; Clarke, L.; Dahe, Q.; Dasgupta, P.; et al. *Climate Change 2014: Synthesis Report. Contribution of Working Groups I, II and III to the Fifth Assessment Report of the Intergovernmental Panel on Climate Change*; IPCC: Geneva, Switzerland, 2014.
2. Costa, J.; Vaz, M.; Escalona, J.; Egipto, R.; Lopes, C.; Medrano, H.; Chaves, M. Modern viticulture in southern Europe: Vulnerabilities and strategies for adaptation to water scarcity. *Agric. Water Manag.* **2016**, *164*, 5–18. [CrossRef]
3. UN-Water. *2018 UN World Water Development Report, Nature-Based Solutions for Water*; UNESCO, Paris, France, 2018.
4. Castellarin, S.D.; Bucchetti, B.; Falginella, L.; Peterlunger, E. Influenza del deficit idrico sulla qualità delle uve: Aspetti fisiologici e molecolari. *Italus Hortus* **2011**, *18*, 63–79.
5. Monaghan, J.M.; Daccache, A.; Vickers, L.H.; Hess, T.M.; Weatherhead, E.K.; Grove, I.G.; Knox, J.W. More 'crop per drop': Constraints and opportunities for precision irrigation in European agriculture. *J. Sci. Food Agric.* **2013**, *93*, 977–980. [CrossRef]
6. Sanchez, L.; Sams, B.; Alsina, M.; Hinds, N.; Klein, L.; Dokoozlian, N. Improving vineyard water use efficiency and yield with variable rate irrigation in California. *Adv. Anim. Biosci.* **2017**, *8*, 574–577. [CrossRef]
7. McClymont, L.; Goodwin, I.; Whitfield, D.; O'Connell, M. Effects of within-block canopy cover variability on water use efficiency of grapevines in the Sunraysia irrigation region, Australia. *Agric. Water Manag.* **2019**, *211*, 10–15. [CrossRef]
8. Matese, A.; Di Gennaro, S.F. Technology in precision viticulture: A state of the art review. *Int. J. Wine Res.* **2015**, *7*, 69–81. [CrossRef]
9. Pádua, L.; Vanko, J.; Hruška, J.; Adão, T.; Sousa, J.J.; Peres, E.; Morais, R. UAS, sensors, and data processing in agroforestry: A review towards practical applications. *Int. J. Remote Sens.* **2017**, *38*, 2349–2391. [CrossRef]
10. Nebiker, S.; Annen, A.; Scherrer, M.; Oesch, D. A light-weight multispectral sensor for micro UAV: Opportunities for very high resolution airborne remote sensing. *Int. Arch. Photogramm. Remote Sens. Spat. Inf. Sci.* **2008**, *37*, 1193–1199.

11. Tsouros, D.C.; Bibi, S.; Sarigiannidis, P.G. A Review on UAV-Based Applications for Precision Agriculture. *Information* **2019**, *10*, 349. [CrossRef]
12. Corti, M.; Cavalli, D.; Cabassi, G.; Vigoni, A.; Degano, L.; Gallina, P.M. Application of a low-cost camera on a UAV to estimate maize nitrogen-related variables. *Precis. Agric.* **2019**, *20*, 675–696. [CrossRef]
13. Noh, H.; Zhang, Q. Shadow effect on multi-spectral image for detection of nitrogen deficiency in corn. *Comput. Electron. Agric.* **2012**, *83*, 52–57. [CrossRef]
14. Pádua, L.; Marques, P.; Hruška, J.; Adão, T.; Bessa, J.; Sousa, A.; Peres, E.; Morais, R.; Sousa, J.J. Vineyard properties extraction combining UAS-based RGB imagery with elevation data. *Int. J. Remote Sens.* **2018**, *39*, 5377–5401. [CrossRef]
15. Poblete-Echeverría, C.; Olmedo, G.; Ingram, B.; Bardeen, M. Detection and segmentation of vine canopy in ultra-high spatial resolution RGB imagery obtained from unmanned aerial vehicle (UAV): A case study in a commercial vineyard. *Remote Sens.* **2017**, *9*, 268. [CrossRef]
16. Marques, P.; Pádua, L.; Adão, T.; Hruška, J.; Peres, E.; Sousa, A.; Sousa, J.J. UAV-based automatic detection and monitoring of chestnut trees. *Remote Sens.* **2019**, *11*, 855. [CrossRef]
17. Li, B.; Xu, X.; Han, J.; Zhang, L.; Bian, C.; Jin, L.; Liu, J. The estimation of crop emergence in potatoes by UAV RGB imagery. *Plant Methods* **2019**, *15*, 15. [CrossRef] [PubMed]
18. Pádua, L.; Adão, T.; Sousa, A.; Peres, E.; Sousa, J.J. Individual Grapevine Analysis in a Multi-Temporal Context Using UAV-Based Multi-Sensor Imagery. *Remote Sens.* **2020**, *12*, 139. [CrossRef]
19. Ortuani, B.; Facchi, A.; Mayer, A.; Bianchi, D.; Bianchi, A.; Brancadoro, L. Assessing the Effectiveness of Variable-Rate Drip Irrigation on Water Use Efficiency in a Vineyard in Northern Italy. *Water* **2019**, *11*, 1964. [CrossRef]
20. Ronchetti, G.; Pagliari, D.; Sona, G. DTM generation through UAV survey with a fisheye camera on a vineyard. *Int. Arch. Photogramm. Remote Sens. Spat. Inf. Sci.* **2018**, *XLII-2*, 983–989. doi:10.5194/isprs-archives-XLII-2-983-2018. [CrossRef]
21. Pádua, L.; Marques, P.; Hruška, J.; Adão, T.; Peres, E.; Morais, R.; Sousa, J. Multi-Temporal Vineyard Monitoring through UAV-Based RGB Imagery. *Remote Sens.* **2018**, *10*, 1907. [CrossRef]
22. Rouse J., Jr.; Haas, R.; Schell, J.; Deering, D. Monitoring Vegetation Systems in the Great Plains with ERTS. *NASA Spec. Publ.* **1974**, *351*, 309.
23. Birth, G.S.; McVey, G.R. Measuring the color of growing turf with a reflectance spectrophotometer 1. *Agron. J.* **1968**, *60*, 640–643. [CrossRef]
24. Huete, A.R. A soil-adjusted vegetation index (SAVI). *Remote Sens. Environ.* **1988**, *25*, 295–309. [CrossRef]
25. Kaufman, Y.J.; Tanre, D. Atmospherically resistant vegetation index (ARVI) for EOS-MODIS. *IEEE Trans. Geosci. Remote Sens.* **1992**, *30*, 261–270. [CrossRef]
26. Woebbecke, D.M.; Meyer, G.E.; Von Bargen, K.; Mortensen, D. Color indices for weed identification under various soil, residue, and lighting conditions. *Trans. ASAE* **1995**, *38*, 259–269. [CrossRef]
27. Richardson, A.D.; Jenkins, J.P.; Braswell, B.H.; Hollinger, D.Y.; Ollinger, S.V.; Smith, M.L. Use of digital webcam images to track spring green-up in a deciduous broadleaf forest. *Oecologia* **2007**, *152*, 323–334. [CrossRef]
28. Weiss, M.; Baret, F. Using 3D point clouds derived from UAV RGB imagery to describe vineyard 3D macro-structure. *Remote Sens.* **2017**, *9*, 111. [CrossRef]
29. MATLAB. *Version 9.3 (R2017b)*; The MathWorks Inc.: Natick, MA, USA, 2017. Available online: https://www.mathworks.com/downloads/ (accessed on 14 December 2018).
30. QGIS Development Team. *QGIS Geographic Information System*. Open Source Geospatial Foundation, 2009. Available online: http://qgis.osgeo.org (accessed on 2 January 2020).
31. MacQueen, J.Some methods for classification and analysis of multivariate observations. In Proceedings of the Fifth Berkeley Symposium on Mathematical Statistics and Probability, Oakland, CA, USA, 1 January 1967; Volume 1, pp. 281–297.
32. Ross, S.M. *Probabilità e statistica per l'ingegneria e le scienze*; Apogeo Editore, Milano, Italy, 2003.
33. Bayes, T. LII. An essay towards solving a problem in the doctrine of chances. By the late Rev. Mr. Bayes, FRS communicated by Mr. Price, in a letter to John Canton, AMFR S. *Philos. Trans. R. Soc. Lond.* **1763**, *53*, 370–418. [CrossRef]

34. Geman, S.; Geman, D. Stochastic relaxation, Gibbs distributions, and the Bayesian restoration of images. *IEEE Trans. Pattern Anal. Mach. Intell.* **1984**, 721–741. [CrossRef]
35. Marino, A.; Marotta, F. Crop rows detection through UAV images. In *Master Degree Final Dissertation, Politecnico di Milano*; 2019. Available online: https://www.politesi.polimi.it/ (accessed on 28 November 2019).
36. BorgognoMondino, E.; Gajetti, M. Preliminary considerations about costs and potential market of remote sensing from UAV in the Italian viticulture context. *Eur. J. Remote Sens.* **2017**, *50*, 310–319. [CrossRef]
37. Cinat, P.; Di Gennaro, S.F.; Berton, A.; Matese, A. Comparison of Unsupervised Algorithms for Vineyard Canopy Segmentation from UAV Multispectral Images. *Remote Sens.* **2019**, *11*, 1023. [CrossRef]
38. de Castro, A.; Jiménez-Brenes, F.; Torres-Sánchez, J.; Peña, J.; Borra-Serrano, I.; López-Granados, F. 3-D characterization of vineyards using a novel UAV imagery-based OBIA procedure for precision viticulture applications. *Remote Sens.* **2018**, *10*, 584. [CrossRef]
39. Dong, X.; Zhang, Z.; Yu, R.; Tian, Q.; Zhu, X. Extraction of Information about Individual Trees from High-Spatial-Resolution UAV-Acquired Images of an Orchard. *Remote Sens.* **2020**, *12*, 133. [CrossRef]
40. Ortuani, B.; Sona, G.; Ronchetti, G.; Mayer, A.; Facchi, A. Integrating Geophysical and Multispectral Data to Delineate Homogeneous Management Zones within a Vineyard in Northern Italy. *Sensors* **2019**, *19*, 3974. [CrossRef] [PubMed]
41. Di Gennaro, S.F.; Dainelli, R.; Palliotti, A.; Toscano, P.; Matese, A. Sentinel-2 Validation for Spatial Variability Assessment in Overhead Trellis System Viticulture Versus UAV and Agronomic Data. *Remote Sens.* **2019**, *11*, 2573. [CrossRef]
42. Khaliq, A.; Comba, L.; Biglia, A.; Ricauda Aimonino, D.; Chiaberge, M.; Gay, P. Comparison of satellite and UAV-based multispectral imagery for vineyard variability assessment. *Remote Sens.* **2019**, *11*, 436. [CrossRef]

8

Towards Predictive Modeling of Sorghum Biomass Yields using Fraction of Absorbed Photosynthetically Active Radiation Derived from Sentinel-2 Satellite Imagery and Supervised Machine Learning Techniques

Ephrem Habyarimana [1,*], Isabelle Piccard [2], Marcello Catellani [1,3], Paolo De Franceschi [1] and Michela Dall'Agata [1]

1 CREA Research Center for Cereal and Industrial Crops, Bologna 40128, Italy; marcello.catellani@enea.it (M.C.); paolo.defranceschi@crea.gov.it (P.D.F.); michela.dallagata@crea.gov.it (M.D.)
2 Vlaamse instelling voor technologisch onderzoek N.V., MOL 2400, Belgium; isabelle.piccard@vito.be
3 Department for Sustainability, Italian National Agency for New Technologies, Energy and Sustainable Economic Development (ENEA), Rotondella (MT) 75026, Italy
* Correspondence: ephrem.habyarimana@crea.gov.it

Abstract: Sorghum crop is grown under tropical and temperate latitudes for several purposes including production of health promoting food from the kernel and forage and biofuels from aboveground biomass. One of the concerns of policy-makers and sorghum growers is to cost-effectively predict biomass yields early during the cropping season to improve biomass and biofuel management. The objective of this study was to investigate if Sentinel-2 satellite images could be used to predict within-season biomass sorghum yields in the Mediterranean region. Thirteen machine learning algorithms were tested on fortnightly Sentinel-2A and Sentinel-2B estimates of the fraction of Absorbed Photosynthetically Active Radiation (fAPAR) in combination with in situ aboveground biomass yields from demonstrative fields in Italy. A gradient boosting algorithm implementing the xgbtree method was the best predictive model as it was satisfactorily implemented anywhere from May to July. The best prediction time was the month of May followed by May–June and May–July. To the best of our knowledge, this work represents the first time Sentinel-2-derived fAPAR is used in sorghum biomass predictive modeling. The results from this study will help farmers improve their sorghum biomass business operations and policy-makers and extension services improve energy planning and avoid energy-related crises.

Keywords: sorghum biomass; prediction modeling; machine learning; fAPAR; Sentinel-2 satellite imagery; big data technology; remote sensing

---

1. Introduction

Sorghum (*Sorghum bicolor* (L.) Moench) is a cereal with a C4 carbon fixation (the Hatch—Slack pathway) cultivated mainly for food, feed, forage, and fuel [1]. Sorghum grain was historically used for human consumption in developing countries but, because it is gluten-free, with low glycemic index and high contents of macronutrients and antioxidants, its utilization as food extended worldwide.

There are several types of sorghum. Grain sorghums are generally shorter (usually having recessive alleles at three of the four Dw genes) than biomass sorghums (having recessive alleles at two Dw genes at most), and have been selected to have the grain as the primary sink for photosynthates.

Biomass sorghum of biofuel production interest was used in this work and includes dual purpose (showing high grain and biomass yields), forage, sweet, and biomass sorghum types [2–5]. The sweet sorghum type translocates photosynthates to the seeds and stem; their stems are juicy (d recessive to D) instead of dry and sweet (x recessive to X) instead of nonsweet [6]. Sweet sorghums are high biomass and sugar yielding crops, and were traditionally bred for syrup or molasses production. Forage sorghum's main characteristics include digestible fiber and low lignin content, while biomass and dual purpose sorghums include high biomass yielding genotypes with high contents of structural carbohydrates, which are being developed for feedstock production [5]. Sorghum can therefore supply several products including starch, soluble sugars, structural carbohydrates, and organic matter for energy production purposes. Several countries worldwide, including in higher latitudes [7], are increasingly developing dedicated biomass sorghums in response to the pressing issue for nations to get independent of foreign energy sources and to cut carbon emissions into the atmosphere [7,8]. As a biofuel-dedicated biomass business, growing sorghum will have to meet critical requirements of high and cost-effective productivity of biobased commodities.

Crop yield forecasting is one of the most important strategies in agriculture, which enables sustainable development and helps avoid famines and shortages in several commodities [9–12]. In industrialized countries, crop yield forecasting provides data to governmental structures, companies, and farmers, which results in strategic advantages such as the rationalization of policy adjustments, price predictions and stabilization, efficient agricultural trade, and simplification of business operations particularly through planning harvest and delivery of the produce, better deployments of machineries and logistics, and a better management at the end user level (e.g., bioreactor owner).

Conventionally, and particularly in developed countries, the information on crop production is collected and disseminated through field surveys and censuses, but this system is rather costly and associated with significant uncertainties [13]. Using satellite imagery resulted in a superior solution [14–16]. Remote sensing data has been used for many years to build operational crop yield forecasting systems like the FAO's Global Information and Early Warning System (GIEWS). The use of remote sensing satellite data for crop yield forecasting is further motivated by wide coverage, near-real time delivery of data and products, and the ability to provide vegetation indicators at low cost. Many studies have shown that forecasting models based on remote sensing data can give similar or better performance comparing to the more sophisticated physiological crop growth models [13,17–19].

The use of remote sensing parameters as proxies for biomass yields was documented in previous works. Normalized difference vegetation index (NDVI), leaf area index (LAI), and fAPAR (fraction of absorbed photosynthetically active radiation) are among the most frequently used parameters [14,16,20]. Recently, the use of biophysical parameters, such as fAPAR, gained more attention relative to using vegetation indices [21,22]. Biophysical parameters reflect the state of the crops more adequately and thus could be better suited for predicting crop yield and production [15,23,24]. fAPAR is defined as the fraction of radiation absorbed by the green vegetation elements in the 400 to 700 nm spectral domain under specified illumination conditions [25]. It is directly linked to photosynthesis, and therefore expresses a canopy's energy absorption capacity [26]. fAPAR values range from 0 to 1, indicating, respectively, bare soil and fully crop covered soil.

Most of the aforementioned studies were focused on common field crops like corn, wheat, barley or soybeans [16,20,27–30]. A few studies dealt with remote sensing-based yield monitoring and prediction [16] in sorghum. For instance, Shafian et al. [31] described the use of unmanned aerial vehicle-based remote sensing to investigate sorghum crop physiological properties. Yang et al. [32] combined airborne digital videography with ground sampling, regression analysis, and image processing to map spatial sorghum grain yield variability within fields and across the cropping season. Johnson [16] presented a comprehensive assessment of the correlations between commonly used Moderate Resolution Imaging Spectroradiometer (MODIS) products and field crop yields, including sorghum, and used these correlations for biomass yield estimation and prediction.

In most studies, however, remote sensing-based biomass yield estimation or prediction makes use of low- or medium-resolution satellite images from sensors such as SPOT-VEGETATION [14,15,23,24] or MODIS [16]. These satellite products have a coarser spatial resolution (250 to 1000 m) compared to the data collected from the two Sentinel-2 satellites in this work (10-m spatial resolution). With the launch of the Sentinel-2 constellation of satellites the overpass frequency (five days and locally even two to three days) the temporal resolution is nearly as good as for SPOT-VEGETATION and MODIS satellites (one to two days). The high spatial resolution of the Sentinel-2 images is an important asset when monitoring crops in agricultural regions characterized by many small fields. To our knowledge no previous studies assessed the efficiency of high resolution Sentinel-2-derived fAPAR data in predicting within-season biomass sorghum yields, and this paper is therefore aimed at addressing this gap.

Deriving yield information from satellite imagery has shown promising results but this technology is not extensively applied across farmers and crop species worldwide [27–29]. In this work, we developed models for within-season prediction of annual and perennial sorghum biomass yields in Emilia-Romagna, Italy, based on fAPAR measurements from Sentinel 2A and Sentinel 2B satellite images on 42 mostly full-fledged commercial sorghum fields. We used machine learning algorithms to create yield prediction equations. These equations can be implemented in decision support systems to allow farmers and/or farming stakeholders to predict biomass yields from sorghum fields of interest early on in the cropping season. This information is very helpful to efficiently schedule fleets of harvesting machinery, transport vehicles, and storage facilities. The fAPAR-derived predictive models for biomass yields can also be implemented by extension services and policy-makers for several purposes, including the possibility to anticipate potential biomass availability and plan ahead, to avoid specific crises such as fuel shortage.

2. Materials and Methods

2.1. Trial Set-Up

Forty-two demonstration trials were run in this work, 23 and 19 of which were evaluated in 2017 and 2018, respectively. In 2017, the experimental sites were located in Conselice, Nonantola, Mirandola, and Anzola dell'Emilia, in the Italian region of Emilia Romagna (Table 1, Figure 1), while in 2018 the sites were established in Anzola, Mirandola, and Conselice (Figure 1, Table 1). The experimental sites were strategically selected to maximize extension impact by conducting most of the trials in the farmers' fields. The experimental fields in Mirandola and Conselice belonged to respective two big farming cooperatives with more than 2000 members each. In Nonantola, the fields belonged to individual farmers, while in Anzola the fields were established in the experimental station of the Council for Agricultural Research and Economics (CREA). The fields were generally of big size relative to plot sizes commonly used under standard experimental settings [1] in order to serve the purpose of demonstrative pilots with the objective of transferring into the production environment the technology of sorghum crop monitoring using satellite imagery. The fields areas ranged from 0.06 ha to 50.00 ha, with a mean and median of 5.70 ha and 1.10 ha, respectively. All the fields were planted with biomass sorghums including biomass per se (high tonnage), sweet, forage, and dual purpose types. One-grain sorghum trials were established in Anzola in 2017, but it was not included in this work in virtue of a different kind of experiment management and a diverse market of the grain sorghum produce relative to biomass sorghum. Thirty-five out of the 42 trials were sown with a single genotype of *Sorghum bicolor* (annual), while the 17IT_mat was sown with a diversity panel of 228 biomass *Sorghum bicolor* genotypes, and six (15R17, 16R17, 16R18, 15R18, 17R18, and 17US_mat) of the trials installed in Anzola were made up of a diversity panel consisting of advanced perennial interploid biomass hybrids deriving from *S. bicolor* × *S. halepense* (SB × SH) crosses. The original SB × SH materials originated from The Land Institute (Salina, KS, United States of America). *S. bicolor* × *S. halepense* breeding strategy was amply detailed in Piper and Kulakow [33] and in Habyarimana et al. [1]. The 15R, 16R, 17R, and 17US_mat trials were sown in 2015, 2016, and 2017, respectively, meaning that regrowth-derived biomass was evaluated for the 15R, 16R, and 17R trials, while for the 17US_mat trial, the biomass evaluated in this study was produced from direct sowing.

Table 1. Pilots descriptors: name, location, variety, season, and productivity.

Serial Number	Field/Pilot Name	Variety Name	Variety Type	Area (ha)	Dry Biomass Yield (t ha^{-1})	Cropping Season	Location Name
1	Botte 1	Harmattan	Dual purpose	9.00	14.13	2018	Conselice
2	Saracca 5	Harmattan	Dual purpose	6.50	10.52	2018	Conselice
3	*V. serrata*	Harmattan	Dual purpose	44.87	9.69	2018	Conselice
4	Magnana	P845F	Forage	32.05	11.43	2018	Conselice
5	Cà bianca	P845F	Forage	3.72	11.11	2018	Conselice
6	Gamberina 3	Aralba	Dual purpose	7.86	9.67	2018	Conselice
7	Sagrate	Harmattan	Dual purpose	50.00	8.90	2017	Conselice
8	Prato_Mensa	Harmattan	Dual purpose	3.29	19.10	2017	Conselice
9	Comuna	P845F	Forage	27.86	24.50	2017	Conselice
10	Gamberina_1	Aralba	Dual purpose	7.60	12.80	2017	Conselice
11	Botte	Harmattan	Dual purpose	5.33	23.50	2017	Conselice
12	Carafolo_G	Bulldozer	Biomass	2.00	17.10	2017	Nonantola
13	Cavriani_S	Merlin	Biomass	2.00	19.50	2017	Nonantola
14	Ferrari_R	Bulldozer	Biomass	1.00	3.40	2017	Nonantola
15	Mattioli_R	Bulldozer	Biomass	1.00	4.90	2017	Nonantola
16	Serafini_G	Bulldozer	Biomass	2.00	15.30	2017	Nonantola
17	Zavatti_E	Bulldozer	Biomass	0.89	7.60	2017	Mirandola
18	Grandi_Magonza	Bulldozer	Biomass	0.80	12.80	2017	Mirandola
19	Grandi_Ponte	Bulldozer	Biomass	1.20	9.50	2017	Mirandola
20	Zini_L	Palo Alto	Biomass	2.50	8.00	2017	Mirandola
21	Villa_verdetta	Bulldozer	Biomass	1.00	6.25	2018	Mirandola
22	Cama_grande	Bulldozer	Biomass	4.40	14.94	2018	Mirandola
23	Cama_piccolo	Bulldozer	Biomass	4.00	15.35	2018	Mirandola
24	Golinelli_Raimondo	Bulldozer	Biomass	2.02	10.48	2018	Mirandola
25	Barozzi_Lidia	Bulldozer	Biomass	3.00	7.19	2018	Mirandola
26	Molon_A	Palo Alto	Biomass	5.00	8.30	2017	Mirandola
27	T1_Anzola	Sole	Biomass	0.74	13.00	2017	Anzola
28	T2_Anzola	Trudan	Forage	0.71	19.00	2017	Anzola
29	T3_Anzola	Hannibal	Sweet	0.71	17.00	2017	Anzola
30	T4_Anzola	Harmattan	Dual purpose	0.70	14.00	2017	Anzola
31	T1_Anzola	Bulldozer	Biomass	0.74	17.00	2018	Anzola
32	T2_Anzola	Hannibal	Sweet	0.71	19.00	2018	Anzola
33	T3_Anzola	Tarzan	Biomass	0.71	21.00	2018	Anzola
34	T4_Anzola	Trudan	Forage	0.70	13.00	2018	Anzola
35	T5_Anzola	Harmattan	Dual purpose	0.70	15.00	2018	Anzola
36	15R17	Perennial	Biomass	0.06	2.39	2017	Anzola
37	16R17	Perennial	Biomass	0.15	8.27	2017	Anzola
38	17IT_mat	Bicolor	Biomass	0.17	18.30	2017	Anzola
39	17US_mat	Bicolor	Biomass	0.15	21.00	2017	Anzola
40	16R18	Perennial	Biomass	0.15	14.35	2018	Anzola
41	15R18	Perennial	Biomass	0.06	10.73	2018	Anzola
42	17R18	Perennial	Biomass	0.15	17.47	2018	Anzola

Crop management followed local extension services guidelines and was well described in Habyarimana et al. [1]. Planting density was 26 (0.75 m spacing between rows; 0.052 m spacing of hills within row) plants per square meter for most (35) trials, and 13 (0.75 m spacing between rows; 0.10 m spacing of hills within row) plants per square meter for 15R, 16R, 17R, 17IT_mat, and 17US_mat trials. In terms of weather (Figures S1–S4), summer was generally dry across years and locations as expected. In 2017, all sites had relatively wet spring except in Anzola, while in 2018 spring was relatively wet in Anzola and Conselice, but dry in Mirandola.

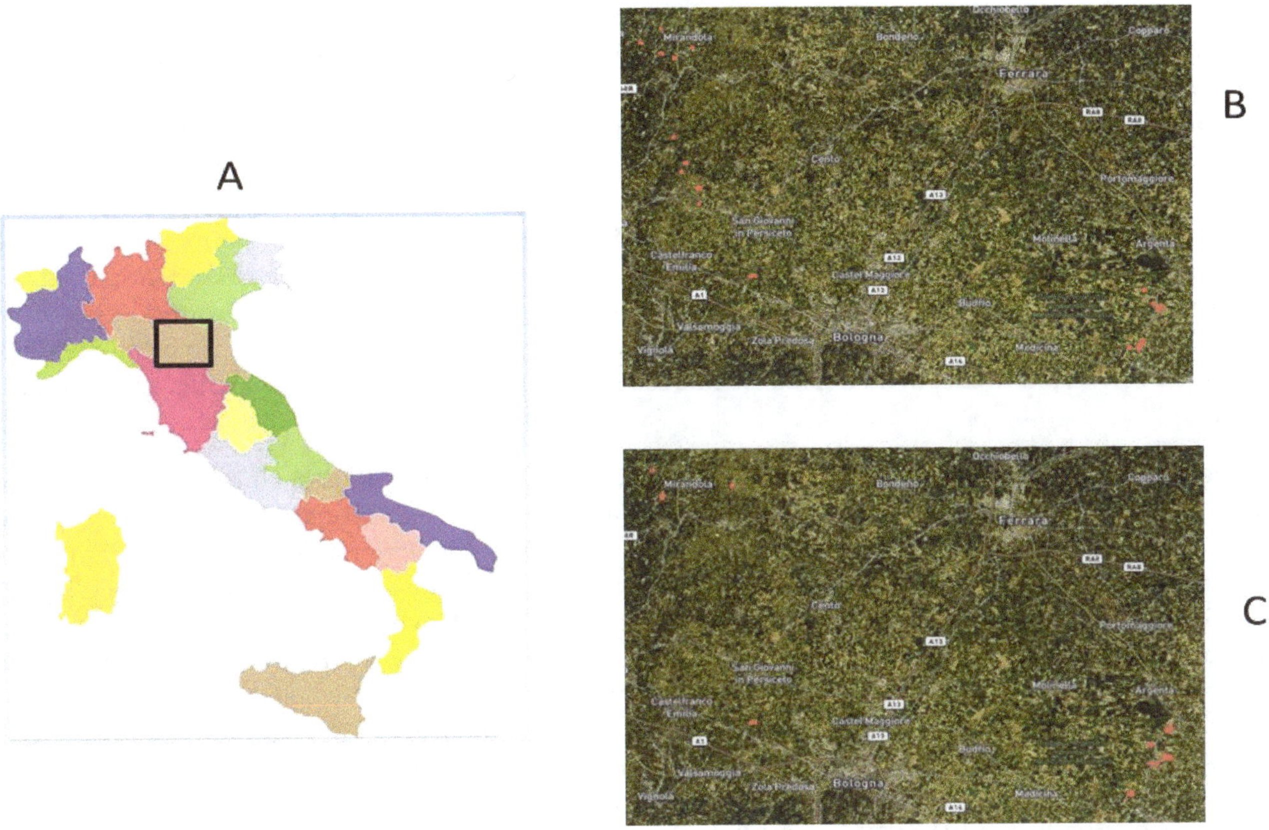

Figure 1. Map of Italy (**A**) with a rectangle inset indicating the geographical location of the experimental sites (red dots) for pilots established in 2017 (**B**) and 2018 (**C**).

2.2. Biomass Data Collection

Trials in Nonantola, Mirandola, and Conselice were harvested at industrial scale from end of August to late November, while all trials in Anzola were harvested end of November using a single-row chopper harvester. Our experience showed that postponing harvest to later times may increase the likelihood for lodging, which may lead to the crop touching the ground, adding grit to the biomass material and possibly reducing biomass quality; delayed harvest also leads to kernel loss and kernel quality deterioration particularly due to molds, insect, and bird damages. The trials were harvested according to two machinery options: forage chopper or swathing the material into windrows and then baling it in large square bales or large round bales. Chopped biomass was weighed immediately at harvest, while baled biomass was weighed when bales were transported to the bioreactor. Chopped and baled biomasses were supplied to private biogas and combustion bioreactors. From each field, a 1kg-composite sample was taken from the sold biomass at the time of shipment to the end user in order to determine the dry mass content in the commercialized produce and calculate the dry biomass yield for each entire field that will be used in modeling. For the diversity panels, samples were taken from each genotype. Fresh samples were weighed and dried at 80 °C to constant weight in a forced air oven. The fresh and dry weights of the samples, and the fresh weight of the entire field's harvest, were used to derive dry mass fraction of the fresh material and dry biomass yield of the entire field. For the diversity panel fields, the final yields integrated the contributions of the component genotypes.

2.3. Satellite Data Acquisition

For this study we used Sentinel-2 optical satellite imagery. The Sentinel-2 mission is based on a constellation of two satellites—Sentinel-2A and Sentinel-2B—both orbiting Earth at an altitude of 786 km, but 180 ° apart to optimize coverage and global revisit times. Swath width, i.e., the image width across the satellite path when scanning the Earth, is 290 km. As a constellation, the revisit time is 5 days. This means that the same spot over the equator is revisited every five days, and even faster at higher latitudes. Sentinel-2 data are acquired on 13 spectral bands in the VNIR (visible and near-infrared) and SWIR (short-wave infrared) range, of which four bands with a spatial resolution of 10 meters (blue, green, red, and near-infrared (NIR)), six bands at 20 meters (three red edge bands, a narrow NIR, and two SWIR bands), and three bands at 60 meters (a coastal aerosol, water vapor, and cirrus band). Spatial resolution refers to the surface area measured on the ground and represented by an individual pixel. Once the Sentinel data are acquired on-board, they are sent to ground and processed by a network of Processing and Archiving Centers. Next, all data products are united, archived, and disseminated online to the users by ESA's Copernicus Space Component (CSC) Ground Segment via the CSC Data Access Coordinated System. To facilitate image transfer and use, the projected Sentinel-2 images are converted to tiles with a fixed size of 100 square kilometers, each of which is approximately 500 MB.

For this study, Sentinel-2A and Sentinel-2B images from tile 32TQQ (including pilots from Conselice) and 32TPQ (including pilots from Anzola, Mirandola, and Nonantola) were downloaded from ESA and processed by Vlaamse Instelling voor Technologisch Onderzoek N.V. (VITO). Processing included atmospheric correction with iCOR [34] and cloud and shadow detection using Sen2COR v2.5.5 (ESA-STEP, ESA, Paris, France). Biophysical parameters fAPAR, fCover, and leaf area index (LAI) were calculated from the top of canopy normalized reflectances following the BV-NET (tool for mapping surface and vegetation variables) method described by Weiss and Baret [35]. The BV-NET methodology is based on neural networks which are trained on a synthetic dataset of ~50,000 simulations using the PROSAIL (PROSPECT and SAIL radiative transfer models) model [36]. The BV-NET version used in this study was calibrated with green, red and near infrared bands, all having a spatial resolution of 10 meters. Sen2Cor and BV-NET are publicly available through ESA's SNAP (Sentinel Application Platform, ESA, Paris, France) toolbox.

Previous studies such as Duveiller et al. [15], López-Lozano et al. [24], and Johnson et al. [16] illustrated the good performance of satellite derived fAPAR for estimating and predicting biomass yields of large field crops, including corn and sugarcane, which, together with sorghum, make-up the world's three economically important C4 crops of the Poaceae family with similar growth habits [37]. We therefore decided to use fAPAR for this study as well. The fAPAR estimates generated with BV-NET from Sentinel-2A and 2B top of canopy reflectances over selected tiles in Emilia-Romagna had a spatial resolution of 10 meters and a temporal resolution of 5 days up to 2–3 days in those areas where the different satellite overpasses overlapped.

For monitoring the sorghum fields in this study "WatchITgrow" (Vlaamse Instelling voor Technologisch Onderzoek N.V., MOL, Belgium) was used. WatchITgrow is a web-based application for crop monitoring developed by VITO. It provides information on crop growth and development as well as possible anomalies derived from Sentinel-2 satellite images and weather data, and it allows the user to store all kinds of collected field data, such as planting and harvest dates and development stages, but also information on crop treatments such as fertilization, spraying, or irrigation. Prior to monitoring, the fields used in this study were geolocalized (Figure 1) using Field GPS (global positioning syatem) application for iPhone with a final field boundary correction using Google Earth. The field polygons were saved as kml files and then imported into WatchITgrow for monitoring. For each field, fAPAR or "greenness" maps were created (see example in Figure 2), and a growth curve was built, showing the evolution of the fAPAR values throughout the cropping season (see example in Figure 3). To build the growth curve the fAPAR values of all pixels within the field were averaged, thereby accounting for an inside buffer of ten meters (one pixel) in order to avoid capturing signals from neighboring fields or other objects. To correct for artifacts in the resulting fAPAR curve such as abnormally low fAPAR values due to undetected clouds, shadows or haze and to interpolate fAPAR values between subsequent acquisition dates, a Whittaker smoothing filter was applied on the curve [38,39].

Figure 2. Greenness (fAPAR) maps derived from Sentinel-2 satellite imagery for five sorghum fields in Anzola (from left to right: T5-grain sorghum, T4-dual purpose sorghum, T3-sweet sorghum, T2-forage sorghum, T1-biomass sorghum) for a selected number of dates in 2017, as available via WatchITgrow. T5-grain sorghum was not included in this study (refer to Section 2.1 for detail).

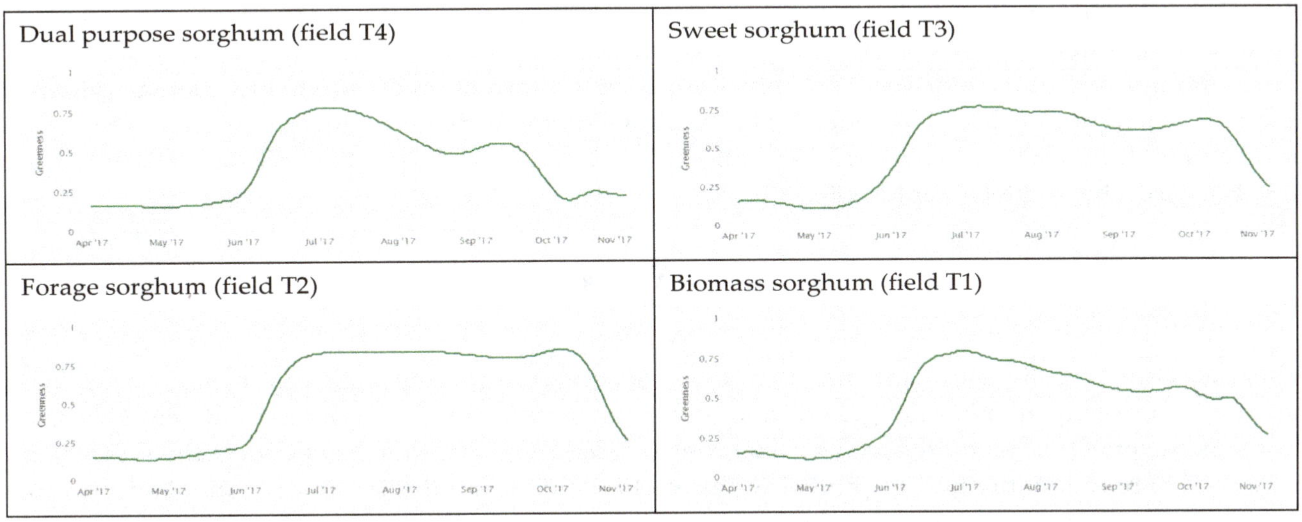

Figure 3. *Cont.*

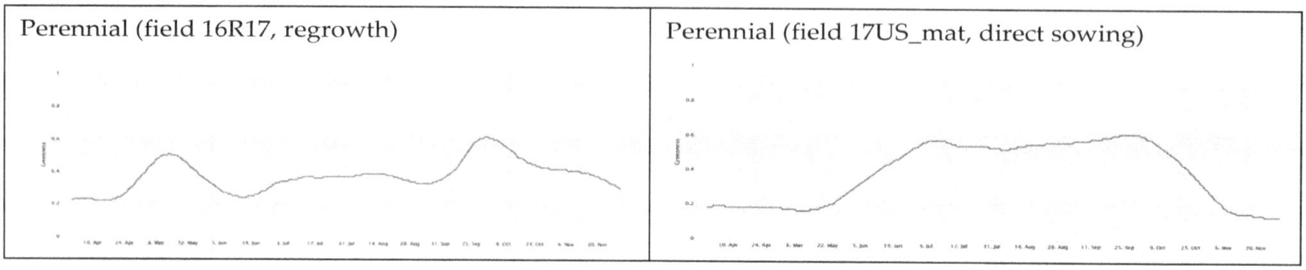

Figure 3. Greenness (fAPAR) graphs derived from Sentinel-2 satellite imagery from 2017 for six sorghum fields in Anzola (T4-dual purpose sorghum, T3-sweet sorghum, T2-forage sorghum, T1-biomass sorghum, and 16R17-perennial, 17US_mat perennial), available via WatchITgrow.

2.4. Modeling Total Aboveground Biomass Yields

Thirteen models were assessed in this study to predict sorghum biomass yields. The models included partial least square discriminant analysis (PLS-DA), principal component analysis discriminant analysis (PCA-DA), neural network (NN), random forest (RF), support vector machine (SVM) with linear classifier (SVML), nonlinear kernel (SVML_G), radial basis kernel (SVM_R), radial basis kernel with polynomial basis kernel (SVM_P), neural network (NNET), eXtreme Gradient Boosting-xgbtree method (GBT), eXtreme Gradient Boosting- xgbDART method (GBD), eXtreme Gradient Boosting-xgbLinear method (GBL), simple linear model (LM), and Neural Network-neuralnet method (NLNET). The simple linear model was used as a benchmark to gauge the performance of the models implemented. The models evaluated in this work were selected based on their robustness as reported in previous studies [40].

The field-based daily interpolated fAPAR estimates extracted from WatchITgrow were converted to fortnightly fAPAR averages. In this study, preference was given to the use of fortnightly fAPAR data as major morphophysiological changes in crops also occur fortnightly [41]. In addition pilots established in experimental stations and in farmers' fields were fortnightly visited at or close to the times the fAPAR images used in this work were acquired. This time management was also favorable and accommodated the busy schedules of farmers and scientists.

Six fortnightly fAPAR values registered from May to July—here referred to as six "days of year" (DOY), that is, DOY 135 and 150 in May, 165 and 180 in June, and 195 and 210 in July—were used as regressor variables in successive predictive modeling of sorghum biomass yields. May, June, and July are important months concerning the predictive modeling, mimicking the 1 to 2 months required to release yield predictions before harvest [42]; taking into account that biomass sorghum in the Mediterranean region is harvested from August to November.

The research questions addressed in this work are (1) how accurately can we predict the yield of a biomass sorghum field based of Sentinel-2-derived fAPAR profile early in the cropping season? (2) Which months and/or days-of-year best contribute useful information for predicting biomass yields in commercial sorghum fields? The solutions to the above problems were evaluated by solving the below linear model for n trials or experimental locations ($i = 1, \ldots, n$) and p prediction times or days of year ($j = 1, \ldots, p$). This model is represented by

$$y_i = \mu + \sum_{j=1}^{p} x_{ij}\beta_j + e_i \tag{1}$$

where μ is the overall mean, y_i is the phenotypic observation (biomass yields) from field i, e_i is the residual comprising all other nongenetic and environmental factors, x_{ij} is the days of year covariates, and β_j is the effect of the jth day of year covariate on y_i [43]. Note that it is beyond the scope being presented here to identify and/or predict within-field yield variability for any potential applications. In addition, different sorghum types were combined in this study as they qualified for commercial aboveground biomass production and to mimic farming practices in the region of the study. We also assumed that the test region was homogeneous with respect to climatic conditions.

All statistical analyses were carried out using R software [44]. The predictive models were fitted using the caret R package. In this work, the "one standard error" rule of Breiman et al. [45] was implemented to avoid overfitting, and the caret built-in features were invoked to automatically choose the tuning parameters associated with the best performance of the regression routines. During data preparation, zero-variance regressors were removed and those remaining were centered and scaled in order to avoid regressors with zero or near-zero variance, which often constitute a problem as they behave as second intercepts in predictive models [40]. The dataset was randomly partitioned into training (80% of the entire dataset; 34 observations) and testing set (20% of the entire dataset; eight observations). The training set was used to run a cross-validation experiment to train and assess the models using a 10× repeated 5-random fold cross-validation (CV) iterations, rendering a total of 50 estimates of accuracy and prediction error; a large number of repetitions is expected to compensate for the high variance stemming from a reduced number of folds. Models were validated on the testing set which was an external test (validation) sample set needed so that the model performance can be characterized on data that were not used in the model training. The models were evaluated based on the prediction accuracy, the mean absolute error (MAE), and the mean absolute percentage error (MAPE). The MAE built within the repeated cross validation procedure (model calibration) was used to assess the variability (dependability) of the model performance. On the other hand, the MAE, MAPE, and accuracy obtained on the testing set were used to assess the model predictive ability. The MAPE allows us to compare the prediction of different dependent variables in different scales. The MAE measured the average magnitude of the errors in the set of predictions of biophysical variable values produced in this work, without considering their direction. It represented the average over the test sample of the absolute differences between prediction and actual observation where all individual differences had equal weight. The MAE was chosen for the model verification because it provides an unambiguous measure of the magnitude of the average error and is therefore more appropriate than the Root Mean Square Error (RMSE) for dimensioned evaluations of average model performance error [46]. The distribution of the 50 MAE estimates from the optimal cross-validated models was characterized using boxplot, while the comparison of mean accuracies across models and across prediction times was performed using Duncan's test [47]. The importance of the regressor variables (useful prediction times) was determined using a 0 to 100 index, with 0 corresponding to no effect and 100 corresponding to the highest magnitude of the regressor's importance. The accuracy was defined as the Pearson correlation coefficient between the predicted and the observed biomass yield values in the testing set [5]. From the computed accuracy, r-squared values can be derived in order to better compare, for each model, the proportion of the variance in the dependent variable that is predictable from the regressors.

3. Results

3.1. fAPAR Index Pattern Across Sorghum Types

Three fAPAR curve and map patterns were consistently observed as illustrated in the above Figures 2 and 3 using data from the 2017 cropping season. In dual purpose and biomass sorghums, a major peak was observed earlier in July followed by a drop and then a weak increase at the beginning of the second half of September. For the sweet, forage, and the perennial sorghum (SB × SH) grown from seeds, the fAPAR increased significantly in early July to reach a plateau from then up to late September/early October, whereas, in October, the curve decreases sharply to reach the minimum value in early November. On the other hand, in perennial sorghum regrown from rhizomes, two fAPAR peaks (smaller peak in mid-May, bigger peak in late September/early October) were observed that were separated by a deep drop extending from June to August.

3.2. Assessment and Validation of the Predictive Models, and Importance of Regressors in Total Biomass Prediction

Thirteen models implemented in this work were assessed using the prediction accuracy (Table 2) in the validation set, the mean absolute error (MAE) and mean absolute percentage error (MAPE) metrics (Table 2, Figure 4) produced during the repeated cross-validation, and during the validation stage in the independent sample set. A repeated cross-validation iteration was run for each model, resulting in MAE resample vectors, each containing 50 elements. Over the months evaluated, the range and mean accuracy (Table 2) for NNET, RF, SVM-R, PCA-DA, PLS-DA, SVM-P, SVML, SVML-G, GBT, GBD, GBL, LM, and NLNET, were 0.16–0.78 and 0.56, 0.39–0.82 and 0.63, −0.36–0.88 and 0.16, −0.13–0.76 and 0.43, −0.02–0.77 and 0.47, 0.09–0.81 and 0.50, 0.49–0.80 and 0.64, 0.46–0.80 and 0.64, 0.56–0.81 and 0.66, −0.01–0.84 and 0.49, 0.03–0.93 and 0.47, 0.46–0.78 and 0.61, and 0.01–0.79 and 0.47, respectively.

Table 2. Predictive models × prediction time accuracies and May MAE for the training and validation sets.

(1) Model	(2) Accuracy							(3) May_MAE.T	May_MAE.V	
	May	June	July	May–June	June–July	May–July	Mean	t ha^{-1}	t ha^{-1}	%
PLS-DA	0.77	0.49	−0.02	0.69	0.32	0.56	0.47 ab	5.01 bcd	3.62	26.81
PCA-DA	0.76	0.49	−0.13	0.67	0.29	0.52	0.43 ab	5.00 bcd	2.91	21.56
RF	0.82	0.39	0.64	0.68	0.53	0.74	0.63 a	5.05 bcd	2.27	16.81
SVML	0.80	0.49	0.58	0.67	0.59	0.70	0.64 a	4.82 cd	3.74	27.70
SVML-G	0.80	0.49	0.58	0.66	0.61	0.72	0.64 a	4.84 cd	3.74	27.70
SVM-R	0.88	−0.36	0.51	0.02	−0.15	0.08	0.16 b	4.95 bcd	1.87	13.85
SVM-P	0.81	0.49	0.09	0.63	0.42	0.53	0.50 a	4.64 d	6.22	46.07
NNET	0.78	0.56	0.16	0.75	0.38	0.70	0.56 a	11.99 a	12.50	92.59
GBT	0.78	0.56	0.69	0.58	0.81	0.57	0.66 a	5.29 bc	2.68	19.85
GBD	0.84	0.37	−0.01	0.76	0.48	0.51	0.49 a	4.80 d	2.18	16.15
GBL	0.45	0.11	0.89	0.93	0.03	0.43	0.47 ab	5.43 b	3.40	25.19
LM	0.78	0.50	0.56	0.73	0.46	0.65	0.61 ab	4.88 cd	4.53	33.56
NLNET	0.79	0.09	0.33	0.79	0.01	0.79	0.47 ab	5.36 b	2.34	17.33
MEAN	0.77a	0.36b	0.37b	0.66a	0.37b	0.58ab	0.52	5.54	4.00	29.63

(1) PLS-DA, PCA-DA, RF, SVML, SVML-G, SVM-R, SVM-P, NNET, GBT, GBD, GBL, LM, and NLNET, respectively, partial least squares discriminant analysis, principal component analysis discriminant analysis, random forest, Support Vector Machines with Linear Kernel, Support Vector Machines with Linear Kernel grid search, Support Vector Machines with Radial Basis Function Kernel, Support Vector Machines with Polynomial Kernel, neural network, eXtreme Gradient Boosting- xgbtree method, eXtreme Gradient Boosting- xgbDART method, eXtreme Gradient Boosting-xgbLinear method, Linear model, and Neural Network neuralnet method. (2) Accuracy represents the Pearson correlation coefficient between the predicted and the observed values in the validation set. (3) May_MAE.T mean absolute error relative to the optimal prediction model in the month of May using repeated cross-validation in the training set; May MAE.V (MAE and MAPE) magnitude of the error relative to the predicted values in the validation (testing) set; means with the same letter in a same column or row, are not significantly different at the 5% probability level using Duncan's multiple range test.

The mean comparison showed that SVM-R was the least accurate model. The other models showed comparable accuracies, but RF, SVML, SVML-G, SVM-P, NNET, GBT, and GBD showed prediction ability greater than SVM-R. GBT's prediction ability was consistently greater than 0.5 across the prediction times. Apart from GBL, the prediction ability of all models was high (prediction accuracy greater than or equal to 0.76) and/or better in the month of May (Table 2). The mean accuracy across models was high and not significantly different in May, May–June, and May–July. The across-model average accuracy computed in May was significantly superior to the mean accuracy obtained in June, June–July, and July. June, June–July, and July were statistically equally worst times for predicting biomass yields in sorghum under the Mediterranean region.

The range and mean MAE values (in t ha^{-1}) produced for each model during the calibration experiments at one (May) of the best prediction times were 9.3–14.4 and 12.0, 2.4–7.4 and 5.0, 3.1–7.0 and 4.9, 2.8–7.3 and 5.0, 2.7–7.2 and 5.0, 3.1–6.7 and 4.6, 3.1–7.1 and 4.8, 3.1–7.1 and 4.8, 3.19–7.95 and 5.29, 2.64–7.34 and 4.80, 2.77–10.21 and 5.43, 2.82–7.09 and 4.88, and 2.18–7.63 and 5.36, for NNET, RF, SVM-R, PCA-DA, PLS-DA, SVM-P, SVML, SVM-G, GBT, GBD, GBL, LM, and NLNET, respectively (Figure 4, Table 2). The average MAE values during the training process were statistically higher in

NNET followed by NLNET and GBL. Prediction error (MAE) was lower in SVM-P and GBD, while it was not statistically different in PLS-DA, PCA-DA, RF, SVML, SVML-G, SVM-R, GBT, and LM (Figure 4, Table 2). The MAE values (in t ha^{-1}) calculated using the validation (testing) set and the best prediction time (May) were, in increasing order, 1.87 (13.85%), 2.18 (16.15%), 2.27 (16.81%), 2.34 (17.33%), 2.68 (19.85%), 2.91 (21.56%), 3.40 (25.19%), 3.62 (26.81%), 3.74 (27.70%), 3.74 (27.7%), 4.53 (33.56%), 6.22 (46.07%), and 12.5 (92.59%), for SVM-R, GBD, RF, NLNET, GBT, PCA-DA, GBL, PLS-DA, SVML, SVML-G, LM, SVM-P, and NNET algorithms, respectively.

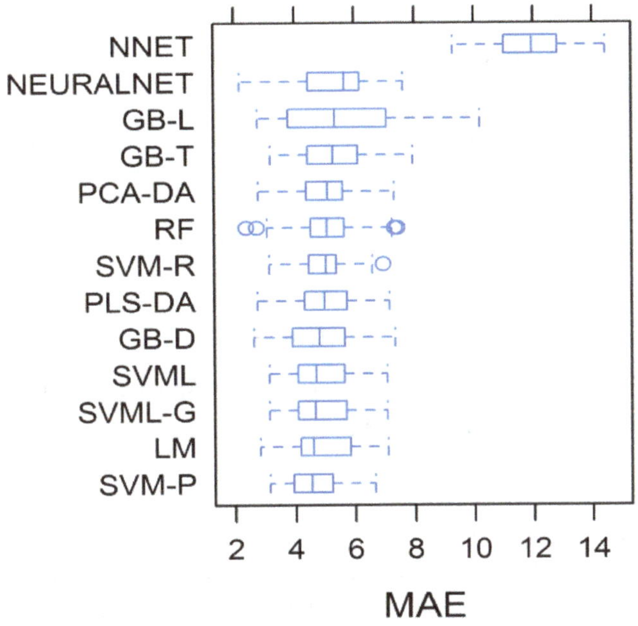

Figure 4. Visualization of models MAE (t ha^{-1}) dispersion using boxplot approach and fAPAR acquired in May. PLS-DA, PCA-DA, RF, SVML, SVML-G, SVM-R, SVM-P, NNET, GBT, GBD, GBL, LM, and NLNET, respectively, partial least squares discriminant analysis, principal component analysis discriminant analysis, random forest, Support Vector Machines with Linear Kernel, Support Vector Machines with Linear Kernel grid search, Support Vector Machines with Radial Basis Function Kernel, Support Vector Machines with Polynomial Kernel, neural network, eXtreme Gradient Boosting xgbtree method, eXtreme Gradient Boosting xgbDART method, eXtreme Gradient Boosting xgbLinear method, Linear model, and Neural Network neuralnet method.

Spearman's rank correlation coefficient (Spearman's rho) between model accuracy and MAE values (t ha^{-1} and %) corresponding to the testing set was −0.40. The Spearman's rho method assesses how well the relationship between two variables can be described using a monotonic function between ordered sets that preserves or reverses the given order [48]. The Spearman's rho approach was selected to account for the small size of the samples whose pairwise statistical dependences could not be correctly assessed with parametric approaches that have to be implemented on normally distributed data. Indeed the Shapiro–Wilk test of normality for the vectors of model accuracies and MAE values, was very highly significant ($p < 0.001$), meaning that we couldn't assume the normality.

Over the May to July prediction time interval, six days of year corresponding to fortnightly fAPAR indices, were used as regressors in this work. Among these regressors, the most important times to predict the aboveground sorghum biomass yields were investigated using the GBT algorithm as this model showed high and dependable performance that was insensitive to the prediction times across the cropping season. The model showed that the day of year 150 was the most important (index = 100) followed by DoY 165 (index = 80), DoY 135 (index = 30), DoY 195 (index = 20), and DoY 210 (index = 10) (Figure 5). The day of year 180 was associated with no importance in terms of fAPAR-based prediction of the aboveground biomass yields in sorghum under the Mediterranean environment.

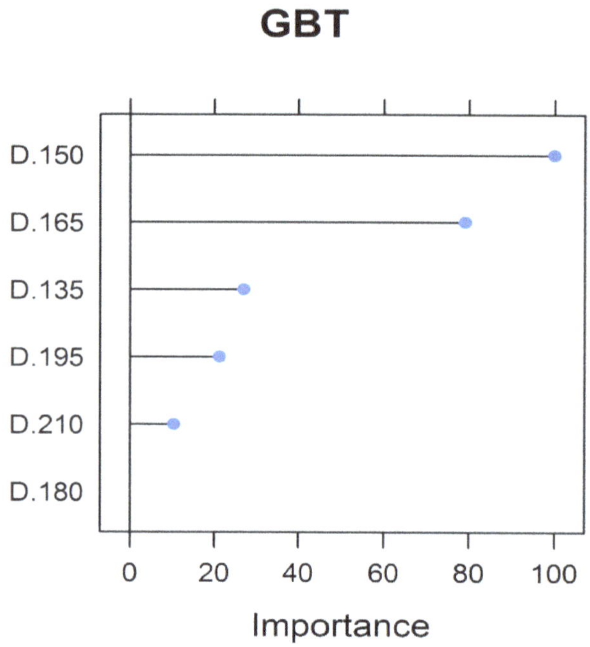

Figure 5. Relative importance of regressors (day of year, D) on sorghum biomass yields in 2017 and 2018, using eXtreme Gradient Boosting xgbtree (GBT) method.

4. Discussion

The fAPAR biophysical variable used in this work was derived from satellite imagery, which is part of Earth Observation's big data. Big data technology (BDT) is a new technological paradigm that is driving entire economies, including low-tech industries such as agriculture where it is implemented under the banner of precision farming (PF) [49]. In this work, BDT was built on geocoded maps of agricultural experiment fields and the real-time monitoring of sorghum crops on commercial farms in order to assess the possibility to monitor sorghum growth and development, with the ultimate aim of predicting the aboveground biomass yields. Early prediction of biomass production has positive implications including increased efficiency in biomass, biofuel, and farming resource management [50], and avoidance of energy crises. Forty-two sorghum pilot trials were evaluated in this work using fAPAR and different sorghum varieties belonging to four biomass producing sorghum types of dual purpose, sweet, forage, and biomass per se. Combining different types of biomass producing sorghums in this study was motivated by the need to mimic farming practice in the Mediterranean region. In this region, farmers, farming cooperatives, and third-party biomass harvesting and biodigesting companies manage the above-mentioned sorghum types indiscriminately on a regular basis. It made therefore sense not to discriminate the biomass producing sorghum types as sources of variation in the models implemented in this work. Similar investigations were reported in previous studies working on different crop species [16,51]. Important regressors of interest were identified and used in the predictive algorithms as suggested in literature [14,51].

The fAPAR index produced unique curves and maps that discriminated between the types of sorghums evaluated in this work. The fAPAR profile paralleled the evolution of leaf senescence across sorghum types [52] under the Mediterranean environment. The fAPAR curves presented in this work were purposely derived from sorghum fields established side by side in the same location in Anzola dell'Emilia. These pilots were sown and harvested on the same dates and managed identically, which allows a coherent comparison. The above-described shapes of the curves were generally similar also across locations in this study, though with slight discrepancies for some sorghum types. All sorghum trials reported in this work were conducted under a rainfed regime. Given that Mediterranean region is characterized by a semiarid climate wherein summer crops rely heavily on winter soil-stored moisture and experience postanthesis drought stress, it can be inferred that the fAPAR in dual purpose and biomass sorghum types did not rise during the reproduction growth stage probably due to a combined effect of sink demand and soil water scarcity in dual purpose, and mostly soil water scarcity in biomass sorghum. Postanthesis drought stress in sorghum under the Mediterranean environment was amply described by Habyarimana et al. [52–55]. The fAPAR profile in the sweet, forage, and SB × SH grown from seed reflects the reduced importance of the sink and the delayed leaf senescence in these types. In these sorghum types, a slow fAPAR increase toward the harvest can be explained by the precipitations registered in early fall in most locations (Figures S1 and S2), which stimulated the growth of axillary tillers in annual Sorghum bicolor [54,55] and the growth of axillary tillers and ramets in perennial SB × SH sorghum [54–57]. The above explanation holds also in the case of the SB × SH regrown from rhizomes. In these plants, the deep fAPAR drop from mid-June (anthesis) to early fall corresponds to the observed dry summers (Figures S1 and S2) and testifies to the increased susceptibility to drought stress in these plants. The conclusions drawn on fAPAR profile held particularly for fields established in the same location. Therefore, further investigations with replications in time and space are in order before any generalization is made.

In this work, high levels of model prediction accuracy ($r \geq 0.70$ or $r^2 \geq 0.50$) were obtained for 12 out of the 13 models deployed at the best prediction time (May). The models were therefore able to explain 50% of the variability that existed in the sorghum biomass yield data, while the remaining variance can be related to other factors non accounted for in this study such as the heterogeneity of external environmental and anthropogenic factors including rainfall distribution, soil types, and planting/tilling practices that could lead to different yield responses across the farms. The modeling performance metrics achieved in this work are nonetheless comparable to previous findings. For instance, Battude et al. [58], Shafian et al. [31], and Panda et al. [30] came across similar accuracy in their work on maize biomass and grain yields, and sorghum yields, respectively. On the other hand, the accuracy realized in this work was greater or equal to the values reported in Gao et al. [51], Diouf et al. [14], and the optimal values in sorghum as presented in Johnson [16]. Linear and nonlinear models performed comparably in terms of accuracy and mean absolute error implying that the relationship between fAPAR and biomass yield was mainly linear, which was expected and also supported by previous findings [14]. At the best prediction time (month of May), the correlation between the model accuracies and the MAE values was negative, denoting the expected inverse relationship between the two metrics of model prediction performance.

The simple linear model was implemented in this work to serve as a benchmark with respect to the most complex models requiring parameters optimization. Since thirteen models were implemented in this study, it is interesting to select the best algorithms. As biomass sorghum in the Mediterranean region is harvested from end of August to late November, it can be interesting to be able to predict the biomass production from May to July, allowing the farmer to know the amount to be produced one to six months ahead of harvest [42,51]. SVML, SVML-G, GBT, and LM showed good prediction accuracy ($r \geq 0.50$) across the evaluated prediction times, with MAE values (%) of 27.70, 27.70, 19.85, and 33.56, respectively. The GBT model was therefore the best algorithm as it performed consistently well ($r \geq 0.60$) across the prediction times, and was associated with low prediction error. The GBT model can therefore be recommended for sorghum biomass yield prediction using Sentinel-2-derived fAPAR as biophysical variable under the Mediterranean region. This model can be deployed anywhere from May to July without significant loss function.

In terms of biomass yields prediction times, June, July, and June–July were the worst times. May, May–June, and May–July showed comparable average accuracies, but accuracy in May was generally high ($r \geq 0.70$) across models except GBL. The month of May can therefore be recommended as the best time to predict sorghum biomass yields in the Mediterranean region. In this work, several types of sorghum were used, including high tonnage, sweet, forage, and dual purpose types. The suitability of the month of May for sorghum biomass yields prediction can be partly explained by the fact that in early sorghum growth stages, particularly in the period of time around the fast growth stage, the four sorghum types exhibit similar levels of growth and development. Furthermore, sorghum crop as currently grown in the Mediterranean region, reaches the fast growth stage generally in the month of May, meaning that predictions run in May are carried out on populations of sorghum types that are mostly at the same stage of growth and development. Overall, the days of year 150 and 165 were the most important regressors followed by days of year 135, 195, and 210 in decreasing order. The two regressors acquired in May (DoY 150 and 135) had important direct effects on the sorghum biomass, which justifies the good prediction accuracies obtained in this month. On the other hand, the two regressors corresponding to the month of July showed poor importance on biomass yields, while one of the two regressors corresponding to the month of June had meaningless effect on biomass yields, all of which explains the poor prediction accuracies obtained in June, July, and June–July (Table 2). In the Mediterranean region, sorghum is sown mid-to-late April. Therefore, being able to perform accurate sorghum biomass yields prediction in May, i.e., up to six months ahead of harvesting is a remarkable opportunity for the farmer and farming cooperatives that can use this information for several business-related purposes. They can efficiently organize the biomass business operations including the rational mobilization of the fleets of harvesting machinery, transport vehicles, and storage facilities. The predictive models developed in this work can also be used by extension services and policy-makers for strategic purposes. Obtaining the information on potential within-season biomass availability early on before actual harvest will help assess alternative means for energy supply internally, import or export, which is expected to help avoid specific crises such as fuel shortage. The findings in this work are limited in scope to one province in Italy, within the Mediterranean region. The prediction equations produced in this work can therefore be safely used in analogous modeling experiments in other Mediterranean areas. However, for these equations to be extended to modeling activity at a global level, the training populations of farms would require updates with inclusion of data accounting for sampling additional latitudes and longitudes relevant for sorghum cultivation.

5. Conclusions

The importance of sorghum as food, feed, and biofuel crop was amply described in several scientific literatures. Biomass sorghum demonstrated higher yields with better energy balance relative to major crops of agroindustrial interest. As dedicated biomass sorghum crops are steadily increasing and precision farming is driving agricultural economies worldwide, the harnessing satellite technology is well-poised to bring about agricultural advantages including cutting farming operational costs. Sentinel-2-derived fraction of absorbed photosynthetically active radiation was found to satisfactorily explain primary productivity and was used in this study as biophysical variable in the predictive modeling of aboveground biomass yields in annual and perennial sorghums. Across month combinations from May to July and the thirteen machine learning prediction algorithms used in this work, the gradient boosting machine learning algorithm implementing xgbtree was identified as the best predictive model. The best prediction time for sorghum biomass was particularly the month of May, followed by May–June and May–July using fortnightly fAPAR indices. To the best of our knowledge, the present work represents the first time Sentinel-2-derived fAPAR is used in predictive modeling of sorghum biomass yields. The outcome from this study is important and can serve several purposes including farmers being able to improve their sorghum biomass business operations. Policy-makers and extension services will also benefit from the findings in this work allowing them early on within season information on potential biomass availability, which is critical to wider energy planning and avoiding energy-related crises.

Author Contributions: Conceptualization, E.H.; Data Curation, E.H. and I.P.; Formal Analysis, E.H.; Funding Acquisition, E.H. and I.P.; Investigation, E.H.; Methodology, E.H.; Project Administration, E.H.; Writing—Original Draft, E.H.; Writing—Review & Editing, I.P., M.C., P.D.F., and M.D.

Acknowledgments: Part of this work was supported (beneficiary: first author) by the project Data-driven Bioeconomy (www.databio.eu), GA number: 732064 (H2020-ICT-2016-1—innovation action), and the project Risorse GeneticheVegetali (RGV/FAO) 2014e2016 of the Ministero delle PoliticheAgricole, Alimentari e Forestali, Rome. The authors would like to thank the anonymous reviewers and editor for their informative remarks that contributed to improving the quality of this paper.

References

1. Habyarimana, E.; Lorenzoni, C.; Redaelli, R.; Alfieri, M.; Amaducci, S.; Cox, S. Towards a perennial biomass sorghum crop: A comparative investigation of biomass yields and overwintering of *Sorghum bicolor* x *S. halepense* lines relative to long term *S. bicolor* trials in northern Italy. *Biomass Bioenergy* **2018**, *111*, 187–195. [CrossRef]
2. Damasceno, C.M.B.; Schaffert, R.E.; Duweikat, I. *Mining Genetic Diversity of Sorghum as a Bioenergy Feedstock*; Springer: New York, NY, USA, 2014; pp. 81–106.
3. Hoffmann, L., Jr.; Rooney, W.L. Cytoplasm has no effect on the yield and quality of biomass sorghum hybrids. *JSBS* **2013**, *3*, 129–134. [CrossRef]
4. Prakasham, R.S.; Nagaiah, D.; Vinutha, K.S.; Uma, A.; Chiranjeevi, T.; Umakanth, A.V. Sorghum biomass: A novel renewable carbon source for industrial bioproducts. *Biofuels* **2014**, *5*, 159–174. [CrossRef]
5. Habyarimana, E. Genomic prediction for yield improvement and safeguarding of genetic diversity in CIMMYT spring wheat (*Triticum aestivum* L.). *Aust. J. Crop Sci.* **2016**, *10*, 127–136.
6. Rooney, W.L. Genetics and cytogenetics. In *Sorghum: Origin, History, Technology, and Production*; Smith, C.W., Frederiksen, R.A., Eds.; John Wiley & Sons: New York, NY, USA, 2000; pp. 261–307.
7. El Bassam, N. *Handbook of Bioenergy Crops: A Complete Reference to Species, Development and Applications*; Earthscan Ltd.: London, UK, 2010; pp. 45–477.
8. Stefaniak, T.R.; Dahlberg, J.A.; Bean, B.W.; Dighe, N.; Wolfrum, E.J.; Rooney, W.L. Variation in biomass composition components among forage, biomass, sorghum-sudangrass, and sweet sorghum types. *Crop Sci.* **2012**, *52*, 1949–1954. [CrossRef]
9. Kussul, N.; Sokolov, B.V.; Zyelyk, Y.I.; Zelentsov, V.A.; Skakun, S.V.; Shelestov, A.Y. Disaster risk assessment based on heterogeneous geospatial information. *J. Autom. Inform. Sci.* **2010**, *42*, 32–45. [CrossRef]
10. Kussul, N.; Shelestov, A.; Skakun, S. *Flood Monitoring from SAR Data Use of Satellite and In-Situ Data to Improve Sustainability*; Springer: Dordrecht, The Netherlands, 2011; pp. 19–29.
11. Skakun, S.; Kussul, N.; Kussul, O.; Shelestov, A. Quantitative estimation of drought risk in Ukraine using satellite data. In Proceedings of the IEEE International Geoscience and Remote Sensing Symposium (IGARSS), Quebec City, QC, Canada, 13–18 July 2014; pp. 5091–5094.
12. Skakun, S.; Kussul, N.; Shelestov, A.; Kussul, O. The use of satellite data for agriculture drought risk quantification in Ukraine. *Geomat. Nat. Hazards Risk* **2016**, *7*, 901–917. [CrossRef]
13. Gallego, J.; Kravchenko, A.N.; Kussul, N.N.; Skakun, S.V.; Shelestov, A.Y.; Grypych, Y.A. Efficiency assessment of different approaches to crop classification based on satellite and ground observations. *J. Autom. Inform. Sci.* **2012**, *44*, 67–80. [CrossRef]
14. Diouf, A.A.; Brandt, M.; Verger, A.; El Jarroudi, M.; Djaby, B.; Fensholt, R.; Ndione, J.A.; Tychon, B. Fodder Biomass Monitoring in Sahelian Rangelands Using Phenological Metrics from FAPAR time series. *Remote Sens.* **2015**, *7*, 9122–9148. [CrossRef]
15. Duveiller, G.; López-Lozano, R.; Baruth, B. Enhanced Processing of 1-km Spatial Resolution fAPAR Time Series for Sugarcane Yield Forecasting and Monitoring. *Remote Sens.* **2013**, *5*, 1091–1116. [CrossRef]
16. Johnson, D.M. A comprehensive assessment of the correlations between field crop yields and commonly used MODIS products. *Int. J. Appl. Earth Obs. Geoinf.* **2016**, *52*, 65–81. [CrossRef]
17. Kogan, F.; Kussul, N.; Adamenko, T.; Skakun, S.; Kravchenko, O.; Kryvobok, O.; Shelestov, A.; Kolotii, A.; Kussul, O.; Lavrenyuk, A. Winter wheat yield forecasting in Ukraine based on Earth observation, meteorological data and biophysical models. *Int. J. Appl. Earth Obs. Geoinf.* **2013**, *23*, 192–203. [CrossRef]

18. Kogan, F.; Kussul, N.; Adamenko, T.; Skakun, S.; Kravchenko, O.; Kryvobok, O.; Shelestov, A.; Kolotii, A.; Kussul, O.; Lavrenyuk, A. Winter wheat yield forecasting: A comparative analysis of results of regression and biophysical models. *Int. J. Autom. Inform. Sci.* **2013**, *45*, 68–81. [CrossRef]
19. Kowalik, W.; Dabrowska-Zielinska, K.; Meroni, M.; Raczka, T.U.; de Wit, A. Yield estimation using SPOTVEGETATION products: A case study of wheat in European countries. *Int. J. Autom. Inform. Sci.* **2014**, *32*, 228–239.
20. Kross, A.; McNairn, H.; Lapen, D.; Sunohara, M.; Champagne, C. Assessment of RapidEye vegetation indices for estimation of leaf areaindex and biomass in corn and soybean crops. *Int. J. Appl. Earth Obs. Geoinf.* **2015**, *34*, 235–248. [CrossRef]
21. Camacho, F.; Cernicharo, J.; Lacaze, R.; Baret, F.; Weiss, M. GEOV1: LAI, FAPAR Essential Climate Variables and FCOVER global time series capitalizing over existing products. Part 2: Validation and intercomparison with reference products. *Remote Sens. Environ.* **2013**, *137*, 310–329. [CrossRef]
22. Shelestov, A.; Kolotii, A.; Camacho, F.; Skakun, S.; Kussul, O.; Lavrenuik, M. Mapping of biophysical parameters based on high resolution EO imagery for JECAM test site in Ukraine. In Proceedings of the 2015 IEEE International Geoscience and Remote Sensing Symposium (IGARSS), Milan, Italy, 26–31 July 2015.
23. Kussul, N.; Kolotii, A.; Skakun, S.; Shelestov, A.; Kussul, O.; Oliynuk, T. Efficiency estimation of different satellite data usage for winter wheat yield forecasting in Ukraine. In Proceedings of the 2014 IEEE International Geoscience and Remote Sensing Symposium (IGARSS), Quebec City, QC, Canada, 13–18 July 2014; pp. 5080–5082.
24. López-Lozano, R.; Duveiller, G.; Seguini, L.; Meroni, M.; García-Condado, S.; Hooker, J.; Leo, O.; Baruth, B. Towards regional grain yield forecasting with 1km-resolution EO biophysical products: Strengths and limitations at pan-European level. *Agric. For. Meteorol.* **2015**, *206*, 12–32. [CrossRef]
25. Baret, F.; Weiss, M.; Lacaze, R.; Camacho, F.; Makhmara, H.; Pacholcyzk, P.; Smets, B. Geov1: LAI and FAPAR essential climate variables and FCOVER global time series capitalizing over existing products. Part1: Principles of development and production. *Remote Sens. Environ.* **2013**, *137*, 299–309. [CrossRef]
26. Fensholt, R.; Sandholt, I.; Rasmussen, M.S.; Stisen, S.; Diouf, A. Evaluation of satellite based primary production modelling in the semi-arid Sahel. *Remote Sens. Environ.* **2006**, *105*, 173–188. [CrossRef]
27. Tucker, C.J.; Holben, B.N.; Elgin, J.H.; McMurtrey, J.E. Relationship of spectral data to grain yield variation. *Photogramm. Eng. Remote Sens.* **1980**, *46*, 657–666.
28. Barnett, T.L.; Thompson, D.R. The use of large-area spectral data in wheatyield estimation. *Remote Sens. Environ.* **1982**, *12*, 509–518. [CrossRef]
29. Hatfield, J.L. Remote sensing estimators of potential and actual crop yield. *Remote Sens. Environ.* **1996**, *13*, 301–311. [CrossRef]
30. Panda, S.S.; Ames, D.P.; Panigrahi, S. Application of vegetation indices for agricultural crop yield prediction using neural network techniques. *Remote Sens.* **2010**, *2*, 673–696. [CrossRef]
31. Shafian, S.; Rajan, N.; Schnell, R.; Bagavathiannan, M.; Valasek, J.; Shi, Y.; Olsenholler, J. Unmanned aerial systems-based remote sensing for monitoring sorghum growth and development. *PLoS ONE* **2018**, *13*, e0196605. [CrossRef] [PubMed]
32. Yang, C.; Everitt, J.H.; Bradford, J.M.; Escobar, D.E. Mapping grain sorghum growth and yield variations using airborne multispectral digital imagery. *Trans. ASAE* **2000**, *43*, 1927–1938. [CrossRef]
33. Piper, J.K.; Kulakow, P.A. Seed yield and biomass allocation in Sorghum bicolor and F1 and backcross generations of *S. bicolor* x *S. halepense* hybrids. *Can. J. Bot.* **1994**, *72*, 468–474. [CrossRef]
34. De Keukelaere, L.; Sterckx, S.; Adriaensen, S.; Knaeps, E.; Reusen, I.; Giardino, C.; Bresciani, M.; Hunter, P.; Neil, C.; Van der Zande, D.; et al. Atmospheric correction of Landsat-8/OLI and Sentinel-2/MSI data using iCOR algorithm: Validation for coastal and inland waters. *Eur. J. Remote Sens.* **2018**, *51*, 525–542. [CrossRef]
35. Weiss, M.; Baret, F. ATBD S2ToolBox Level 2 Products: LAI, FAPAR, FCOVER (Version 1.1). 2016. Available online: http://step.esa.int/docs/extra/ATBD_S2ToolBox_L2B_V1.1.pdf (accessed on 7 November 2018).
36. Jacquemoud, S.; Verhoef, W.; Baret, F.; Bacour, C.; Zarco-Tejada, P.J.; Asner, G.P.; François, C.; Ustin, S.L. PROSPECT + SAIL models: A review of use for vegetation characterization. *Remote Sens. Environ.* **2009**, *113*, S56–S66. [CrossRef]
37. Sage, R.F. A portrait of the C4 photosynthetic family on the 50th anniversary of its discovery: Species number, evolutionary lineages, and Hall of Fame. *J. Exp. Bot.* **2016**, *67*, 4039–4056. [CrossRef]
38. Eilers, P.H.C. A perfect smoother. *Anal. Chem.* **2003**, *75*, 3631–3636. [CrossRef]

39. Atzberger, C.; Eilers, P.H.C. A smoothed 1-km resolution NDVI time series (1998–2008) for vegetation studies in South America. *Int. J. Digit. Earth* **2010**, *4*, 365–386. [CrossRef]
40. Kuhn, M. Building predictive models in R using the caret Package. *J. Stat. Softw.* **2008**, *28*, 1–26. [CrossRef]
41. Benor, D.; Baxter, M. *Training and Visit Extension*; The World Bank: Washington, DC, USA, 1984; pp. 21–212.
42. Hoefsloot, P.; Ines, A.V.; van Dam, J.; Duveiller, G.; Kayitakire, F.; Hansen, J. Combining crop models and remote sensing for yield prediction: Concepts, applications and challenges for heterogeneous smallholder environments. In *JRC Scientific and Policy Reports*; Report of CCFAS-JRC Workshop at Joint Research Centre; Joint Research Centre of the European Commission: Ispra, VA, Italy, 2012; pp. 7–41.
43. Meuwissen, T.H.E.; Hayes, B.J.; Goddard, M.E. Prediction of total genetic value using genome-wide dense marker maps. *Genetics* **2001**, *157*, 1819–1829. [PubMed]
44. R Core Team. *R: A Language and Environment for Statistical Computing*; R Foundation for Statistical Computing: Vienna, Austria, 2013.
45. Breiman, L.; Friedman, J.H.; Olshen, R.A.; Stone, C.J. *Classification and Regression Trees*; Wadsworth Inc.: Blelmont, CA, USA, 1984.
46. Willmott, C.J.; Matsuura, K. Advantages of the mean absolute error (MAE) over the root mean square error (RMSE) in assessing average model performance. *Clim. Res.* **2005**, *30*, 79–82. [CrossRef]
47. Duncan, D.B. Multiple range and multiple F tests. *Biometrics* **1955**, *11*, 1–42. [CrossRef]
48. Spearman, C. The proof and measurement of association between two things. *Am. J. Psychol.* **1904**, *15*, 72–101. [CrossRef]
49. Schellberg, J.; Hill, M.J.; Gerhards, R.; Rothmund, M.; Braun, M. Precision agriculture on grassland: Applications, perspectives and constraints. *Eur. J. Agron.* **2008**, *29*, 59–71. [CrossRef]
50. Segarra, E. *Precision Agriculture Initiative for Texas High Plains*; Annual Comprehensive Report; Texas A&M University Research and Extension Center: Lubbock, TX, USA, 2002.
51. Gao, F.; Anderson, M.; Daughtry, C.; Johnson, D. Assessing the Variability of Corn and Soybean Yields in Central Iowa Using High Spatiotemporal Resolution Multi-Satellite Imagery. *Remote Sens.* **2018**, *10*, 1489. [CrossRef]
52. Habyarimana, E.; Lorenzoni, C.; Busconi, M. Search for new stay-green sources in *Sorghum bicolor* (L.) Moench. *Maydica* **2010**, *55*, 187–194.
53. Habyarimana, E.; Laureti, D.; Di Fonzo, N.; Lorenzoni, C. Biomass production and drought resistance at the seedling stage and in field conditions in sorghum. *Maydica* **2002**, *47*, 303–309.
54. Habyarimana, E.; Laureti, D.; De Ninno, M.; Lorenzoni, C. Performances of biomass sorghum [*Sorghum bicolor* (L.) Moench] under different water regimes in Mediterranean region. *Ind. Crop Prod.* **2004**, *20*, 23–28. [CrossRef]
55. Habyarimana, E.; Bonardi, P.; Laureti, D.; Di Bari, V.; Cosentino, S.; Lorenzoni, C. Multilocational evaluation of biomass sorghum hybrids under two stand densities and variable water supply in Italy. *Ind. Crop Prod.* **2004**, *20*, 3–9. [CrossRef]
56. Cox, T.S.; Van Tassel, D.L.; Cox, C.M.; DeHaan, L.R. Progress in breeding perennial grains. *Crop. Pasture Sci.* **2010**, *61*, 513–521. [CrossRef]
57. Nabukalu, P.; Cox, T.S. Response to selection in the initial stages of a perennial sorghum breeding program. *Euphytica* **2016**, *209*, 103–111. [CrossRef]
58. Battude, M.; Al Bitar, A.; Morin, D.; Cros, J.; Huc, M.; Sicre, C.M.; Le Dante, V.; Demarez, V. Estimating maize biomass and yield over large areas using high spatial and temporal resolution Sentinel-2 like remote sensing data. *Remote Sens. Environ.* **2016**, *184*, 668–681. [CrossRef]

9

Performance Characterization of the UAV Chemical Application based on CFD Simulation

Hang Zhu [1,*], Hongze Li [1], Cui Zhang [1], Junxing Li [2] and Huihui Zhang [3,*]

1 School of Mechanical and Aerospace Engineering, Jilin University, Changchun 130022, China; lihz18@mails.jlu.edu.cn (H.L.); zhangcui@jlu.edu.cn (C.Z.)
2 Jilin Academy of Agricultural Machinery, Changchun 130062, China; junxingli_jl@163.com
3 Water Management and Systems Research Unit, USDA-ARS, Fort Collins, CO 80526, USA
* Correspondence: zhuhang@jlu.edu.cn (H.Z.); Huihui.Zhang@ars.usda.gov (H.Z.)

Abstract: Battery-powered multi-rotor UAVs (Unmanned Aerial Vehicles) have been employed as chemical applicators in agriculture for small fields in China. Major challenges in spraying include reducing the influence of environmental factors and appropriate chemical use. Therefore, the objective of this research was to obtain the law of droplet drift and deposition by CFD (Computational Fluid Dynamics), a universal method to solve the fluid problem using a discretization mathematical method. DPM (Discrete Phase Model) was taken to simulate the motion of droplet particles since it is an appropriate way to simulate discrete phase in flow field and can track particle trajectory. The figure of deposition concentration and trace of droplet drift was obtained by controlling the variables of wind speed, pressure, and spray height. The droplet drifting models influenced by different factors were established by least square method after analysis of drift quantity to get the equation of drift quantity and safe distance. The relationship model, Y_i(m), between three dependent variables, wind speed X_w(m s^{-1}), pressure X_p(MPa) and spray height X_h(m), are listed as follows: The edge drift distance model was $Y_1 = 0.887X_w + 0.550X_p + 1.552X_h - 3.906$ and the correlation coefficient (R^2) was 0.837; the center drift distance model was $Y_2 = 0.167X_w + 0.085X_p + 0.308X_h - 0.667$ and the correlation coefficient (R^2) was 0.774; the overlap width model was $Y_3 = 0.692x_w + 0.529x_p + 1.469x_h - 3.374$ and the correlation coefficient (R^2) was 0.795. For the three models, the coefficients of the three variables were all positive, indicating that the three factors were all positively correlated with edge drift distance, center drift distance, and overlap width. The results of this study can provide theoretical support for improving the spray quality of UAV and reducing the drift of droplets.

Keywords: UAV chemical application; droplet drift; flat-fan atomizer; simulation analysis; control variables

---

1. Introduction

China is a large agricultural country with the most serious occurrence of crop diseases and pests in the world, and has the largest use of pesticides [1]. Diseases and pests are important factors in agricultural production, and chemical pesticides are the main means to prevent and control crop diseases and pests in China [2,3]. The traditional method of pesticide application is manual control of ground application machinery, but this is strictly limited by the terrain [4]. With the development and implementation of new aerial application technology, the use of unmanned aerial vehicles (UAV) for aerial pesticide application is an inevitable trend for the intelligent development of green agriculture [5,6]. Therefore, it is critical to analyze and evaluate various performance parameters of UAV pesticide application technology. The performance evaluation of a UAV pesticide application system mainly involves an analysis of droplet drift and deposition characteristics under different working conditions [7].

In recent years, research concerning drift deposition characteristics and nozzle performance of UAVs used for the protection of small plants has increased. Due to aerial operation environment and the influence of natural airflow, the drifting and deposition problems of spraying is complicated [8–10]. Chen et al. [11] measured the wind field distribution under the rotor of a multi-rotor electric UAV using a UAV rotor wind field measurement system. The results showed that the vertical wind speed had an impact on the droplet deposition in the effective spray area. Qiu et al. [12] studied the relationship between spray deposition and flight height for unmanned helicopters. The results showed the significant effects of application height on deposition concentration. Shen et al. [13] obtained the flow field characteristics of multi-rotor UAV at different speeds by simulating the flow field of multi-rotor UAV through CFD. Nuyttens et al. [14] established a CFD three-dimensional spray drift model, which considered droplet characteristics, meteorological conditions, chemical characteristics, canopy structure and crop characteristics, etc. They carried out field experiments, verifying that their CFD model was useful for reducing spray drift in the field. Teske et al. [15–17] developed the AGDISP model based on the droplet trajectory model, which covered aircraft models, aircraft vortices, nozzle types, weather factors, and more. Luo et al. [18] carried out gas–solid two-phase flow field simulations for three types of nozzles, acquiring data to assist in the selection of nozzles for specific applications. Chen et al. [19] carried out multi-nozzle atomization field simulations using the UDF method. Their results showed that there was interference between multiple nozzles, and the number and position of nozzles affected the overall atomization effect. There have been many analyses of UAV performance as agricultural aerial sprayers, but relatively fewer analyses concerned with the effect of nozzle characteristics. Thus, this study aimed to simulate and analyze the droplet drift and deposition law of a flat fan nozzle under different working conditions and explore the droplet drift and deposition phenomenon under the influence of different factors.

2. Materials and Methods

2.1. Geometric Model Building

According to the actual spray situation (Figure 1), a cuboid model is established in software ANSYS ICEM CFD15.0 (NASDAQ: ANSS, Canonsburg, PA, USA). The length and width of the simulation calculation area was set as 10 m and 4 m to simulate the spraying area, the height was set as 0.8 m, 1 m and 1.5 m to simulate the spraying heights, for total grid area of 228,000. The left side of the cuboid is an application target area, and the right side is a droplet drift (off-target) area. Grids in the target area are encrypted to have a better spatial resolution. The two nozzles are positioned directly above the center point of the target area (Figure 2).

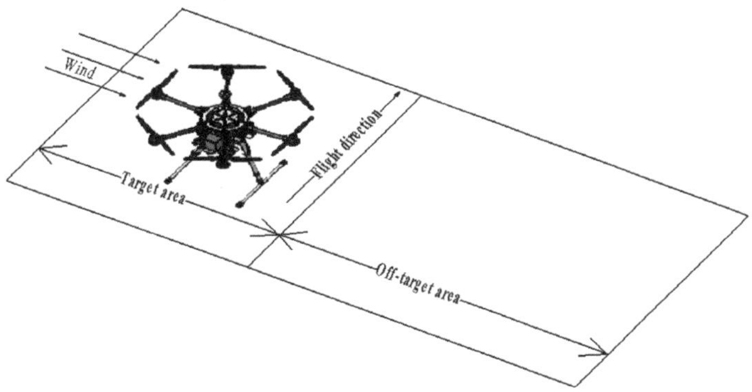

Figure 1. The sketch of UAV, wind and flight direction is shown in diagram to simulate the actual spray picture. The target area is set under the UAV and the off-target area is set down wind relative to the UAV.

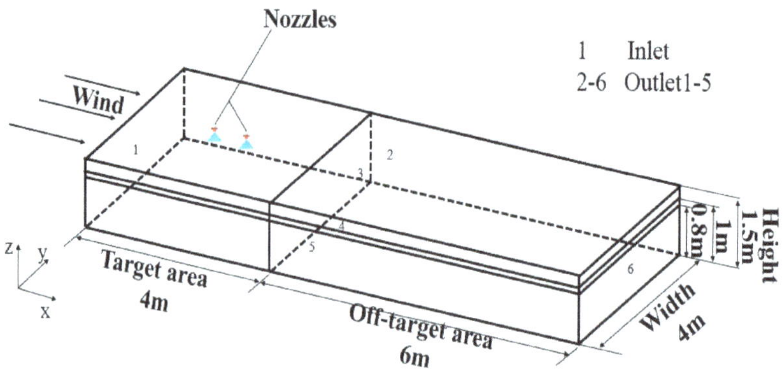

Figure 2. Computational domain and boundary setup are shown. The two nozzles are positioned on top of the domain.

The simulation analysis mainly obtains the droplet deposition rule and its influencing factor indexes. The boundary conditions are set as follows: The left face of the cuboid is the velocity inlet; the right face is the droplet receiving area pressure outlet; the other four surfaces are all pressure outlets; and the outlets are set as boundary escapes. According to the actual spraying operation height of the plant protection UAV, the sprayer height was changed to 0.8 m, 1 m and 1.5 m, respectively, by adjusting the position of the nozzles. The parameter settings are shown in Table 1.

Table 1. Parameters of the flat fan atomizer used to simulate XR8002 nozzle.

Parameters (Unit)	Value
X-Center (m)	1.75/2.25
Y-Center (m)	2
Z-Center (m)	1.494
X-Virtual Center (m)	1.75/2.25
Y-Virtual Center (m)	2
Z-Virtual Center (m)	1.5
X-Fan Normal Vector	0
Y-Fan Normal Vector	−1
Z-Fan Normal Vector	1
Flow Rate (kg s^{-1})	0.01316
Spray Half Angle (deg)	40
Orifice Width (m)	0.00091
Flat Fan Sheet Constant	3
Atomizer Dispersion Angle (deg)	6

In the simulation analysis, the continuous phase substance is air and the discrete uses the parameters of liquid water to simulate chemical. In the steady-state calculation mode, the standard k-ε model (ANSYS, 15.0) [20,21] is selected to simulate the turbulent wind flow. Its transport equations are shown in Equations (1) and (2).

$$\frac{\partial(\rho k)}{\partial t} + \frac{\partial(\rho k u_i)}{\partial x_i} = \frac{\partial}{\partial x_j}[(\mu + \frac{\mu_t}{\sigma_k})\frac{\partial k}{\partial x_j}] + G_k + G_b - \rho\varepsilon - Y_M + S_k \quad (1)$$

$$\frac{\partial(\rho\varepsilon)}{\partial t} + \frac{\partial(\rho\varepsilon u_i)}{\partial x_i} = \frac{\partial}{\partial x_j}[(\mu + \frac{\mu_t}{\sigma_\varepsilon})\frac{\partial \varepsilon}{\partial x_j}] + C_{1\varepsilon}\frac{\varepsilon}{k}(G_k + C_{3\varepsilon}G_b) - C_{2\varepsilon}\rho\frac{\varepsilon^2}{k} + S_\varepsilon \quad (2)$$

where k is the turbulence kinetic energy, ε is the turbulence dissipation rate, μ is the dynamic viscosity, μ_t is the turbulence viscosity, G_k is generated by turbulent kinetic energy caused by the average velocity gradient, and G_b is generated by turbulent kinetic energy caused by buoyancy. Y_M is a pulsating expansion term in compressible turbulence, $C_{1\varepsilon}$, $C_{2\varepsilon}$ and $C_{3\varepsilon}$ are empirical constants, Prandtl numbers of σ_k and σ_ε correspond to turbulent kinetic energy k and turbulent dissipation rate respectively, S_k and S_ε is a user-defined source item.

The variation of air density under standard atmospheric pressure and normal temperature is less than 5%, it is regarded as an incompressible fluid, the pressure-based solver type is selected. All the simulations are based on the transient calculation. The convergence criterion is set to 10^{-5}, which means that the converged solutions are reached when the residuals of several significant variables are equal to or less than 10^{-5}.

2.2. Design of Experiment

As is shown in Figure 3, the spray effect is seriously affected by wind speed. The control variates method was used to solve the problem including multivariate by changing one of the factors. This study used different wind speeds, 0 m s^{-1}, 1 m s^{-1}, 3 m s^{-1} and 5 m s^{-1}, to seek the law of droplet drift, and different particle mass flow rates, 0.01083 kg s^{-1}, 0.01316 kg s^{-1} and 0.01516 kg s^{-1}, to control the spray pressure at 0.2 MPa, 0.3 MPa and 0.4 MPa, respectively. The overall control variates of the parameters are shown in Table 2.

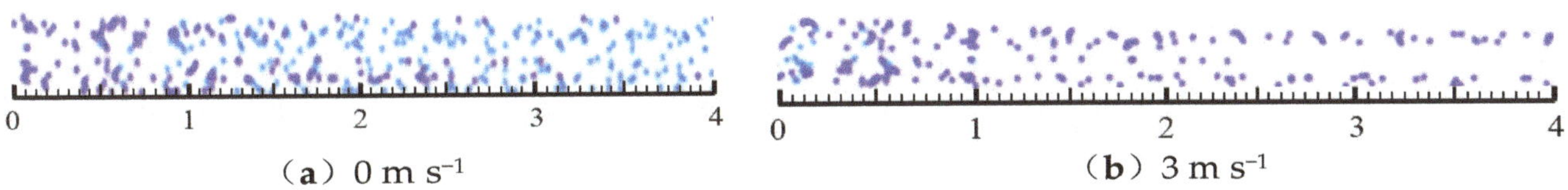

Figure 3. Droplets deposition in target area at different wind speed (0 m s^{-1} and 3 m s^{-1}). (**a**) The distribution of droplets is ideal at 0 m s^{-1}, (**b**) The droplet was influenced by wind clearly and the distribution of droplets is not satisfied at 3 m s^{-1}.

Table 2. Total variates of spray height (0.8, 1, and 1.5 m), pressure (0.2, 0.3, and 0.4 MPa) and wind speed (0, 1, 3, and 5 m s^{-1}).

Spray Height(m)		0.8			1			1.5	
Pressure (MPa)	0.2	0.3	0.4	0.2	0.3	0.4	0.2	0.3	0.4
Wind Speed (m s^{-1})									
0	A_{11}	A_{12}	A_{13}	B_{11}	B_{12}	B_{13}	C_{11}	C_{12}	C_{13}
1	A_{21}	A_{22}	A_{23}	B_{21}	B_{22}	B_{23}	C_{21}	C_{22}	C_{23}
3	A_{31}	A_{32}	A_{33}	B_{31}	B_{32}	B_{33}	C_{31}	C_{32}	C_{33}
5	A_{41}	A_{42}	A_{43}	B_{41}	B_{42}	B_{43}	C_{41}	C_{42}	C_{43}

The ability of DPM has been shown to accurately simulate particle dispersion and deposition [22–24]. In this study, the flat fan atomizer model of the DPM was selected to simulate the XR8002 nozzle of Teejet Company (Wheaton, IL, USA, 60187). In the DPM model, Euler method is used to describe the continuous phase. Navier-Stokes equation [25] is used to obtain velocity and other parameters. The discrete phase is described by Lagrange method, and its movement is obtained by integrating the motion equations of a large number of particles. Therefore, this model is called Euler-Lagrange model, and its transport equation can be expressed as [26]

$$\frac{d_{u_p}}{dt} = \frac{18\mu C_D R_e}{24\rho_p d_p{}^2}(u - u_p) + \frac{g_x(\rho_p - \rho)}{\rho_p} + \frac{1}{2}\frac{\rho}{\rho_p}\frac{d}{dt}(u - u_p) \tag{3}$$

where u is the continuous phase velocity, u_p is the velocity of particle, ρ_p is the density of particle, d_p is the particle diameter, g_x is the acceleration of gravity, R_e is the relative Reynolds number, and C_D is the drag coefficient.

2.3. Method of Analysis

In order to show the influence of three factors (wind speed, spray pressure, and spray height) on the droplet drift distance, deposition center drift distance, and overlap width, the multivariate curve fitting was carried out by the least square method [27,28]. The equation is listed as follows:

$$y_1 = a_0 + a_1 x_w + a_2 x_p + a_3 x_h \tag{4}$$

$$y_2 = a_4 + a_5 x_w + a_6 x_p + a_7 x_h \tag{5}$$

$$y_3 = a_8 + a_9 x_w + a_{10} x_p + a_{11} x_h \tag{6}$$

Satisfy

$$\sum_{i=1}^{n} (y - y_i)^2 = \min \sum_{i=1}^{n} (y - y_i)^2 \tag{7}$$

The model obtained by the least square method can be verified by Equation (8),

$$y_o = a_0 - y_n + y_f \tag{8}$$

where y_o is the overlap width influenced by wind, a_o is the overlap width in nature, y_n is the drift distance of near tuyere edge, and y_f is the drift distance of far tuyere edge. Their relationship can be portrayed in Figure 4.

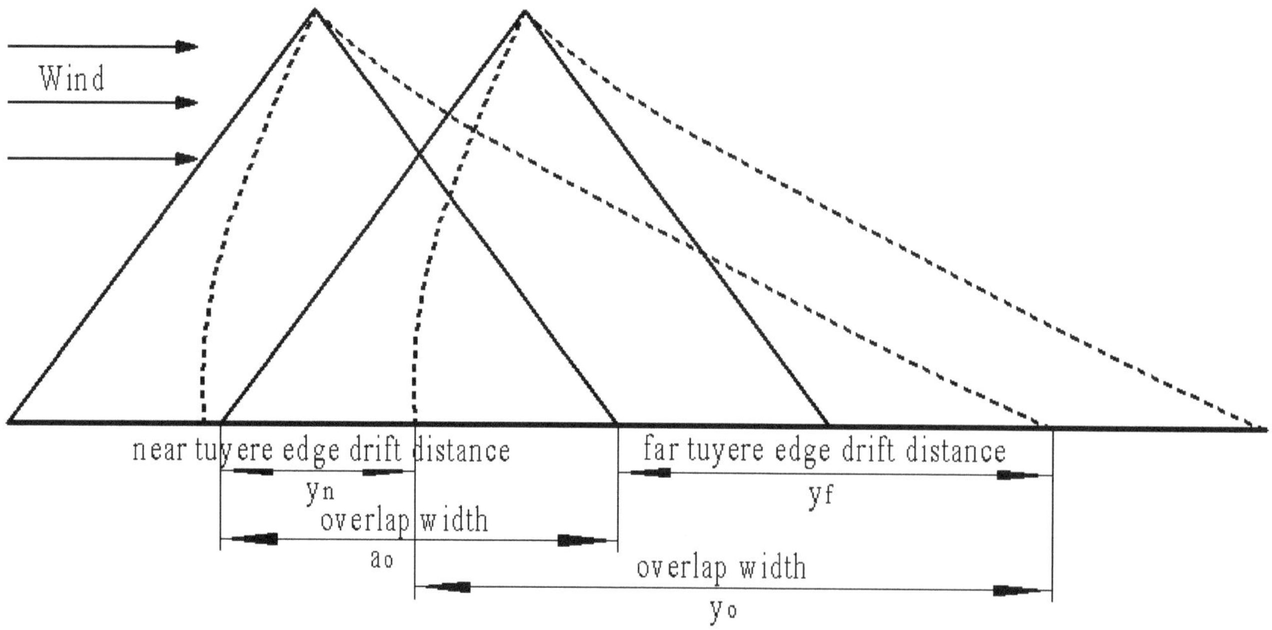

Figure 4. The natural curve of droplets is shown by solid lines and the drift curve influenced by wind is shown by dotted lines, a_0 is overlap width, y_2 is the drift distance of near tuyere edge, y_3 is the drift distance of far tuyere edge, and y_4 is the overlap width influenced by wind.

3. Results and Discussion

3.1. Simulation of Influence of Different Factors on Droplet Drift

3.1.1. Influence of Wind Speed on Droplet Drift

Inlet pressure (0.3 MPa) and spray height (1.5 m) were set to constant and the effect of ambient wind speeds of 0 m s^{-1}, 1 m s^{-1}, 3 m s^{-1}, and 5 m s^{-1} were explored. The droplet deposition density at different target positions is plotted in Figure 5

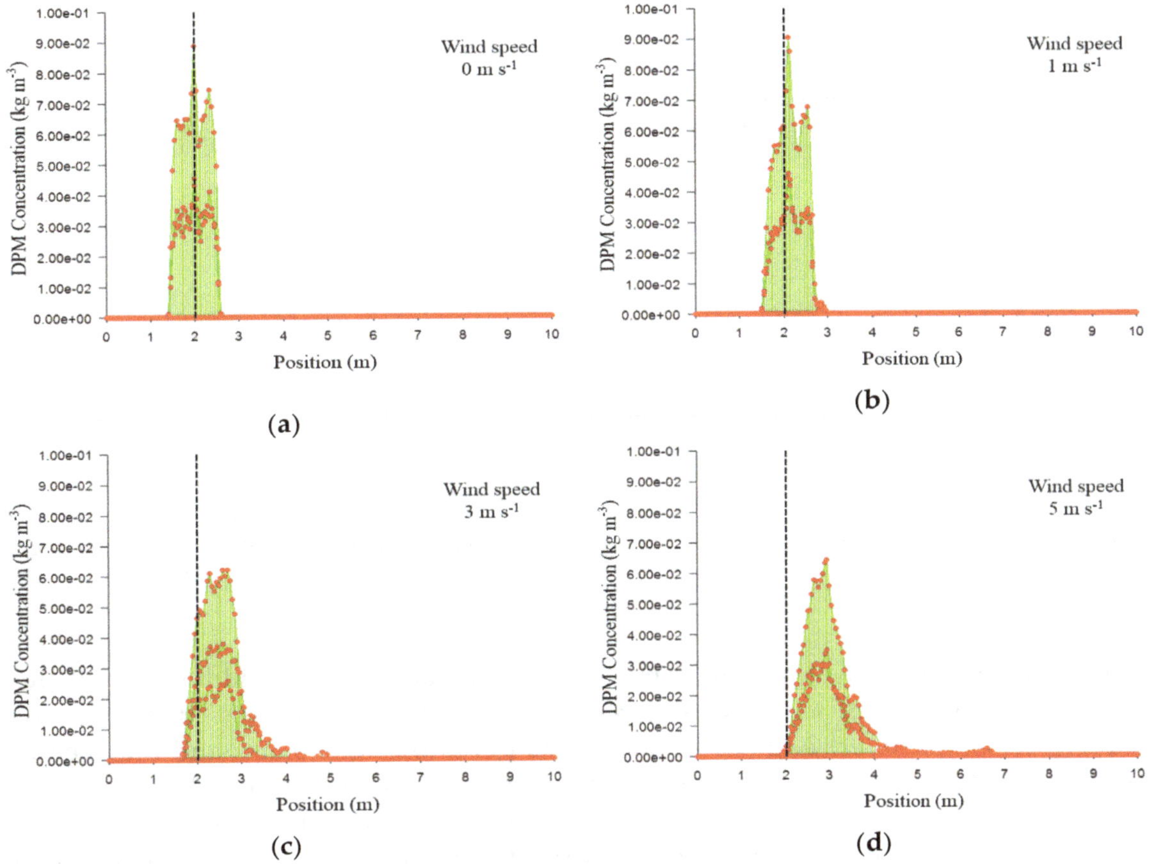

Figure 5. Concentration of droplets deposition at different wind speeds, (**a**) 0 m s^{-1}, (**b**) 1 m s^{-1}, (**c**) 3 m s^{-1}, and (**d**) 5 m s^{-1}. X-axis indicates a different position of target area and off-target area. Y-axis shows DPM concentration (kg m^{-3}) of droplets deposition at different wind speeds, 0 m s^{-1}, 1 m s^{-1}, 3 m s^{-1}, and 5 m s^{-1}. The point means concentration of different position, the dotted line means the droplet concentration deposition center, and the solid line means the biggest concentration of corresponding point.

It is clear that droplet deposition concentration at different target locations varied with wind speed (Figure 5). When the wind speed was 0 m s^{-1}, the droplets did not drift. The droplet concentration deposition center point was 2 m and the drift distance was 0. When the wind speed was 1 m s^{-1}, the droplet had a minor drift phenomenon. The droplet concentration deposition center point was 2.05 m and the drift distance was 0.05 m. When the wind speed was 3 m s^{-1}, the droplet had obvious drift phenomenon. The droplet concentration deposition center point was 2.35 m and the drift distance was 0.35 m. When the wind speed was 5 m s^{-1}, the droplets had a large amount of drift. The droplet concentration deposition center point was 2.85 m and the drift distance was 0.85 m. The three main performance indicators to evaluate drift characteristics, drift distance of deposition center, near tuyere edge, and far tuyere edge, and the relationships between these parameters and wind speed are shown in Table 3 and Figure 6.

Table 3. Droplet drift distance of deposition center, near tuyere edge, far tuyere edge, and overlap width at different wind speed, 0 m s^{-1}, 1 m s^{-1}, 3 m s^{-1} and 5 m s^{-1}.

Wind Speed (m s^{-1})	Deposition Center (m)	Near Tuyere Edge (m)	Far Tuyere Edge (m)	Overlap Width (m)
0	0	0	0	0.20
1	0.05	0.05	0.05	0.65
3	0.45	0.25	2.40	2.35
5	0.85	0.55	4.60	3.95

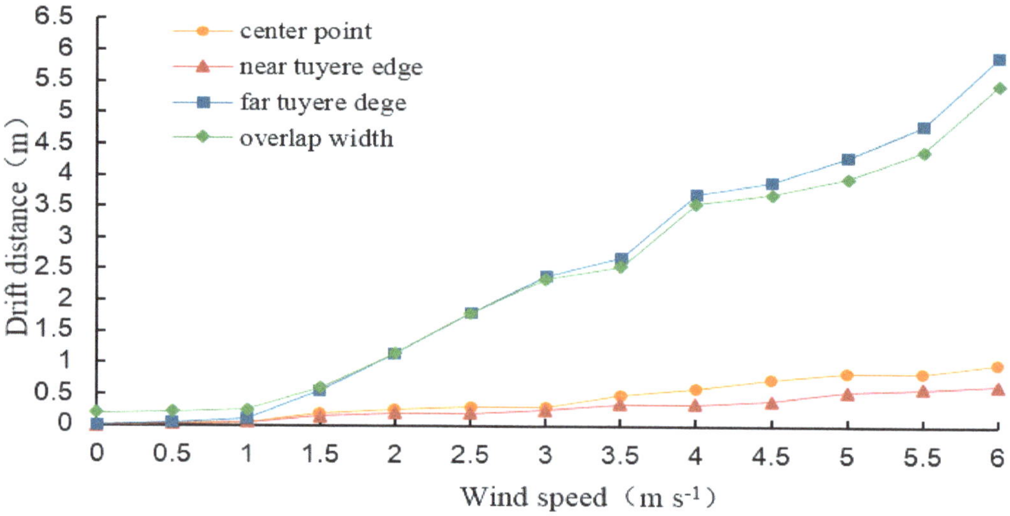

Figure 6. The relationship between wind speed and drift distance (center point, near tuyere edge, far tuyere edge, and overlap width).

The least square method was used to fit the drift distance and overlap width under different wind speeds. The center point drift model was $y_1 = 0.176x_w + 0.091$, near tuyere edge drift model was $y_2 = 0.110x_w - 0.038$, far tuyere edge drift model was $y_3 = 1.016x_w - 0.636$, and the overlap width model was $y_4 = 0.901x_w - 0.382$. In these models, x_w was wind speed (m s^{-1}) and y was distance (m). It can be seen that the wind speed was positively correlated with drift distance and overlap width. For every 1 m s^{-1} increase in wind speed, drift distance of droplet center point was increased by 0.176 m, near tuyere edge was increased by 0.11 m, far tuyere edge was increased by 1.016 m, and overlap width was increased by 0.901 m. The accuracy can be verified by the model obtained above with a_0 representing the droplet overlap width in the natural state. The simulation showed that when $a_0 = 0.2$, $y_4 - (a_0 - y_2 + y_3) = 0.005x + 0.0156$ was about 0, which proved that the model fits the actual situation.

3.1.2. Influence of Inlet Pressure on Droplet Drift

Wind speed (3 m s^{-1}) and spray height (1.5 m) were set to constant and the pressure of 0.2 MPa, 0.3 MPa and 0.4 MPa was explored. The droplet deposition density at different target positions is plotted in Figure 7.

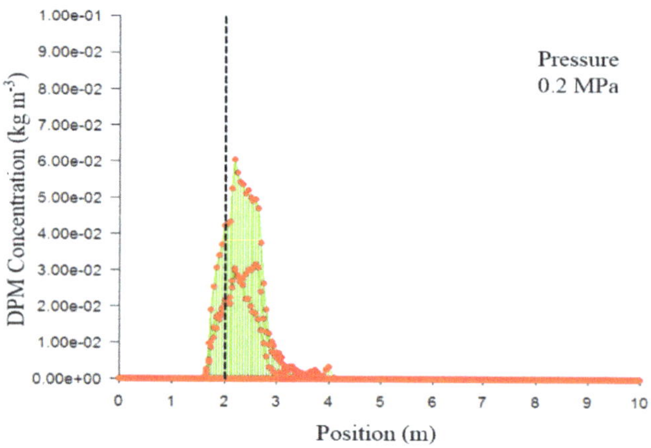

(a)

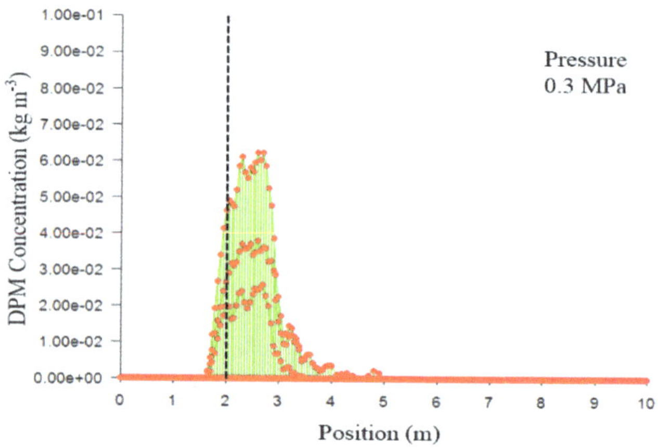

(b)

Figure 7. *Cont.*

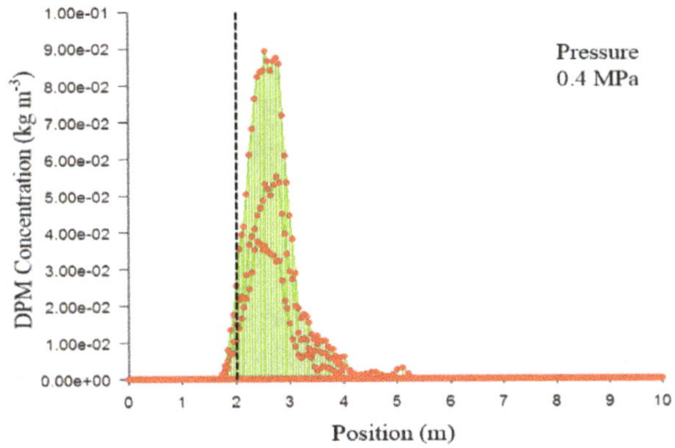

(c)

Figure 7. Concentration of droplets deposition under different pressures, (**a**) 0.2 MPa, (**b**) 0.3 MPa, and (**c**) 0.4 MPa. X-axis indicates different position of target area and off-target area, Y-axis shows DPM Concentration (kg m^{-3}) of droplets deposition under different pressure, 0.2 MPa, 0.3 MPa, and 0.4 MPa. The point means concentration of different position, the dotted line means the droplet concentration deposition center, and the solid line means the biggest concentration of corresponding point.

It is clear that droplet deposition concentration at different target locations varies under the influence of pressure (Figure 7). When the pressure was 0.2 MPa, the droplet concentration deposition center point is 2.35 m and the drift distance is 0.35 m. When the pressure was 0.3 MPa, the droplet concentration deposition center point is 2.45 m and the drift distance is 0.45 m. When the pressure was 0.4 MPa, the droplet concentration deposition center point is 2.6 m and the drift distance is 0.6 m. The drift distance of deposition center, the edge of near tuyere, and the edge of far tuyere and the relationships between these parameters and pressure are shown in Table 4 and Figure 8.

Table 4. Droplet drift distance of deposition center, near tuyere edge, far tuyere edge, and overlap width under different pressure, 0.2 MPa, 0.3 MPa, and 0.4 MPa.

Pressure (MPa)	Deposition Center (m)	Near Tuyere Edge (m)	Far Tuyere Edge (m)	Overlap Width (m)
0.2	0.60	0.25	1.70	1.55
0.3	0.45	0.25	2.40	2.35
0.4	0.65	0.35	2.85	2.70

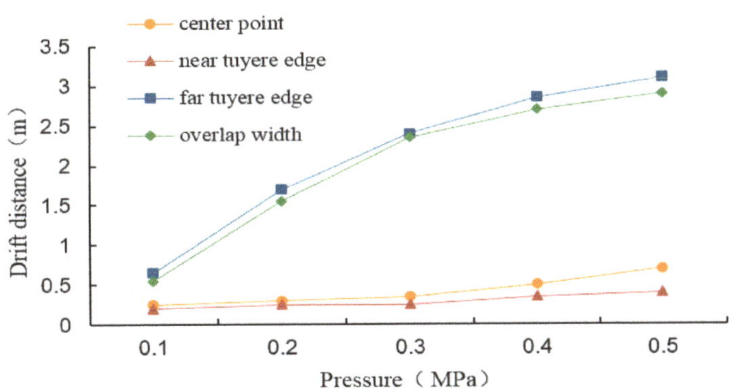

Figure 8. The relationship between pressure and drift distance (center point, near tuyere edge, far tuyere edge and overlap width).

The least square method was used to fit the drift distance and overlap width under different wind speeds. The center point drift model is $y_5 = 0.110x_p + 0.090$, the near tuyere edge drift model is $y_6 = 0.050x_p + 0.140$, the far tuyere edge drift model is $y_7 = 0.605x_p + 0.325$, and the overlap width model is $y_8 = 0.601x_p + 0.600$. In the models, x is inlet pressure (MPa) and y is distance (m). It can be seen that inlet pressure is positively correlated with droplet drift distance and overlap width. For every 0.1 MPa increased in the inlet pressure, the drift distance of deposition center point was increased by 0.11 m, the edge of near tuyere was increased by 0.05 m, the edge of far tuyere was increased by 0.605 m, and the overlap width was increased by 0.6 m.

3.1.3. Influence of Spray Height on Droplet Drift

Wind speed (3 m s^{-1}) and pressure (0.3 MPa) were set to constant and the spray height of 0.8 m, 1 m, and 1.5 m was explored. The droplet deposition density at different target positions is plotted in Figure 9.

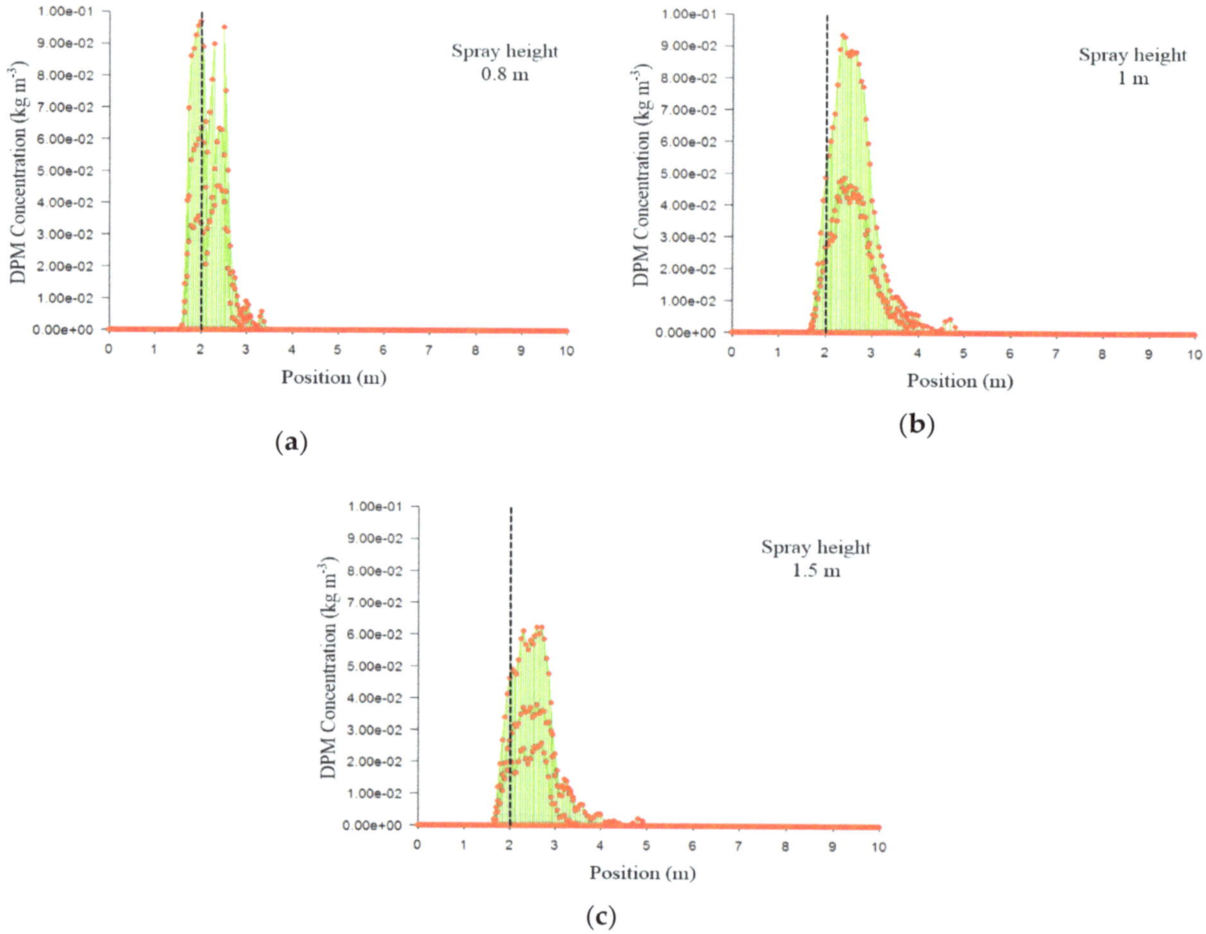

Figure 9. Concentration of droplets deposition in different spray heights, (**a**) 0.8 m, (**b**) 1 m, and (**c**) 1.5 m. X-axis means different position of target area and off-target area, Y-axis means DPM Concentration (kg m^{-3}) of droplets deposition in different spray height, 0.8 m, 1 m, and 1.5 m. The point means concentration of different position, the dotted line means the droplet concentration deposition center and the solid line means the biggest concentration of corresponding point.

It is clear that droplet deposition concentration at different target locations varied with spray height (Figure 9). When the spray height was 0.8 m, the droplet concentration deposition center point is 2.35 m and the drift distance is 0.35 m. When the spray height was 1 m, the droplet concentration deposition center point is 2.45 m and the drift distance is 0.45 m. When the spray height was 1.5 m, the droplet concentration deposition center point is 2.6 m and the drift distance is 0.6 m. The deposition center, near tuyere edge, and far tuyere edge, and the relationship between these parameters and spray height are shown in Table 5 and Figure 10.

Table 5. Droplet drift distance of deposition center, near tuyere edge, far tuyere edge, and overlap width in different spray height, 0.8 m, 1 m, and 1.5 m.

Spray Height (m)	Deposition Center (m)	Near Tuyere Edge (m)	Far Tuyere Edge (m)	Overlap Width (m)
0.8	0.35	0.15	1.55	1.35
1	0.45	0.25	2.40	2.60
1.5	0.60	0.35	3.25	2.75

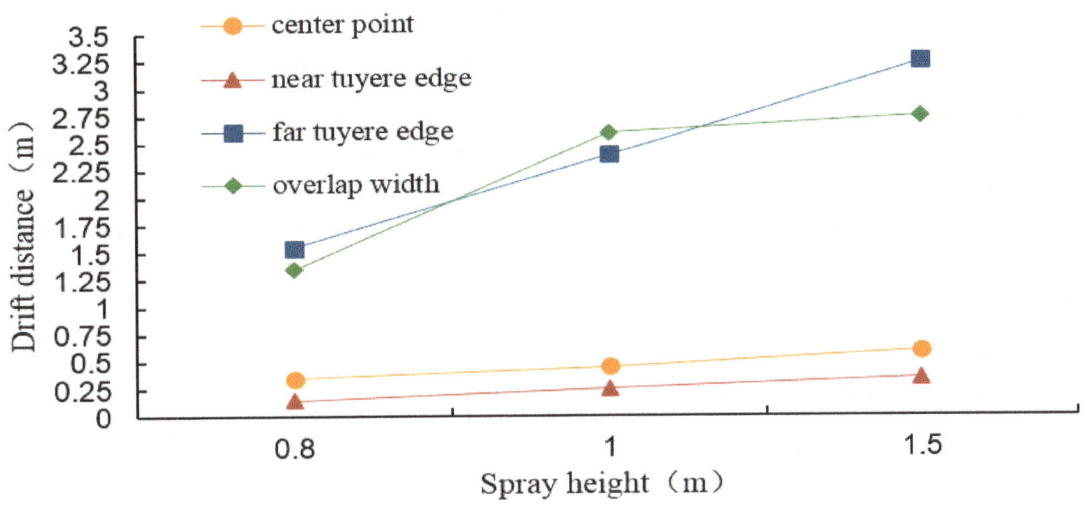

Figure 10. The relationship between spray height and drift distance (center point, near tuyere edge, far tuyere edge, and overlap width).

Again, the least square method was used to fit the drift distance and overlap width under different spray heights. The center point drift model is $y_9 = 0.235x_h + 0.125$, the near tuyere drift model is $y_{10} = 0.115x_h + 0.902$, the far tuyere drift model is $y_{11} = 1.654x_h - 0.081$, and the overlap width is $y_{12} = 1.670x_h + 0.789$. In the models, x is the spray height (m) and y is the distance (m). It can be seen that the spray height is positively correlated with droplet drift distance and overlap width, i.e., for every 1m increase in spray height, the drift distance of deposition center point was increased by 0.235 m, the drift distance of near tuyere increased by 0.115 m, the drift distance of far tuyere was increased by 1.654 m, and the overlap width was increased by 1.67 m.

3.2. *Drift Distance Analysis of Fitting Regression Results*

The least square method was taken to fit a curve of drift, with y_1 (edge drift distance), y_2 (deposition center drift distance), and y_3 (overlap width) as dependent variables, and x_w (wind speed), x_p (inlet pressure), and x_h (spray height) as independent variables, and C (constant term). The corresponding equations of multivariate linear function groups can be explained as follows:

$$\begin{pmatrix} 27 & 81 & 81 & 29.7 \\ 81 & 315 & 243 & 89.1 \\ 89.1 & 243 & 288 & 89.1 \\ 29.7 & 89.1 & 89.1 & 35 \end{pmatrix} \begin{pmatrix} a_0 \\ a_1 \\ a_2 \\ a_3 \end{pmatrix} = \begin{pmatrix} 57.025 \\ 243.925 \\ 180.975 \\ 66.36 \end{pmatrix} \tag{9}$$

$$\begin{pmatrix} 27 & 81 & 81 & 29.7 \\ 81 & 315 & 243 & 89.1 \\ 89.1 & 243 & 288 & 89.1 \\ 29.7 & 89.1 & 89.1 & 35 \end{pmatrix} \begin{pmatrix} a_0 \\ a_1 \\ a_2 \\ a_3 \end{pmatrix} = \begin{pmatrix} 11.5 \\ 46.5 \\ 36.025 \\ 13.37 \end{pmatrix} \tag{10}$$

$$\begin{pmatrix} 27 & 81 & 81 & 29.7 \\ 81 & 315 & 243 & 89.1 \\ 89.1 & 243 & 288 & 89.1 \\ 29.7 & 89.1 & 89.1 & 35 \end{pmatrix}\begin{pmatrix} a_0 \\ a_1 \\ a_2 \\ a_3 \end{pmatrix} = \begin{pmatrix} 51.475 \\ 204.275 \\ 172.5 \\ 60.06 \end{pmatrix} \tag{11}$$

and solved by MATLAB 2015a®(MathWorks, MA, USA),

$$(a_0\ a_1\ a_2\ a_3)^T = (-3.096\ \ 0.887\ \ 0.550\ \ 1.552)^T \tag{12}$$

$$(a_0\ a_1\ a_2\ a_3)^T = (-0.667\ \ 0.167\ \ 0.085\ \ 0.308)^T \tag{13}$$

$$(a_0\ a_1\ a_2\ a_3)^T = (-3.374\ \ 0.692\ \ 0.529\ \ 1.469)^T \tag{14}$$

Table 6 shows the variance and regression analysis of the influence of three factors on edge drift distance, deposition center drift distance and overlap width. According to the analysis results (Table 6), the influence of wind speed on edge drift distance, deposition center drift distance and overlap width is very significant, pressure and spray height is significant. The influence of three factors on edge drift distance, deposition center drift distance, and overlap width is also significant, so a linear equation can be established.

Table 6. The variance and regression analysis of the influence of three factors on edge drift distance, deposition center drift distance, and overlap width analysis.

Dependent Variable	Source of Difference	Regression Coefficient	T-Distribution Value	Significance	95% Confidence Interval		R	R^2
					Lower Limit	Upper Limit		
Y_1	X_w	0.887	9.941	**	0.702	1.071	0.915	0.837
	X_p	0.550	3.083	*	0.181	0.919		
	X_h	1.552	3.137	*	0.529	2.576		
	C	−3.096	−4.752		−5.606	−2.206		
Y_2	X_w	0.167	8.196	**	0.125	0.209	0.880	0.774
	X_p	0.085	2.083	*	0.001	0.169		
	X_h	0.308	2.728	*	0.074	0.541		
	C	−0.667	−3.558		−1.054	−0.279		
Y_3	X_w	0.692	8.303	**	0.520	0.865	0.892	0.795
	X_p	0.529	3.173	*	0.184	0.874		
	X_h	1.469	3.176	*	0.512	2.426		
	C	−3.374	−4.391		−4.963	−1.785		

The dependent variables (Y_1, Y_2, Y_3) are edge drift distance, deposition center drift distance, and overlap width. The different sources (X_w, X_p, X_h, C) are wind speed, inlet pressure, spray height and constant term. The significances are the results of significance analysis and the number of stars means the degree of influence by independent variables. The two stars (**) means the influence is very significant, one star (*) is significant, and no star is not significant.

The regression coefficients of the three variables in the regression equation of edge drift distance are 0.887, 0.550, and 1.552, and the constant term (C) is −3.096. Therefore, the relationship model between edge drift distance Y_1 and wind speed X_w (m s^{-1}), pressure X_p (MPa) and spray height X_h (m) is

$$Y_1 = 0.887X_w + 0.550X_p + 1.552X_h - 3.906\ (R^2 = 0.837) \tag{15}$$

In this model (15), the coefficients of the three variables are all positive, indicating that the three factors are all positively correlated with droplet drift distance of far tuyere edge. At the same time, this model also provides a reference for safe distance. This model is the drift distance of the droplet far from far tuyere edge, which is also the farthest distance to which the droplet can drift.

The regression coefficients of the three variables in the regression equation of center drift distance are 0.167, 0.085, and 0.308, and the C is −0.667. Therefore, the relationship model between drift distance Y_2 of deposition center and wind speed X_w (m s^{-1}), pressure X_p (MPa) and spray height X_h (m) is

$$Y_2 = 0.167X_w + 0.085X_p + 0.308X_h - 0.667\ (R^2 = 0.774) \tag{16}$$

In this model (16), the coefficients of the three variables are all positive, indicating that the three factors are positively correlated with the drift distance of droplet deposition center. At the same time, this model also provides a reference for the selection of spray deposition center. This model is the drift distance of droplet deposition center, which is also the point where droplet deposition concentration is the largest.

The regression coefficients of the three variables in the regression equation of overlap width are 0.692, 0.529, and 1.469, and the C is −3.374. Therefore, the relationship model between overlap width Y_3 and wind speed X_w (m s^{-1}), pressure X_p (MPa), and spray height X_h (m) is

$$Y_3 = 0.692x_w + 0.529x_p + 1.469x_h - 3.374\ (R^2 = 0.795) \tag{17}$$

In this model (17), the coefficients of the three variables are all positive, indicating that the three factors are positively correlated with the overlap width. At the same time, this model also provides a reference for the selection of nozzle distance and how to get the best droplet overlapping effect.

3.3. *Analysis of Droplet Drift Curve Characteristic*

Under the influence of different wind speeds, the droplet drift curve of XR8002 (spray height 1.5 m; inlet pressure 0.3 MPa) is shown in Figure 11.

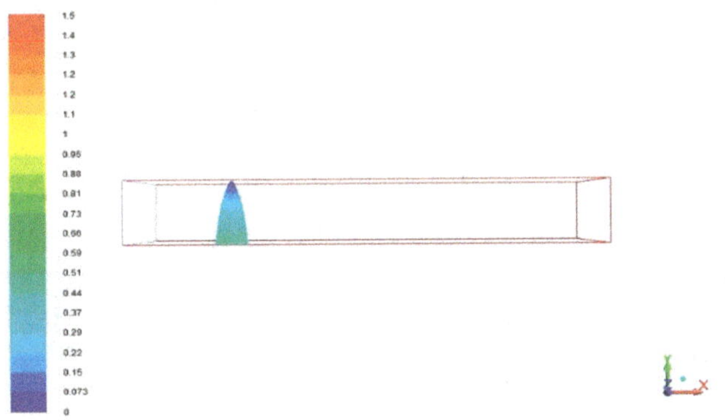

(**a**) 0 m s^{-1}

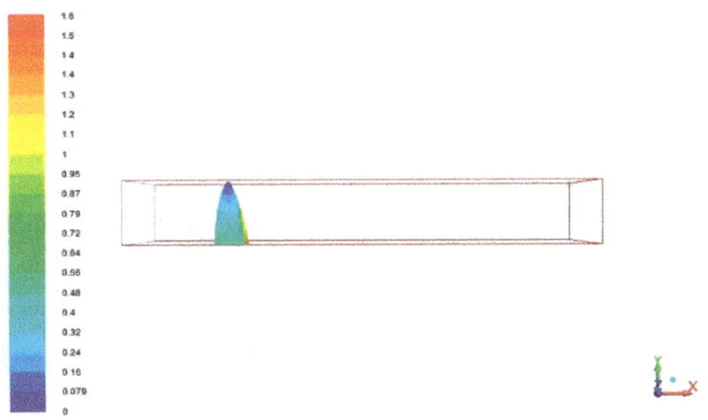

(**b**) 1 m s^{-1}

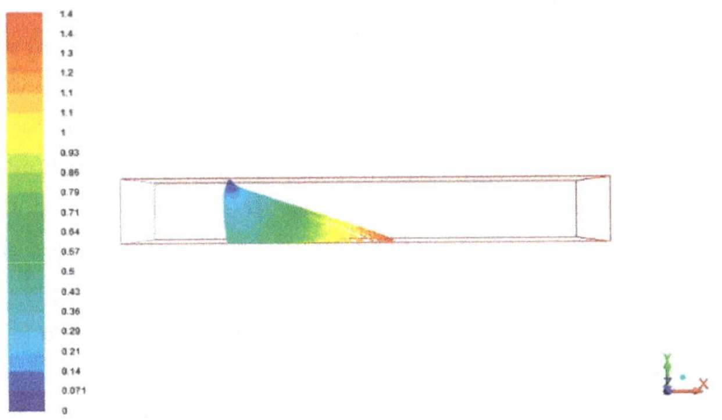

(**c**) 3 m s^{-1}

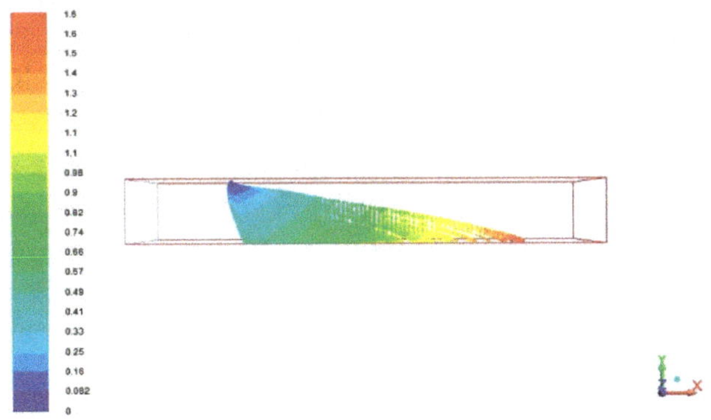

(**d**) 5 m s^{-1}

Figure 11. Trace of droplet drift curve at wind speed 0, 1, 3, and 5 m s^{-1} when spray height was 1.5 m and inlet pressure was 3 MPa. (**a**) The droplets have no drift at 0 m s^{-1}, (**b**) The droplets hardly drift at 1 m s^{-1}, (**c**) The droplets have obvious drift at 3 m s^{-1}, (**d**) The droplets have serious drift at 5 m s^{-1}.

As can be seen from the above Figure 11, when the wind speed was 1 m s^{-1}, the droplets hardly drift. When the wind speed exceeded 1 m s^{-1}, the droplets drift slightly. When the wind speed exceeded 3 m s^{-1}, the droplets will obviously drift. It is also the reason why UAV could not work in high wind speed. The analysis of droplet drift curves at the wind speeds of 3 m s^{-1} and 5 m s^{-1} were also made in Figure 12.

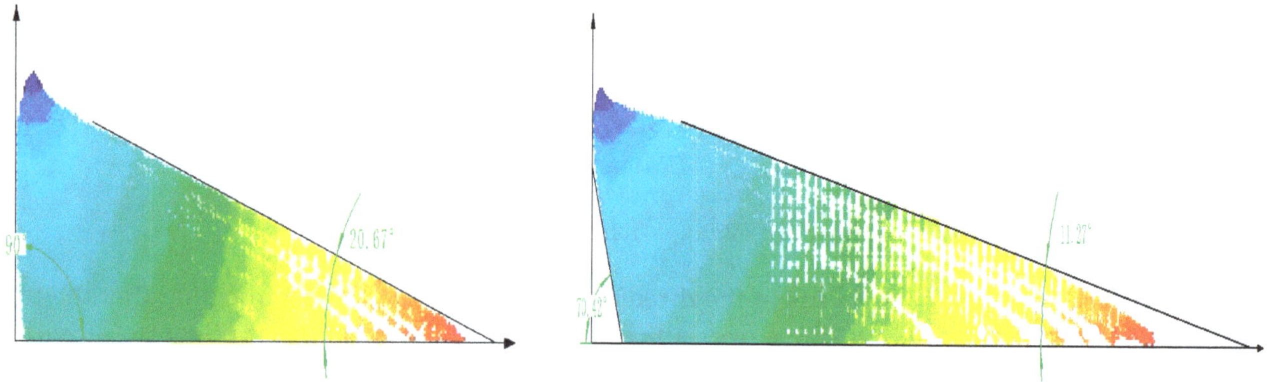

(**a**) 3 m s^{-1} (**b**) 5 m s^{-1}

Figure 12. Droplet drift curve of different wind speed (3 and 5 m s^{-1}), X-axis means deposition area of 0 to 10m, and Y-axis means spray height. (**a**) The droplet angle was 90° and 20.67° at 3 m s^{-1}, (**b**) The droplet angle was 70.42° and 11.27° at 5 m s^{-1}.

As is shown in Figure 12, both sides of the droplet drift curve were approximately a straight line. When the wind speed was 3 m s^{-1}, the droplet angle on the left side was 90°and the right side was 20.67°. When the wind speed was 5 m s^{-1}, the droplet angle on the left side was 70.42°and the right side was 11.27°, which showed that the inclination angle of the straight line was negatively related to the wind speed. Results displayed in Figure 12 show trends matching the previously observed behavior of a drift distance increase with a droplet angle decrease. The droplet angle also could be used to estimate the influence of wind, and the drift distance.

4. Conclusions

In this study, CFD simulation method was used to acquire droplet deposition distribution and drift under the influence of different factors. DPM model was used to simulate droplet drift from a double XR8002 nozzle at various wind speeds (0, 1, 3, 5 m s^{-1}), nozzle pressures (0.2, 0.3, 0.4 MPa) and spray heights (0.8, 1, 1.5 m). The information obtained from these simulations provided valuable insight into the characteristic of spray drift. General analysis suggest that the droplet drift curves were influenced by the three factors. Additionally, the influence coefficients of the three factors on the droplet drift distance were calculated. On the basis of analysis of the variance and regression results aimed at the edge drift distance, the center drift distance, and the overlap width, the three models were established. The expressions of three models are important on guiding significance to the practice. The analysis of the droplet drift curves showed that the droplet angle is closely related to drift. Results made from these simulations have provided a tool which can be used to ensure future UAV chemical application can be designed to maximize efficacy, reduce waste, and minimize damage to organisms not being targeted.

Author Contributions: Formal analysis, C.Z.; resources, J.L.; supervision, H.Z. (Hang Zhu); writing—original draft, H.L.; writing—review & editing, H.Z. (Huihui Zhang). All authors have read and approved the final manuscript.

References

1. Han, H.Q.; Gao, H.J.; Yang, H.T.; Zhang, X.; Hu, T.; Zhou, Q.Q. Analysis on decoupling between total grain production and pesticide usage in China. *Subtrop. Agric. Res.* **2018**, *14*, 91–98.

2. Cheng, X.B. Current situation and development countermeasures of agricultural extension in China. *Agric. Sci. Technol.* **2018**, *23*, 255–259.
3. Wu, K.M.; Lu, Y.H.; Wang, Z.Y. Advance in integrated pest management of crops in China. *Chin. Bull. Entomol.* **2009**, *46*, 831–836.
4. Wang, G.; Gong, Y.; Zhang, X.; Chen, X.; Liu, D.J.; Chen, W.; Miao, Y.Y. The test of droplet deposition distribution for different kinds of pesticide application equipment in corn field. *J. Agric. Mech. Res.* **2017**, *39*, 177–182.
5. Meng, Y.H.; Zhou, G.Q.; Wu, C.B. Discussion on application and promotion of agricultural plant protection unmanned aerial vehicle in China. *China Plant Prot.* **2014**, *34*, 33–39.
6. Shang, C.Y.; Cai, J.F.; Huang, S.J.; Pan, H.L.; Zhong, F.L. Applications status and prospect analysis of agricultural UAVs in China. *J. Anhui Agric. Sci.* **2017**, *45*, 193–195.
7. Wang, X.N.; He, X.K.; Wang, C.L.; Wang, Z.C.; Li, L.L.; Wang, S.L.; Janes, B.; Andreas, H.; Wang, Z.G. Spray drift characteristics of fuel powered single-rotor UAV for plant protection. *Trans. Chin. Soc. Agric. Eng.* **2017**, *33*, 117–123.
8. Bradley, K.F. Role of atmospheric stability in drift and deposition of aerially applied sprays-preliminary results. *ASAE* **2004**, *58*, 742–755.
9. Kirk, L.W.; Teske, M.E.; Thistle, H.W. What about upwind buffer zones for aerial applications? *J. Agric. Saf. Health* **2002**, *8*, 333–336. [CrossRef]
10. Ivan, W.K.; Bradley, K.F.; Wesley, C.H. Aerial methods for increasing spray deposits on wheat heads. *ASAE* **2004**, *58*, 716–729.
11. Chen, S.D.; Lan, Y.B.; Bradley, K.F. Effect of wind Field below Rotor on Distribution of Aerial Spraying Droplet Deposition by Using Multi-rotor UAV. *Trans. Chin. Soc. Agric. Mach.* **2017**, *48*, 105–113.
12. Qiu, B.J.; Wang, L.W.; Cai, D.L.; Wu, J.H.; Ding, G.R.; Guan, X.P. Effects of flight altitude and speed of unmanned helicopter on spray deposition uniform. *Trans. Chin. Soc. Agric. Mach.* **2013**, *29*, 25–32.
13. Shen, A.; Zhou, S.D.; Wang, M. Simulation and analysis of multi-rotor UAV flow field. *Trans. Chin. Soc. Agric. Mach.* **2018**, *36*, 29–33.
14. Nuyttens, D.; De Schampheleire, M.; Baetens, K.; Brusselman, E.; Dekeyser, D.; Verboven, P. Drift from field crop sprayers using an integrated approach: Results of a 5-year study. *Trans. ASABE* **2011**, *54*, 403–408. [CrossRef]
15. Teske, M.E.; Bird, S.L.; Esterly, D.M.; Curbishley, T.B.; Ray, S.L.; Perry, S.G. AgDrift: A model for estimating near-field spray drift from aerial applications. *Environ. Toxicol. Chem.* **2002**, *21*, 659–671. [CrossRef] [PubMed]
16. Teske, M.E.; Thistle, H.W.; Riley, C.M.; Hewitt, A.J. Laboratory measurements on the sensitivity of evaporation rate of droplets inside a spray cloud. *Trans. ASABE* **2017**, *60*, 361–366.
17. Bilanin, A.J.; Teske, M.E.; Barry, J.W.; Ekblad, R.B. AGDISP: The aircraft spray dispersion model, code development and experimental validation. *Trans. ASABE* **1989**, *32*, 327–334. [CrossRef]
18. Luo, J.; Zeng, G.H.; Li, B.K. Simulation analysis of two-phase flow field inside pass of nozzle base on FLUENT. *Modul. Mach. Tool Autom. Manuf. Tech.* **2016**, *10*, 44–47.
19. Chen, X.; Ge, S.C.; Zhang, Z.W.; Jing, D.J. Numerical simulation and analysis of multi-nozzle interference base on fluent. *Chin. J. Environ. Eng.* **2014**, *8*, 2503–2508.
20. Sun, G.X.; Wang, X.C.; Ding, W.M.; Zhang, Y. Simulation and analysis on characteristics of droplet deposition base on CFD discrete phase model. *Trans. Chin. Soc. Agric. Eng.* **2012**, *28*, 13–19.
21. Zhang, S.F.; Song, L.L. Simulation of Spray characteristics of pressure swirl nozzle. *J. Anhui Agric. Sci.* **2009**, *37*, 8098–8100.
22. Adeniyi, A.A.; Morvan, H.P.; Simmons, K.A. A coupled Euler-Lagrange CFD modelling of droplets-to-film. *Aeronaut. J.* **2017**, *121*, 1897–1918. [CrossRef]
23. Ryan, S.D.; Gerber, A.G.; Holloway, A.G.L. A computational study on spray dispersal in the wake of an aircraft. *Trans. ASABE* **2013**, *56*, 847–868.
24. Zhang, B.; Tang, Q.; Chen, L.P.; Xu, M. Numerical simulation of wake vortices of crop spraying aircraft close to the ground. *Biosyst. Eng.* **2016**, *145*, 52–64. [CrossRef]
25. Yang, F.B.; Xue, X.Y.; Cai, C. Numerical Simulation and analysis on spray drift movement of multirotor plant protection unmanned aerial vehicle. *Energies* **2018**, *11*, 2399. [CrossRef]
26. Roberts, T.W.; Murman, E.M. Solution method for a hovering helicopter rotor using the Euler equations. In Proceedings of the 23rd Aerospace Sciences Meeting, Reno, NV, USA, 14–17 January 1985.

27. Jin, Y.H.; Jae, S.R.; Kyu, H.K. New least squares method with geometric conservation law (GC-LSM) for compressible flow computation in meshless method. *Comput. Fluids* **2018**, *172*, 122–146.
28. Salazar-Campoy, M.M.; Morales, R.D.; Najera-Bastida, A.; Calderon-Ramos, I.; Cedillo-Hernandez, V.; Delgado-Pureco, J.C. A physical model to study the effects of nozzle design on dispersed two-phase flows in a slab mold casting ultra-low-carbon steels. *Metall. Mater. Trans. B* **2018**, *49*, 812–830. [CrossRef]

10

Automatic Grapevine Trunk Detection on UAV-Based Point Cloud

Juan M. Jurado [1,*], Luís Pádua [2,3], Francisco R. Feito [1] and Joaquim J. Sousa [2,3]

1 Computer Graphics and Geomatics Group of Jaén, University of Jaén, 23071 Jaén, Spain; ffeito@ujaen.es
2 Engineering Department, School of Science and Technology, University of Trás-os-Montes e Alto Douro, 5000-801 Vila Real, Portugal; luispadua@utad.pt (L.P.); jjsousa@utad.pt (J.J.S.)
3 Centre for Robotics in Industry and Intelligent Systems (CRIIS), INESC Technology and Science (INESC-TEC), 4200-465 Porto, Portugal
* Correspondence: jjurado@ujaen.es

Abstract: The optimisation of vineyards management requires efficient and automated methods able to identify individual plants. In the last few years, Unmanned Aerial Vehicles (UAVs) have become one of the main sources of remote sensing information for Precision Viticulture (PV) applications. In fact, high resolution UAV-based imagery offers a unique capability for modelling plant's structure making possible the recognition of significant geometrical features in photogrammetric point clouds. Despite the proliferation of innovative technologies in viticulture, the identification of individual grapevines relies on image-based segmentation techniques. In that way, grapevine and non-grapevine features are separated and individual plants are estimated usually considering a fixed distance between them. In this study, an automatic method for grapevine trunk detection, using 3D point cloud data, is presented. The proposed method focuses on the recognition of key geometrical parameters to ensure the existence of every plant in the 3D model. The method was tested in different commercial vineyards and to push it to its limit a vineyard characterised by several missing plants along the vine rows, irregular distances between plants and occluded trunks by dense vegetation in some areas, was also used. The proposed method represents a disruption in relation to the state of the art, and is able to identify individual trunks, posts and missing plants based on the interpretation and analysis of a 3D point cloud. Moreover, a validation process was carried out allowing concluding that the method has a high performance, especially when it is applied to 3D point clouds generated in phases in which the leaves are not yet very dense (January to May). However, if correct flight parametrizations are set, the method remains effective throughout the entire vegetative cycle.

Keywords: grapevine detection; precision viticulture; 3D vineyard structure; photogrammetry

---

1. Introduction

The monitoring and management of agricultural crops, particularly with regard to nutrient level, water stress, diseases and pests, and phenological status, are vital for successful agricultural operations [1]. Traditionally, these activities are carried out through visual examinations of the crops, or by analysing plants and soil, which are time-consuming and invasive approaches [2]. Considering the fact that it is necessary to maximise yield and resources, while reducing environmental impacts, mainly by optimising the use of water and significantly reducing fertilisers and pesticides [3]. This can only be achieved by obtaining data that allow the intelligent and sustainable management of agricultural parcels [4]. It will then be possible, in a rational and economical way, to resort differentiated and localised actions with regard to the use of water and nutrients, and to control the soil and vegetation cover, as well as the plant's phytosanitary status.

The technological advances in recent years, have enabled the miniaturisation of electronic components and a significantly reduction in prices, taking Precision Agriculture (PA) to another level. For example, the advent of Unmanned Aerial Vehicles (UAVs) capable of capturing aerial high resolution data using different kind of sensors (RGB, multi and hyperspectral, thermal and LiDAR), together with new photogrammetric processing methods, allow the computation of diverse outcomes such as orthophoto mosaics, vegetation indices and 3D point clouds [5]. UAVs are a popular tool in PA and the obtained aerial imagery is turned into information which can be used to optimise crop inputs through variable rate applications [6–8].

In a short period of time PA approaches and practices have become very popular and were introduced in all agricultural sectors [9]. The vine and wine sector is among those that have most benefited of precision farming techniques, applied to optimise vineyard performance [10]. Thus, the Precision Viticulture (PV) concept was introduced and can be defined as a particular field of PA, whose purpose is maximising grape yield and quality while minimising environmental impacts and risks [11]. Therefore, it is possible to avoid unnecessary treatments, which can be harmful and polluting, and reduce costs [12]. The ability of UAVs to obtain high spatiotemporal resolution and geocoded images from different sensors, make them a powerful tool for a better understanding of the spatial and multi-temporal heterogeneity of vineyard plots, allowing the estimation of parameters that directly affect its state. Thus, individual grapevine identification and location is of great importance to precisely assess the vineyard status estimating different parameters per individual plant [13]. However, there are many features in vineyards that make these scenarios very complex to develop automatic methods for trunk individual detection and location [14]. Therefore, segmentation methods that consist of processes of dividing input data into several disjoint areas that maintain the unique and homogeneous features from surrounding have to be employed.

Regarding vineyard vegetation detection, several methods were already proposed based on different approaches using the photogrammetric outcomes from UAV-based imagery by applying image processing techniques, machine learning methods and by filtering dense 3D point clouds and Digital Elevation Models (DEMs) [14–18]. Those methods are capable of distinguishing grapevine from non-grapevine vegetation and to extract different vineyard macro properties such as the number of vine rows, row spacing, width and height, potential missing plants and vineyard vigour maps.

The outcomes resulting from photogrammetric processing applied to UAV-based imagery can be used to estimate individual geometrical and biophysical grapevine parameters, providing a plant-specific application for PV [19]. In this scope some studies can be found in the literature. De Castro et al. [20] developed an Object-Based Image Analysis (OBIA) method applied to high-resolution vineyard Digital Surface Models (DSMs) to estimate grapevine vegetation. Then, the individual position of grapevines were marked, assuming a constant space between plants. This way, missing plants were also estimated and some geometrical parameters were estimated. In a different study, proposed by Matese and Di Genaro [21], missing plants detection was assessed in a semi-automatic procedure by filtering the DSM and by manually placing small polygons, representing individual plants and, then, analysing the number of pixels intercepted by each polygon by using a five-classes approach based on quantiles to verify the probability of a missing plant presence. A binary multivariate-logistic regression model was used by Primicerio et al. [22] for the individual detection of grapevines, including missing grapevines, in orthophoto mosaics. In the referred studies it is highlighted that the integration of other sensors data could allow the extraction of single plant vigour, health and water status. In this regard, Pádua et al. [13] performed an individual grapevine estimation for site-specific management in a multi-temporal context, helping winegrowers to fully explore the potential of the high-resolution data provided by UAVs and to combine data resultant from the different imagery sensors for a more precise decision support and a quick vineyard inspection. More recently, several studies have explored 3D point clouds resulting from UAV-based imagery photogrammetry processing to identify vineyards. Point cloud models consist of large datasets of points representing the surface of visible objects and can be derived from UAV-based imagery by photogrammetry and computer vision algorithms such as, for example, Structure from Motion (SfM). Alternatively, 3D point clouds can be directly provided by Light Detection and Ranging Systems (LiDAR). Comba et al. [23] proposed an unsupervised algorithm for vineyard detection and vine-rows features evaluation, based on 3D point-cloud maps processing. However, as final result, only the vineyards and local evaluation of vine rows orientation were retrieved. Comba et al. [24] applied a multivariate linear regression model to crop canopy descriptors derived from the 3D point cloud, to estimate vineyard's Leaf Area Index (LAI). Marie Weiss and Frédéric Baret [17], applied a SfM algorithm to extract 3D dense point cloud over the vineyard and used the terrain altitude, extracted from the dense point cloud, to get the 2D distribution of height of the vineyard. Then, a threshold on the height was applied to separate the rows. Mesas-Carrascosa et al. [25] used 3D point clouds generated using photogrammetric techniques to RGB images acquired by UAV to derive vineyard canopy information. Additionally, to the geometry, each 3D point also stored the colour which was used to discriminate between vegetation and bare soil. Aboutalebi et al. [26] used UAV-based 3D information to monitor and assess vineyard plant's condition. Different aspects of 3D point cloud were used to estimate height, volume, surface area, and projected surface area of plant's canopy. Then biomass information was used to assess its relationship with in situ LAI. Other studies, such as that by Moreno et al. [27], used terrestrial LiDAR sensors to reconstructed vineyard crops. Although accurate, these methods are time-consuming and very expensive.

As it can be concluded, through the studies previously presented, there are many groups of researchers who are dedicated to the development of methods to extract useful information from vineyards. Although it is considered by everyone of fundamental importance the detection and location of individual plants, there are no methods capable of making it fully automatic. Indeed, the various methods found in the literature, are able to estimate the position of trunks, but using prior knowledge related to the number of plants per row and the distance between plants. Therefore, a fully automatic method able to detect and locate grapevine trunks is desirable and would have the potential to create base maps for most PV studies.

In this article, we present an innovative and fully automatic method able to detect and locate individual grapevine trunks, by exploring 3D point clouds derived from photogrammetric processing of UAV-based RGB imagery. The proposed method proved to be effective even when applied to complex vineyards plots. It is able to distinguish posts from trunks and to mark missing plants.

2. Materials and Methods

2.1. Study Area

To develop the method proposed in this manuscript, several commercial vineyards (Figure 1a), in the northern region of Portugal, were selected to its application. Commercial vineyards usually present the great advantage of a proper management, where the best practices are applied to improve yield and quality. Therefore, these vineyards present well treated rows with a regular vegetative wall, facilitating the individual grapevine detection. Then, in the scope of this study, a complex and challenging vineyard plot was analysed (Figure 1b, 41°17′08.1″ N, 7°44′09.9″ W, 472 m altitude). The main purpose of using this vineyard, located in the campus of the University of Trás-os-Montes e Alto Douro (UTAD), Vila Real, Portugal, in the Douro demarcated region, was to push the method's application to its limit. The plot has an area of 3200 m^2 and is composed of 55 rows, and a double guyot trained system is used. The selection of this vineyard is based on the different levels of vigour and missing plants along the vine rows, providing a diverse variety of cases that are hard to be found in commercial vineyards.

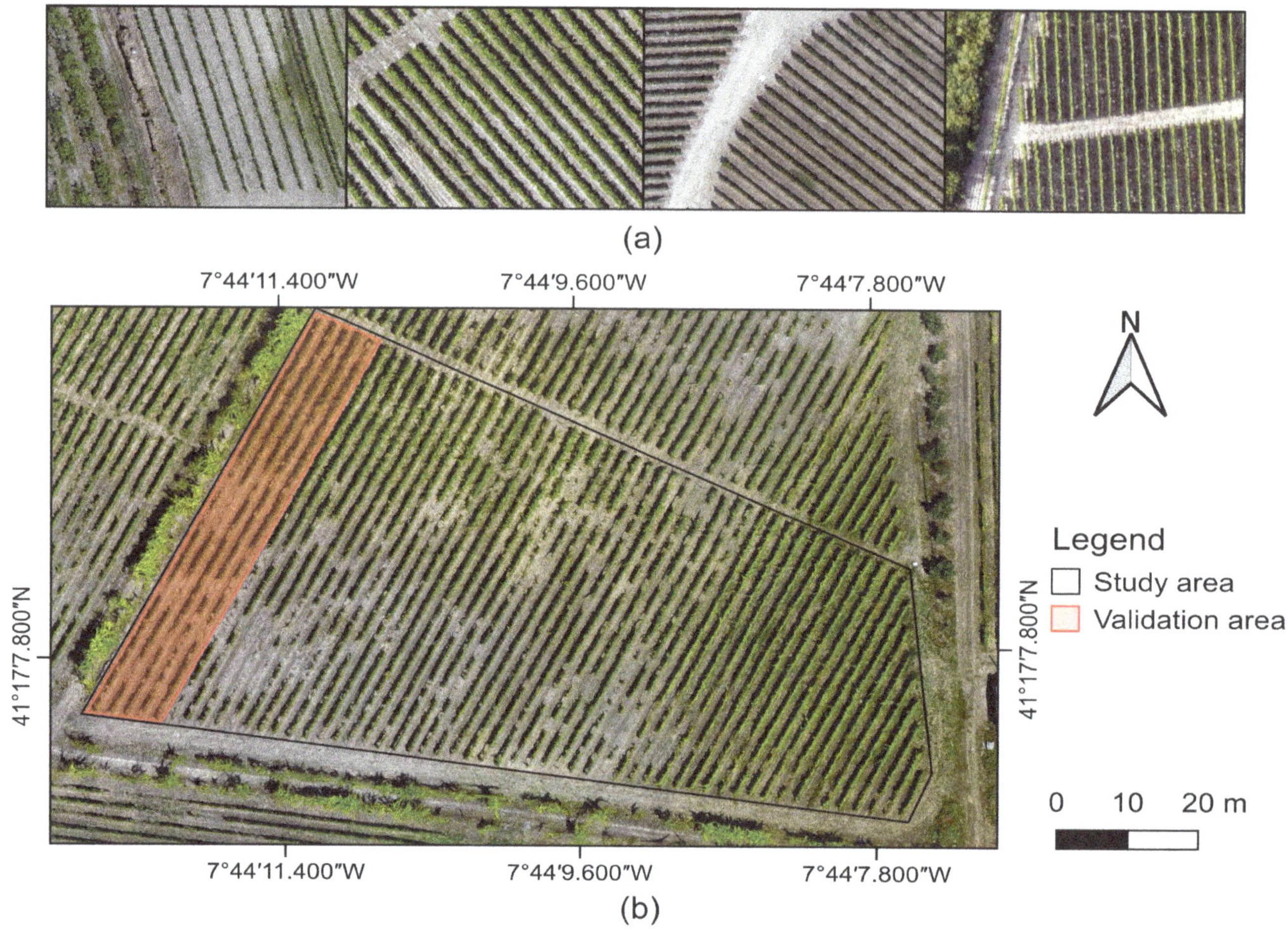

Figure 1. General overview of the study areas: (**a**) some examples of commercial vineyards; and (**b**) vineyard plot used for validation and to assess the limits of the proposed methodology. Coordinates in WGS84 (EPSG:4326).

2.2. *UAV-Based Data Acquisition*

Aerial RGB imagery acquisition was performed using the multi-rotor UAV DJI Phantom 4 (DJI, Shenzhen, China). Its native camera was used for RGB imagery acquisition, FCC 3 model, a CMOS sensor with 12.4 MP resolution mounted in a 3-axis electronic gimbal.

Different flights were conducted over distinct areas using a single-grid configuration and a flight height varying between 30 m (June and July flight campaigns) and 50 m (flights from January to May), from the UAV take-off position and with an imagery frontal overlap of 90% and 80% side overlap. The missions were planned and executed using DroneDeploy (California, CA, USA) in an Android smartphone. Regarding the most complex test site (Figure 1b), the flight was performed on 30 July 2019 and a total of 228 images were acquired. The whole flight campaign was carried in 13 min, five minutes for UAV assembly/disassembly operations and mission uploading, while the duration of the flight was eight minutes. The camera can be facing down, i.e., in the nadir direction, in all the flights conducted during the season preparation period. In the remaining flights, conducted in the growing and/or harvesting preparation period, the camera was used with an inclination angle of 65°, relative to the nadir direction. This choice was done due to the absence or presence of leaves, capable of obstructing trunk detection (respectively, Phase 1 and Phases 2 and 3, of Figure 2).

Figure 2. The three moments of the vegetative cycle that influence camera parameterization and flight height. Phase 1 (beginning of the wine campaign), where the influence of leaves is negligible—camera can be used facing down and the flight altitude can be higher; Phase 2 (critical phase of phenological development); and Phase 3 (preparation of the vintage and estimating production)—In these phases the camera should be used with an angle and the flight height must be low (20–30 m).

2.3. *Proposed Method*

The main steps of the automatic grapevine trunk detection method based on a geometric segmentation on point cloud data are presented in Figure 3.

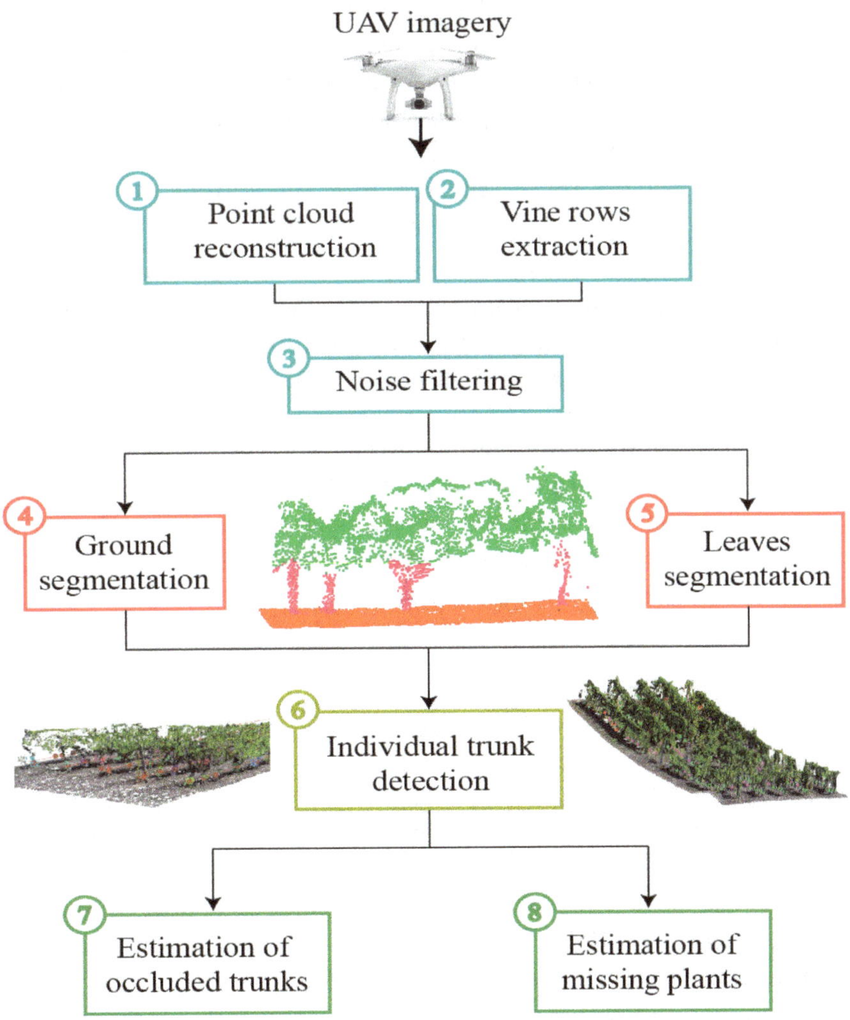

Figure 3. The flowchart diagram of the proposed methodology.

Firstly, the RGB aerial imagery are used to generate a 3D dense point cloud which was geometrically corrected using ground control points (GCPs). Then, a noise reduction is applied to remove many points which usually are close to the plant body. This step is important to achieve a more accurate trunk detection on areas with dense vegetation. Secondly, the location of vine rows is obtained in the form of lines using the method proposed in Pádua et al. [14]. In this way, the search area is confined to the vine rows only, optimising the whole process. Consequently, 3D points close to these lines were selected from the rest of the point cloud. Thirdly, a geometric segmentation is carried out in order to remove 3D points representing the ground and leaves. Thus, points belonging to trunks are isolated and spatial clustering can be performed. Finally, the 3D position for each grapevine is determined and, therefore, the number of existing and missing plants is calculated. The method was developed and implemented in C++ using the Point Cloud Library (PCL) [28].

2.3.1. SFM Reconstruction and Noise Removal

SfM techniques [29] are widely used for 3D reconstruction of multiple scenarios of the real-world [30]. These image-based methods are able to identify and match key points between overlapping images. In contrast to LiDAR-based solutions [27], the application of SfM enables collecting the fit-for-purpose data to model the geometry from some viewpoints. In general, plant modelling poses some challenges due to irregular surfaces, occlusion and varying illumination. In this regard, some considerations should be taken into account to process data correctly. For this study, 3D dense point clouds were generated over vineyard plots, by considering the following processing options: (1) a high overlapping rate ($\geq$80%), (2) a valid key point must be visible in at least three images and (3) the image scale is set to 1/2 in order to increase recognisable key points per image. The photogrammetric processing of the acquired RGB imagery was performed using Pix4Dmapper Pro (Pix4D SA, Lausanne, Switzerland). A 3D dense point cloud was generated using the multi-scale half-image size, a high point density and a minimum of three matches per image. It was exported in polygon file format (PLY). Moreover, raster products were also computed after 3D point cloud interpolation using Inverse Distance Weighting (IDW).

Noisy points that inevitably surround the vegetation areas have to be removed in order to make an accurate geometric segmentation on the 3D model. The point cloud is filtered applying a noise filter which is provided by PCL, this method is based on the computation of distance between neighbours [31]. For each 3D point, the mean distance from it to all its neighbours is computed. Thus, all points whose mean distances are outside an interval defined by the global distances mean and standard deviation are considered to be outliers. The neighbour search was developed by considering a specific radius which is related to the point cloud density. The several tests performed allowed to conclude that 0.05 m should be used to increase method's and results' quality. By applying this noise filter to the 3D point cloud, most of erroneous 3D points in the lower parts of the grapevines were removed, allowing a better recognition of the trunks.

2.3.2. Vine Rows Extraction

In order to reduce the research area, the method proposed by Pádua et al. [14] is first applied. In this way, the vector lines representing individual axis of vine rows are identified. In short, the identification of the lines is based on the use of a crop surface model—computed from the subtraction of the DEM to the Digital Surface Model (DSM)—in combination with the green percentage index (G%) [32], computed using the red, green and blue bands of the orthophoto mosaic. Then, grapevine vegetation is estimated from the threshold and concatenation of both raster products— the Canopy Surface Model (CSM), according to a height range and the G%, by using the Otsu's method [33]. After this procedure, a binary image is generated with a set of clusters of pixels mostly representing the grapevine vegetation. In this way, the vine rows and its central lines are estimated, considering the orientation of the most representative clusters.

Thus, the estimated vine rows axis (central lines) are used as a virtual guide to create a buffer allowing identifying points which compose the plant geometry and the surrounding soil (see Figure 4a). Since the vegetative wall width of grapevine plants usually varies between 30 and 50 cm, it was decided to create a 60 cm buffer to selected 3D points to be analysed. As it is shown in Figure 4b, the green points represent the selected points considering 30 cm width to each side the vine row.

(a) (b)

Figure 4. Example of a segmentation of vine rows: (**a**) visualisation of the 3D vine rows generated as presented in Pádua et al. [14]; and (**b**) 3D points selection in the point cloud, using a 60 cm buffer.

2.3.3. Ground and Leaves Segmentation

The vine rows extraction method enables removing 3D points, which were outside of 3D buffers. However, in addition to points, which potentially may define the trunk's geometry, there are other 3D points in the vine row space to be discarded. In order to isolate trunk points, it is necessary to remove those 3D points representing leaves, which usually appear in the upper section of the point cloud and ground points, which are located in the lower part of the 3D model. For this purpose, only geometric and spatial features as well as the point colour were considered, and the following three steps strategy was applied: (1) spatial subdivision of the vine row buffers, based on height thresholds, (2) ground removal (3) leaves removal. This process is fully automatic and no human intervention is required. In effect, the method has the ability to be applied even in vineyard plots with irregular slope, distinct density of plant foliage and voids (missing plants) along the vine row.

However, and before the application of the this procedure, 3D buffers need to be divided in different segments (Figure 5). This subdivision is determined based on the buffer's length and the terrain's slope. In this task, it is crucial to apply height thresholds, mainly important if the terrain slope varies. In flat terrains this step could be avoided, still to keep the method as general as possible, it was decided to include this step. If the terrain's slope is irregular, a higher number of segments will be required to allow a better fit. By default, the segment length is set to 1 m since it proved to fit most scenarios.

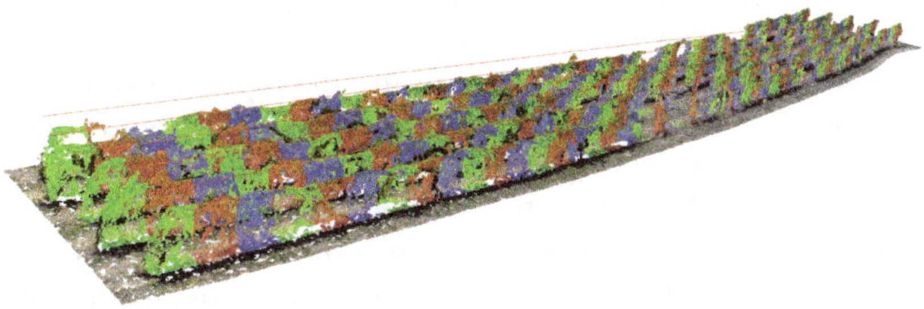

Figure 5. Subdivision of the vine row into n segments.

The leaves and ground removal operation is carried out by considering the vine row buffers segmentation. In fact, the following geometric operations were developed for each segment. Firstly, 3D points with the highest and lowest heights are detected. Then, the terrain's slope is fitted by changing the orientation of a cutting plane for n iterations. Thus, ground points which are under this plane were automatically discarded. Figure 6 shows the main iterations of this step.

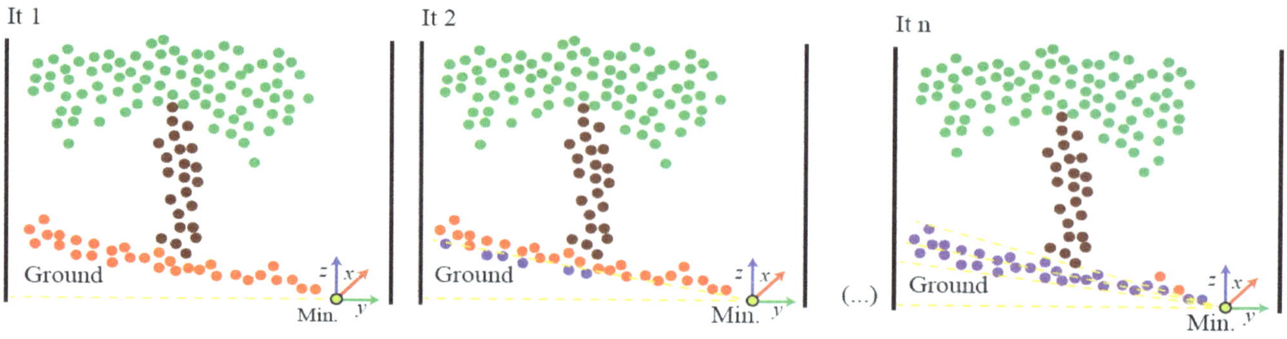

Figure 6. Cutting plane adjustment for the ground removal in segment n_1.

Initially, this 3D plane is fixed by the point with the minimum height and the up vector (**v** = 0,0,1), whose direction is perpendicular to the horizontal plane. Then, for each iteration, the plane is rotated 5° around the x-axis. This geometric transformation was performed by applying the rotation matrix showed in Equation (1). The stopping criterion was determined by several 3D points, selected in each iteration, i.e., those which were under the cutting plane influence. The plan inclination angle is set to the value allowing a proper fit of 3D plane to the terrain slope. Finally, to determine the location of points relative to the cutting plane, the point-normal equation for the cutting plane was applied. If the result is lower than zero, the point is under the plane (Equation (2)), so it is classified as ground and automatically removed in the point cloud.

$$R_x = \begin{pmatrix} 1 & 0 & 0 \\ 0 & \cos(\alpha) & -\sin(\alpha) \\ 0 & \sin(\alpha) & \cos(\alpha) \end{pmatrix} \tag{1}$$

$$A(x - x1) + B(y - y_1) + C(z - z_1) < 0 \tag{2}$$

where A, B and C represent the coefficients of the normal vector, (x, y, z) represent the coordinates of the point on the plane and (x_1, y_1, z_1) represent the coordinates of 3D points.

Regarding the leaves removal, the same inclination angle used for the ground cutting plane was applied. To fix the position of this upper plane, an offset value was calculated considering points with the maximum and minimum height for each segment. This procedure is illustrated in Figure 7.

After ground and leaves points removal, the next step consists in the identification of individual trunks. This is done based on the application of a spatial segmentation especially developed for that purpose. In this regard, the trunk is considered to be a 3D geometric shape, and a clustering method sharing the following features was developed: (1) the Euclidean distance between the 3D points must be lower than 50 cm (this value is estimated considering the typical distance between grapevines, which is never of this order); and (2) the minimum number of points for clustering is set to five. According to these constraints, a correct limitation of the growing region was determined for each cluster. Therefore, n groups of points were segmented for each vine row.

The value h represents the vertical height of plants in the segment, while the value f represents the offset obtained by applying Equation (3). According to this setting, the top plane was adapted by the geometric features for each segment. In this case, upper points to the cutting planes, which were considered vegetation, were removed in the point cloud. However, some outlying points could not be correctly filtered by the cutting plane. For this reason, another filter was applied based on the point colour. A threshold for the green channel was fixed in order to remove vegetation points characterised by the green channel higher than 120 as well as red and blue channel lower than 80. This combination was considered adequate and derived from the many tests performed. Lower values would cause the removal of points belonging to the trunks.

$$f = |Z_{Max} - Z_{Min}|x0.6 \quad (3)$$

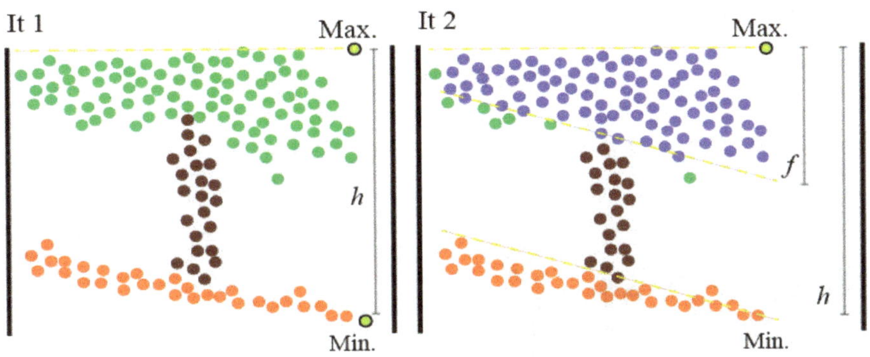

Figure 7. Cutting plane adjustment for the leaves removal.

2.3.4. Trunk Detection

After ground and leaves points removal, the next step consists in the identification of individual trunks. This is done based on the application of a spatial segmentation developed for that purpose. In this regard, the trunk is considered to be a 3D geometric shape, and a clustering method sharing the following features was developed: (1) the Euclidean distance between the 3D points must be lower than 50 cm (this value is estimated considering the typical distance between grapevines, which is never of this order); and (2) the minimum number of points for clustering is set to five. According to these constraints, a correct limitation of the growing region was determined for each cluster. Therefore, k groups of points were segmented for each vine row.

Regarding the results of spatial clustering, two optimisations are considered to improve the final results. The first optimisation focuses on solving errors related to the spatial segmentation. In areas characterised by dense vegetation, where the trunk was partially occluded by leaves, the trunk's 3D reconstruction is, in general, generated with a lower detail. In these cases, the trunk is potentially composed by a few real points and many noisy points. Consequently, an inaccurate segmentation is achieved, which causes false positives in those regions. This issue is overtaken by testing the angle (α) between two vectors: (1) the direction vector of the vine row axis; (2) the direction formed by two consecutive centroids of clusters. A maximum deviation of 20° is allowed. This value proved to be adequate to remove clusters which are not correctly segmented. As illustrated in Figure 8, green points are valid centroids and red points are those centroids which are discarded. The blue arrows depict the correct patch formed by all plants of the row. This test is performed along the vine row in the same direction on the x-axis.

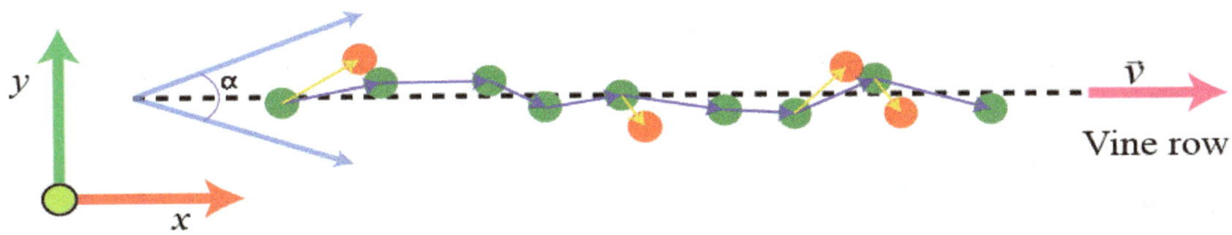

Figure 8. Optimisation of the clustering segmentation procedure.

The second optimisation consists on the automatic recognition of posts, which are considered to be trunks, since geometrically both are very similar. For this purpose, key geometric features were considered to detect posts for each vine row. In Figure 9a, trunks and posts are marked by a green and yellow rectangle, respectively. The main difference is the number of vegetation points around them. As shown in Figure 9b, a neighbour search is developed for each centroid, limited by a radius (r = 80 cm). This value is the most adequate considering the mean size of grapevines in the dataset. To check points which were into this spherical-searching region, Equation (4) is applied. Then, the point colour is also considered in order to detect vegetation points. Vegetation points are characterised by a threshold for the R, G and B channels so, if some of them are located inside the search area, the cluster is considered to be trunk. Likewise, if no vegetation points are found, the cluster is classified as a post.

$$x^2 + y^2 + z^2 - r^2 < 0 \tag{4}$$

where for each cluster, x, y and z represent the coordinates of a 3D point and r the radius of the search region.

Figure 9. Recognition of posts: (**a**) post location in the point cloud; and (**b**) the search of vegetation points around the trunk.

2.3.5. Estimation of Missing Plants and Occluded Trunks

Depending on the epoch of the year in which data are acquired, vegetation may occlude trunks (Figure 2). The proposed method is also prepared to deal with such scenarios, being able to estimate the position of no visible trunks and missing plants. In general, and mostly in recent commercial vineyards, each vine row is formed by plants which are equidistant from each other. However, the presented method estimates the space between plants automatically, which makes its use universal and fully automatic. For testing this feature, the method was applied on a complex vineyard plot presented in Figure 1. Therefore, the method can be used in any vineyard even in those which present challenging features such as the irregular distance between plants, replanted grapevines with different trunk diameter and some plants not visible from aerial images due to a dense foliage (Figure 10).

Figure 10. Cases where the trunks are occluded or cannot properly modelled.

The estimation of missing plants and occluded trunks is useful in order to optimise the results of plant recognition as well as to know the number of voids along the vine rows, which could be occupied by new plants. For this purpose, the resulting data from the individual grapevine detection were used in order to identify areas where no plants were found. Firstly, the distances between consecutive grapevines are calculated and sorted from the lowest to the highest, considering all plants detected in a given vine row.

Then, all distances less than 50 cm are discarded and the top 10% values of the sorted list are used to calculate the average distance (*d*). This value was calculated for each vine row and is also used to highlight areas that can potentially contain missing plants or occluded grapevines. If the distance between two consecutive plants (*D*) is higher than *d*, and the rate d/D is higher than one, the integer part of the quotient represents the number of plants that should be detected in the area. Then, missing plants and/or occluded trunks are marked using the average distance (*d*) as reference. Figure 11 illustrates the representation of two detected grapevines and the area between them where there are missing plants and occluded trunks. The last step consists in determining if the marked point represents a missing plant or an occluded trunk. To this end, a point inclusion test in a three-dimensional cylinder (height of 1 m and a radius of 20 cm) is implemented, considering the same constrains which were used for the trunk detection.

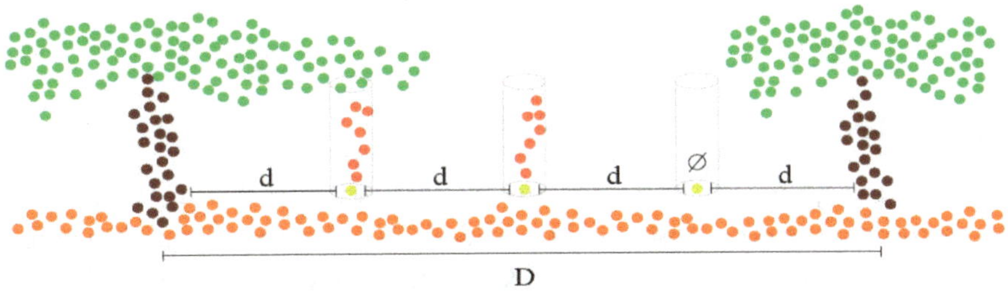

Figure 11. Recognition of missing or occluded plants.

2.4. Validation Process

To analyse the robustness and effectiveness of the proposed method in all its features, five vine rows of the complex study area (polygon highlighted in Figure 1) were used. A field campaign was performed in order to map the real state of vine rows and to determine the location of missing plants. This way, for each vine row, the results provided by the application of the proposed method were compared with ground-truth data allowing computing the overall accuracy of the whole validation area and of each estimation.

Grapevine estimation evaluation was conducted based on the number of correct (true positive—TP) and incorrect (false positive—FP) grapevine estimations and also considering the correct/incorrect estimation of missing plants along the vine rows as, respectively, true negatives (TN) and false negatives (FN). From theses data precision, recall, F1score and the overall accuracy were computed for each vine row and for the whole validation area.

3. Results

3.1. Point Cloud Reconstruction and Processing

The proposed method works perfectly when it is applied to well-maintained commercial vineyards plots, using aerial data acquired as described in Section 2.2. In those cases, the method is able to detect all existing trunks. However, the method was pushed to its limit, being applied to a complex area, arising from the existence of distinct vegetation density/vigour areas, several voids caused by missing plants, and new plants that were replaced defective or dead plants.

As results, a 3D point cloud formed by 26,656,371 points was generated and the time for point cloud densification was 01 h:29 m:23 s. Figure 12 shows the generated point cloud and the virtual lines that represent the vine rows axis. Most of plants could be fully modelled, but there are some grapevines whose trunks were partially occluded by leaves. Moreover, noisy points were produced around the trunks, between the leaves of plants and the ground. These 3D points negatively affect the recognition of trunk's shape. To address this problem, a noise filter was applied and most of noising points could be removed. This process was applied using a kernel size of 0.05 m. In summary, 10.448.046 points were discarded in the generated dense point cloud.

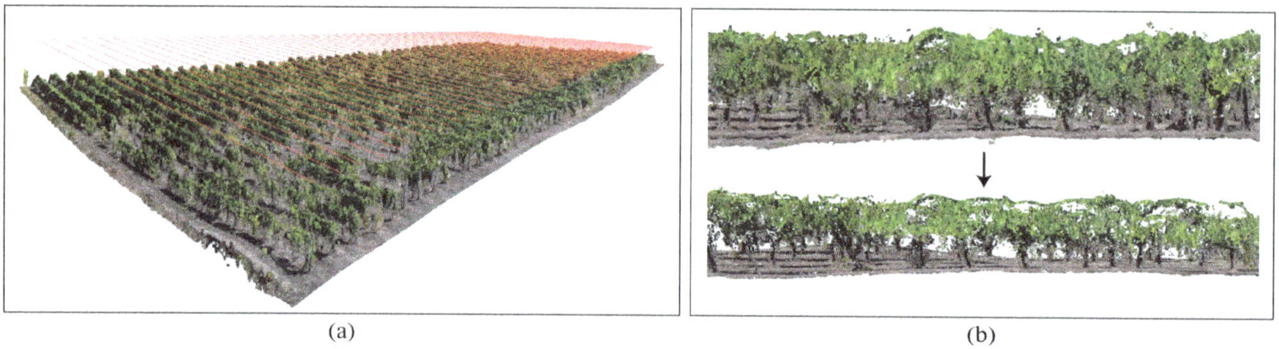

Figure 12. 3D model generated of the complex vineyard plot: (**a**) the reconstruction of study area using all the 3D points; and (**b**) final model, after application of noise filter.

3.2. Individual Grapevine Detection

Once the point cloud is generated and the noisy points filtered out, the 3D model is segmented in order to discard ground and leaves points. Figure 13 shows the results of this step, when the method is applied to the vine rows of the validation area. Consequently, red points (classified as ground and leaves, Figure 13b) are discarded and just trunk points, which will be used as input data for the spatial clustering, remain.

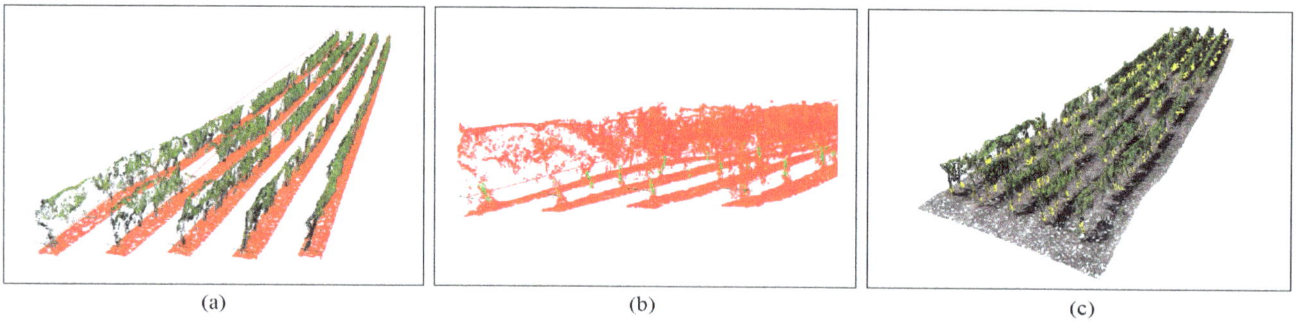

Figure 13. Main steps for individual trunk detection: (**a**) ground points identification and removal, (**b**) vegetation/leaf points identification and removal; and (**c**) trunk detection.

The method was applied to the whole plot and the location of detected and missing plants was analysed in the QGIS software. This output is presented in Figure 14, with the orthophoto mosaic in the background. A total of 1916 grapevines were estimated and 402 plants were classified as being missing.

Regarding the efficiency of the method, the time required for the automatic recognition of each individual grapevine in the whole plantation was 38 s using a PC with CPU (Intel Xeon(R) W-2145) and RAM (64 GB). The low time required for computing the methods makes possible the use of portable devices for on-site processing.

Figure 14. General overview of the results obtained from the application of the proposed method to the whole complex plot. Coordinates in WGS84 (EPSG:4326).

3.3. Grapevine Estimation Accuracy

The results of the grapevine estimation from the application of the proposed method to the five validation vine rows are presented in Table 1.

Table 1. Number of plants and overall accuracy (OA) of the proposed method compared to the ground-truth data of five vine rows. Row total values—maximum possible number of plants in a given vine row—are also provided.

Vine Row	Number of Grapevines			Missing Grapevines			Row Total		
	Obs.	Est.	OA (%)	Obs.	Est.	OA (%)	Obs.	Est.	OA (%)
1	46	43	93.5	12	15	75.0	58	58	100.0
2	39	37	94.9	18	21	83.3	57	58	98.2
3	42	45	92.9	15	12	80.0	57	57	100.0
4	40	46	85.0	17	9	52.9	57	55	96.5
5	49	50	98.0	7	7	100.0	56	57	98.2
Total	216	221	97.7	69	64	92.8	285	285	100.0

Regarding the total number of grapevines presented in the evaluated vine rows (ranging between 39 to 46), an overestimation of five plants is observed. Under detection of grapevines is verified in two rows, differing in three and two grapevines, respectively. The opposite is observed in the other three vine rows, with ten plants being overestimated (respectively, three, six and one). According to the number of missing grapevines (69, in total, ranging between 7 to 18 per row, average of 14 missing plants) the results obtained by the method show an overestimation of six missing grapevines in two vine rows (three in each) and one vine row is in agreement with the ground-truth data. As for the other vine rows the under estimation of missing plants diverges from three to eight plants (64 missing plants in total, with an average of 13 missing plants per row, 93% overall accuracy). As for the total number of possible grapevines in a given vine row (sum of the grapevines and missing grapevines), a mean of 57 plants were estimated, the same number when observing the ground-truth data, being the total also 285 plants, obtaining three rows with the same number of plants, two rows with one plant less, and one row with two plants more. For this specific parameter, the overall accuracy ranges between 97% (underestimation) and 100% considering all vine rows. According to the capability of the proposed method in the automatic detection of grapevines, 157 grapevines (71%) were directly detected from the grapevine trunk, and by the analysis of the point cloud density 64 grapevines were estimated. Moreover, several posts were also detected along the vine rows and therefore, were automatically discarded.

To further validate the spatial accuracy of the methods outputs, each detection was analysed to assess if the estimations were correctly located. By using the ground-truth data, false negatives and false positives results—grapevines classified as being missing and the inverse, missing plants classified as being grapevines—are evaluated. Table 2 presents these results. In total, from the 221 estimated plants 197 plants were correctly detected and classified (89%), as for missing plants, 41 (64%) were correctly classified, the overall accuracy considering all data is approximately 84%.

Table 2. Evaluation of the proposed method in the classification grapevines and missing grapevines for the following parameters: precision, recall, F1score, and overall accuracy (OA). The ground-truth of five vine rows was used. TP: true positive; FP: false positive; TN: true negative; FN: false negative.

Vine Row	TP	FP	TN	FN	Precision	Recall	F1score	O.A. (%)
1	39	4	8	7	0.91	0.85	0.88	81.0
2	32	5	16	5	0.86	0.86	0.86	82.8
3	40	5	7	5	0.89	0.89	0.89	82.5
4	42	4	6	3	0.91	0.93	0.92	87.3
5	45	5	4	3	0.90	0.94	0.92	86.0
Total	198	23	41	23	0.90	0.90	0.90	83.9

Figure 15 shows the 3D location of each plant on the test area. Green points represent the centroids of plants that are directly detected in the 3D model. The centroids' position were then checked against its corresponding vine row in order to identify centroids wrongly classified as trunk (red points). These wrong classifications are caused due to points of vegetation, which are around the trunk and could not be totally removed by the noise filter. Then, posts were distinguished from the grapevine trunks (blue points). Finally, missing plants (pink points) and occluded trunks (yellow points) were estimated. In terms of quantitative data, in the validation area, the proposed method was able to detect 221 plants and 64 missing plants.

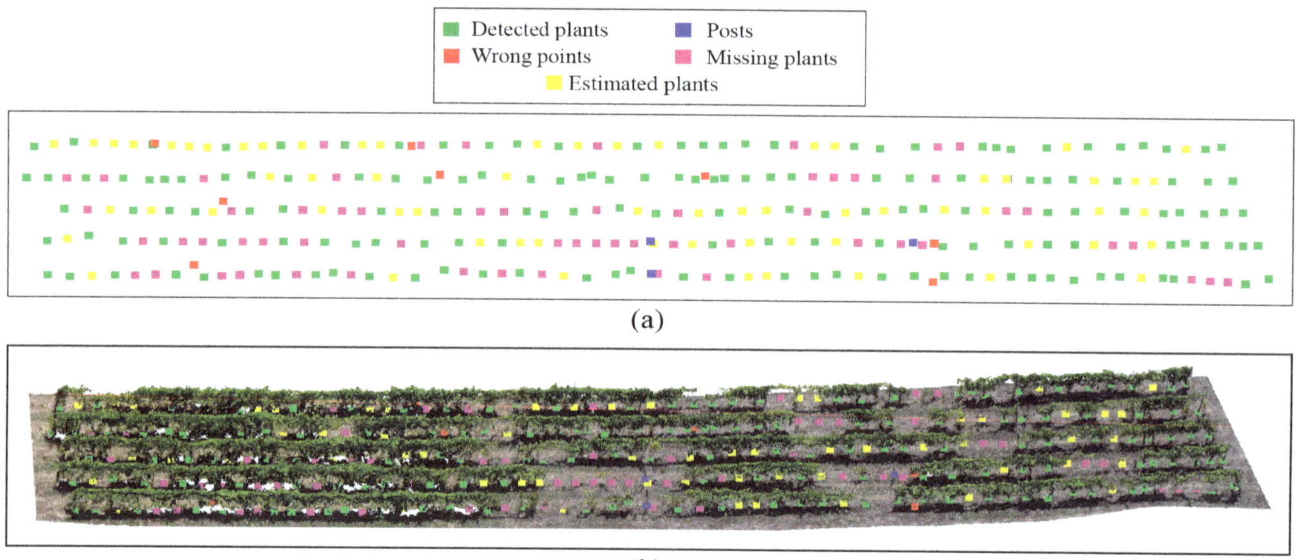

Figure 15. Individual grapevine delineation resulting from the application of the proposed method to the validation area: (**a**) points to represent detected plants (visible and occluded trunks), missing plants, wrong clusters and posts; and (**b**) 3D point cloud and points computed for the plant location.

4. Discussion

4.1. *Point Cloud Reconstruction and Processing*

The generation of point clouds for remote sensing applications was enhanced by the proliferation of innovative UAV-based technologies such as high-resolution cameras and LiDAR systems. By applying photogrammetric techniques, point clouds can be obtained using multiple overlapping images. Other option is the use of LiDAR scanners which provide dense point clouds of natural environments but these are more expensive than digital cameras [34]. LiDAR data discriminate better plant's canopy since it penetrates vegetation [35,36], and, therefore, potentially making the identification of the grapevine trunks easier. Moreover, photogrammetric techniques tend to estimate erroneous points in the cases where some points from the ground are estimated along a post. One of the advantages of the proposed method is that it remains operational even when using point cloud data from other type of sensors, reinforcing that when LiDAR sensors for UAVs become more affordable the method can still be employed. The proposed method is not dependent on any technology, although to get proper results is required enough geometric quality of the 3D model to ensure a partial reconstruction of the trunks at least. In this sense, to avoid noisy points around the trunk and vegetation, the presented solution integrates a noise filter. Hence, the proposed method can be applied using any point cloud data with accurate results.

4.2. *Individual Grapevine Detection*

There are several methods for the automatic detection and parameters extraction on 3D models using crop height models [37], combining terrestrial laser scanner and UAV photogrammetric point clouds [38], fusing RGB and multispectral point clouds to extract individual tree parameters [39], computing 3D vegetation indices in olive groves [40], and obtaining forest structural attributes [41]. However, given the complexity and the unique characteristics of vineyard plots, such methods are not suitable to be applied. Studies focusing on the use of photogrammetric point clouds, generated from UAV-based imagery, were dealing with vineyard detection [23] or detect and describe some of its general properties [17,25,42]. Studies for individual grapevine detection using UAV-based raster outcomes often rely on the coarse position of each vine assuming a mean distance of separation between grapevines along the vine row [13,20,21]. The proposed method addresses all these limitations, using point cloud data and geometrical characteristics, to automatically identify points belonging to grapevine's trunk. Hence, individual plants can be detected and missing plants estimated.

The complex vineyard plot analysed in this study had been used in another study [13] with data acquired in 2018. The whole plot contained 2266 plants and was evaluated by the method with an accuracy of 98%. However, the aerial imagery was acquired in an early phase of the vegetative state [43] with a lower vegetation density, which can help in the detection of missing plants. Furthermore, a constant distance between individual plant was also used. The results presented in Section 2.4 present an overestimation of 2% (Table 1) according to the number of grapevines and an under-estimation of 8% considering the number of missing plants. This fact is related to the presence of vegetation from adjacent grapevines in areas with missing plants that could cover the void. The results are improved by selecting an early period (preferably belonging to phase 2—Figure 2) to conduct grapevine detection, with grapevines with a lower leaf cover, preventing the existence of vegetation in areas with no plants [21]. In Di Gennaro and Matese [44], 3D and 2.5D methods were compared for vineyard biomass and plant detection, but individual grapevine detection was not performed. In that study, as for missing plants, an overestimation was observed in the 3D-based method, false negatives were related to the existence of new plants while some false positives were due to different grapevine canopy thickness. Similar findings were detected in the miss-classifications observed in this study (Table 2). Moreover, proximal sensing approaches for grapevine trunk detection using ground vehicles were also tested by research groups, either using LiDAR [45,46] or depth cameras [47]. However, in contrast to UAVs, such approaches are more expensive—due to the equipment used—and time-consuming, since the vehicles need to go through all vine rows, where some obstacles can also be present in their way.

5. Conclusions

The innovative method presented in this study is proved to be effective for a rapid access of the vineyard status using UAV-based 3D point clouds, with automation levels that allow its applicability in different vineyards, not relying on predefined parameters as the distance among plants. The proposed method is able to detect occluded trunks with reliable accuracy rates and missing plants, where in the vineyard context represents the occurrence of voids along the vine rows. The major contribution of this work is that the approach is fully automatic, not requiring any a prior knowledge of the distance between plants or number of plants per row, as in existing approaches. Moreover, the computational complexity of the proposed technique does not require high-performance computing and is appropriate for use on mobile in-field computing devices.

The applicability of the proposed method can be extended to other types of purposes related to the estimation of biophysical parameters of grapevines, providing a more efficient understanding of data for vineyard management and the validation of the use of UAV-based point clouds. Indeed, its impact is increased in a multi-temporal context. In this way, the estimated canopy of each detected grapevine can be studied to measure its volume which can help in the decision-making process for canopy management operations and, consequently, yield optimisation. To improve data quality and to extend the method capabilities, an in-depth investigation of the flight parameters optimisation (flight height, imagery overlap, camera angle) is required, which can also make possible an automatic detection grape bunches. The results of this study might influence on further research related to individual monitoring of every grapevine, multi-temporal studies and making accurate support decision systems for an optimal vineyard management.

Author Contributions: Conceptualisation, J.M.J., L.P. and J.J.S.; methodology, J.M.J., L.P. and J.J.S.; software, J.M.J.; validation, F.R.F. and J.J.S.; formal analysis, L.P.; investigation, J.M.J. and L.P.; resources, F.R.F. and J.J.S.; data curation, J.M.J. and L.P.; writing—original draft preparation, J.M.J. and L.P.; writing—review and editing, F.R.F. and J.J.S.; visualisation, J.M.J. and L.P.; supervision, F.R.F. and J.J.S.; project administration, J.J.S.; funding acquisition, F.R.F. and J.J.S. All authors have read and agreed to the published version of the manuscript.

References

1. Zhang, C.; Kovacs, J.M. The application of small unmanned aerial systems for precision agriculture: A review. *Precis. Agric.* **2012**, *13*, 693–712. [CrossRef]
2. Mogili, U.R.; Deepak, B. Review on application of drone systems in precision agriculture. *Procedia Comput. Sci.* **2018**, *133*, 502–509. [CrossRef]
3. Bernardes, M.F.F.; Pazin, M.; Pereira, L.C.; Dorta, D.J. Impact of pesticides on environmental and human health. *Toxicol. Stud. Cells Drugs Environ.* **2015**, 195–233.
4. Scott, G.; Rajabifard, A. Sustainable development and geospatial information: A strategic framework for integrating a global policy agenda into national geospatial capabilities. *Geo-Spat. Inf. Sci.* **2017**, *20*, 59–76. [CrossRef]
5. Pádua, L.; Vanko, J.; Hruška, J.; Adão, T.; Sousa, J.J.; Peres, E.; Morais, R. UAS, sensors, and data processing in agroforestry: A review towards practical applications. *Int. J. Remote Sens.* **2017**, *38*, 2349–2391. [CrossRef]
6. Ezenne, G.; Jupp, L.; Mantel, S.; Tanner, J. Current and potential capabilities of UAS for crop water productivity in precision agriculture. *Agric. Water Manag.* **2019**, *218*, 158–164. [CrossRef]
7. Shi, X.; Han, W.; Zhao, T.; Tang, J. Decision support system for variable rate irrigation based on UAV multispectral remote sensing. *Sensors* **2019**, *19*, 2880. [CrossRef] [PubMed]
8. Zhang, M.; Zhou, J.; Sudduth, K.A.; Kitchen, N.R. Estimation of maize yield and effects of variable-rate nitrogen application using UAV-based RGB imagery. *Biosyst. Eng.* **2020**, *189*, 24–35. [CrossRef]

9. Mendes, J.; Pinho, T.M.; Neves dos Santos, F.; Sousa, J.J.; Peres, E.; Boaventura-Cunha, J.; Cunha, M.; Morais, R. Smartphone Applications Targeting Precision Agriculture Practices—A Systematic Review. *Agronomy* **2020**, *10*, 855. [CrossRef]
10. Matese, A.; Di Gennaro, S.F. Technology in precision viticulture: A state of the art review. *Int. J. Wine Res.* **2015**, *7*, 69–81. [CrossRef]
11. Proffitt, A.P.B.; Bramley, R.; Lamb, D.; Winter, E. *Precision Viticulture: A New Era in Vineyard Management and Wine Production*; Winetitles Pty Ltd.: Ashford, SA, USA, 2006.
12. Campos, J.; Llop, J.; Gallart, M.; García-Ruiz, F.; Gras, A.; Salcedo, R.; Gil, E. Development of canopy vigour maps using UAV for site-specific management during vineyard spraying process. *Precis. Agric.* **2019**, *20*, 1136–1156. [CrossRef]
13. Pádua, L.; Adão, T.; Sousa, A.; Peres, E.; Sousa, J.J. Individual Grapevine Analysis in a Multi-Temporal Context Using UAV-Based Multi-Sensor Imagery. *Remote Sens.* **2020**, *12*, 139. [CrossRef]
14. Pádua, L.; Marques, P.; Hruška, J.; Adão, T.; Bessa, J.; Sousa, A.; Peres, E.; Morais, R.; Sousa, J.J. Vineyard properties extraction combining UAS-based RGB imagery with elevation data. *Int. J. Remote Sens.* **2018**, *39*, 5377–5401. [CrossRef]
15. Comba, L.; Gay, P.; Primicerio, J.; Aimonino, D.R. Vineyard detection from unmanned aerial systems images. *Comput. Electron. Agric.* **2015**, *114*, 78–87. [CrossRef]
16. Mathews, A.J.; Jensen, J.L. Visualizing and quantifying vineyard canopy LAI using an unmanned aerial vehicle (UAV) collected high density structure from motion point cloud. *Remote Sens.* **2013**, *5*, 2164–2183. [CrossRef]
17. Weiss, M.; Baret, F. Using 3D point clouds derived from UAV RGB imagery to describe vineyard 3D macro-structure. *Remote Sens.* **2017**, *9*, 111. [CrossRef]
18. Poblete-Echeverría, C.; Olmedo, G.F.; Ingram, B.; Bardeen, M. Detection and segmentation of vine canopy in ultra-high spatial resolution RGB imagery obtained from unmanned aerial vehicle (UAV): A case study in a commercial vineyard. *Remote Sens.* **2017**, *9*, 268. [CrossRef]
19. Caruso, G.; Tozzini, L.; Rallo, G.; Primicerio, J.; Moriondo, M.; Palai, G.; Gucci, R. Estimating biophysical and geometrical parameters of grapevine canopies ('Sangiovese') by an unmanned aerial vehicle (UAV) and VIS-NIR cameras. *Vitis* **2017**, *56*, 63–70.
20. De Castro, A.I.; Jimenez-Brenes, F.M.; Torres-Sánchez, J.; Peña, J.M.; Borra-Serrano, I.; López-Granados, F. 3-D characterization of vineyards using a novel UAV imagery-based OBIA procedure for precision viticulture applications. *Remote Sens.* **2018**, *10*, 584. [CrossRef]
21. Matese, A.; Di Gennaro, S.F. Practical applications of a multisensor uav platform based on multispectral, thermal and rgb high resolution images in precision viticulture. *Agriculture* **2018**, *8*, 116. [CrossRef]
22. Primicerio, J.; Caruso, G.; Comba, L.; Crisci, A.; Gay, P.; Guidoni, S.; Genesio, L.; Ricauda Aimonino, D.; Vaccari, F.P. Individual plant definition and missing plant characterization in vineyards from high-resolution UAV imagery. *Eur. J. Remote Sens.* **2017**, *50*, 179–186. [CrossRef]
23. Comba, L.; Biglia, A.; Aimonino, D.R.; Gay, P. Unsupervised detection of vineyards by 3D point-cloud UAV photogrammetry for precision agriculture. *Comput. Electron. Agric.* **2018**, *155*, 84–95. [CrossRef]
24. Comba, L.; Biglia, A.; Aimonino, D.R.; Tortia, C.; Mania, E.; Guidoni, S.; Gay, P. Leaf Area Index evaluation in vineyards using 3D point clouds from UAV imagery. *Precis. Agric.* **2020**, *21*, 881–896. [CrossRef]
25. Mesas-Carrascosa, F.J.; de Castro, A.I.; Torres-Sánchez, J.; Triviño-Tarradas, P.; Jiménez-Brenes, F.M.; García-Ferrer, A.; López-Granados, F. Classification of 3D point clouds using color vegetation indices for precision viticulture and digitizing applications. *Remote Sens.* **2020**, *12*, 317. [CrossRef]
26. Aboutalebi, M.; Torres-Rua, A.F.; McKee, M.; Kustas, W.P.; Nieto, H.; Alsina, M.M.; White, A.; Prueger, J.H.; McKee, L.; Alfieri, J.; et al. Incorporation of Unmanned Aerial Vehicle (UAV) Point Cloud Products into Remote Sensing Evapotranspiration Models. *Remote Sens.* **2020**, *12*, 50. [CrossRef]
27. Moreno, H.; Valero, C.; Bengochea-Guevara, J.M.; Ribeiro, Á.; Garrido-Izard, M.; Andújar, D. On-Ground Vineyard Reconstruction Using a LiDAR-Based Automated System. *Sensors* **2020**, *20*, 1102. [CrossRef]
28. Rusu, R.B.; Cousins, S. 3d is here: Point cloud library (pcl). In Proceedings of the 2011 IEEE International Conference on Robotics and Automation, Shanghai, China, 9–13 May 2011; IEEE: Piscataway, NJ, USA, 2011; pp. 1–4.
29. Schönberger, J.L.; Frahm, J. Structure-from-Motion Revisited. In Proceedings of the 2016 IEEE Conference on Computer Vision and Pattern Recognition (CVPR), Las Vegas, NV, USA, 26 June–1 July 2016; pp. 4104–4113. [CrossRef]

30. Colomina, I.; Molina, P. Unmanned aerial systems for photogrammetry and remote sensing: A review. *ISPRS J. Photogramm. Remote Sens.* **2014**, *92*, 79–97. [CrossRef]
31. Han, X.F.; Jin, J.S.; Wang, M.J.; Jiang, W.; Gao, L.; Xiao, L. A review of algorithms for filtering the 3D point cloud. *Signal Process. Image Commun.* **2017**, *57*, 103–112. [CrossRef]
32. Richardson, A.D.; Jenkins, J.P.; Braswell, B.H.; Hollinger, D.Y.; Ollinger, S.V.; Smith, M.L. Use of digital webcam images to track spring green-up in a deciduous broadleaf forest. *Oecologia* **2007**, *152*, 323–334. [CrossRef]
33. Otsu, N. A threshold selection method from gray-level histograms. *IEEE Trans. Syst. Man Cybern.* **1979**, *9*, 62–66. [CrossRef]
34. Chen, S.; McDermid, G.J.; Castilla, G.; Linke, J. Measuring vegetation height in linear disturbances in the boreal forest with UAV photogrammetry. *Remote Sens.* **2017**, *9*, 1257. [CrossRef]
35. Lisein, J.; Pierrot-Deseilligny, M.; Bonnet, S.; Lejeune, P. A photogrammetric workflow for the creation of a forest canopy height model from small unmanned aerial system imagery. *Forests* **2013**, *4*, 922–944. [CrossRef]
36. Guimarães, N.; Pádua, L.; Marques, P.; Silva, N.; Peres, E.; Sousa, J.J. Forestry Remote Sensing from Unmanned Aerial Vehicles: A Review Focusing on the Data, Processing and Potentialities. *Remote Sens.* **2020**, *12*, 1046. [CrossRef]
37. Panagiotidis, D.; Abdollahnejad, A.; Surový, P.; Chiteculo, V. Determining tree height and crown diameter from high-resolution UAV imagery. *Int. J. Remote Sens.* **2017**, *38*, 2392–2410. [CrossRef]
38. Tian, J.; Dai, T.; Li, H.; Liao, C.; Teng, W.; Hu, Q.; Ma, W.; Xu, Y. A novel tree height extraction approach for individual trees by combining TLS and UAV image-based point cloud integration. *Forests* **2019**, *10*, 537. [CrossRef]
39. Jurado, J.M.; Ramos, M.; Enríquez, C.; Feito, F. The Impact of Canopy Reflectance on the 3D Structure of Individual Trees in a Mediterranean Forest. *Remote Sens.* **2020**, *12*, 1430. [CrossRef]
40. Jurado, J.M.; Ortega, L.; Cubillas, J.J.; Feito, F. Multispectral mapping on 3D models and multi-temporal monitoring for individual characterization of olive trees. *Remote Sens.* **2020**, *12*, 1106. [CrossRef]
41. Cao, L.; Liu, H.; Fu, X.; Zhang, Z.; Shen, X.; Ruan, H. Comparison of UAV LiDAR and digital aerial photogrammetry point clouds for estimating forest structural attributes in subtropical planted forests. *Forests* **2019**, *10*, 145. [CrossRef]
42. Comba, L.; Zaman, S.; Biglia, A.; Ricauda, A.D.; Dabbene, F.; Gay, P. Semantic interpretation and complexity reduction of 3D point clouds of vineyards. *Biosyst. Eng.* **2020**, *197*, 216–230. [CrossRef]
43. Magalhães, N. *Tratado de Viticultura: A Videira, a Vinha e o Terroir*; Publicações Chaves Ferreira Lisboa: Lisboa, Portugal , 2008; 605p, ISBN 9789899820739.
44. Di Gennaro, S.F.; Matese, A. Evaluation of novel precision viticulture tool for canopy biomass estimation and missing plant detection based on 2.5 D and 3D approaches using RGB images acquired by UAV platform. *Plant Methods* **2020**, *16*, 1–12. [CrossRef]
45. Siebers, M.H.; Edwards, E.J.; Jimenez-Berni, J.A.; Thomas, M.R.; Salim, M.; Walker, R.R. Fast phenomics in vineyards: Development of GRover, the grapevine rover, and LiDAR for assessing grapevine traits in the field. *Sensors* **2018**, *18*, 2924. [CrossRef] [PubMed]
46. Milella, A.; Marani, R.; Petitti, A.; Reina, G. In-field high throughput grapevine phenotyping with a consumer-grade depth camera. *Comput. Electron. Agric.* **2019**, *156*, 293–306. [CrossRef]
47. Mendes, J.; Dos Santos, F.N.; Ferraz, N.; Couto, P.; Morais, R. Vine trunk detector for a reliable robot localization system. In Proceedings of the 2016 International Conference on Autonomous Robot Systems and Competitions (ICARSC), Bragança, Portugal, 4–6 May 2016; IEEE: Piscataway, NJ, USA, 2016; pp. 1–6.

Effect of Large-Scale Cultivated Land Expansion on the Balance of Soil Carbon and Nitrogen in the Tarim Basin

Erqi Xu [1], Hongqi Zhang [1,*] and Yongmei Xu [2,*]

1 Key Laboratory of Land Surface Pattern and Simulation, Institute of Geographic Sciences and Natural Resources Research, Chinese Academy of Sciences, Beijing 100101, China; xueq@igsnrr.ac.cn
2 Institute of Soil and Fertilizer, Xinjiang Academy of Agricultural Sciences, Urumqi 830091, China
* Correspondence: zhanghq@igsnrr.ac.cn (H.Z.); xym1973@163.com (Y.X.)

Abstract: Land reclamation influences the soil carbon and nitrogen cycling, but its scale and time effects on the balance of soil carbon and nitrogen are still uncertain. Taking the Tarim Basin as the study area, the impact of land reclamation on the soil organic carbon (SOC), total nitrogen (TN), and carbon to nitrogen (C:N) ratio was explored by the multiple temporal changes of land use and soil samples. Remote sensing detected that cropland nearly doubled in area from 1978 to 2015. Spatial analysis techniques were used to identify land changes, including the prior land uses and cultivation ages. Using land reclamation history information, a specially designed soil sampling was conducted in 2015 and compared to soil properties in ca. 1978. Results found a decoupling characteristic between the C:N ratio and SOC or TN, indicating that changes in SOC and TN do not correspond directly to changes in the C:N ratio. The land reclamation history coupled with the baseline effect has opposite impacts on the temporal rates of change in SOC, TN and C:N ratios. SOC and TN decreased during the initial stage of conversion to cropland and subsequently recovered with increasing cultivation time. By contrast, the C:N ratio for soils derived from grassland increased at the initial stage but the increase declined when cultivated longer, and the C:N ratio decreased for soils derived from forest and fluctuated with the cultivation time. Lower C:N ratios than the global average and its decreasing trend with increasing reclamation age were found in newly reclaimed croplands from grasslands. Sustainable agricultural management practices are suggested to enhance the accumulation of soil carbon and nitrogen, as well as to increase the C:N ratio to match the nitrogen deposition to a larger carbon sequestration.

Keywords: soil stoichiometry; land use change; soil organic carbon; nitrogen; Tarim Basin

---

1. Introduction

Soil carbon (C) and nitrogen (N) are the main sources of plant mineral nutrition and organic nutrients, which are the two primary factors affecting soil fertility [1,2]. They also play an important role in terrestrial soil C and N pools and global C and N cycles [3,4]. The ratio of soil organic carbon (SOC) and total nitrogen (TN), labeled as the C:N ratio, reliably indicates changes in soil microbial activity, decomposition, mineralization rates of SOC, and the cycle of soil C and N [5,6]. Moreover, the soil C:N ratio can be a good predictor of key parameters in the related C and N cycles, such as the dissolved organic carbon flux [7], nitrate leaching [8,9], and nitrous oxide emissions [10]. Therefore, understanding soil C and N content and their variation can help explore the mechanisms of soil C and N cycling and their coupling effects, and aid in enhancing soil C sequestration and reducing N losses in the ecosystem.

A high soil C:N ratio correlates with low organic matter (SOM) decomposition and mineralization rates while in soils with a low C:N ratio, SOM can be decomposed and mineralized relatively quickly [1]. SOC, TN, and the C:N ratio are influenced by many factors, such as climatic conditions (temperature, moisture), soil properties (soil texture, soil pH), terrain characteristics, and anthropogenic factors (land use and management) [2,11–13]. The soil C:N ratio and its variation are determined by gains (input of SOC and TN) and losses [11]. At the global scale, it has been shown that tundra, natural wetlands, and boreal forest can have a relatively wide range of soil C:N ratios, while croplands have a narrow range [2]. Land reclamation can alter the soil C and N bio-geochemical cycling and spatial-temporal characteristics of soil C:N ratios greatly [14–17]. The conversion from native vegetation to cropland decreases the input of vegetative tissues and increases soil temperatures to accelerate the litter decomposition and soil micro-organisms in the decomposition of soil C and N [18]. Reclamation activities can further cause C and N losses and change the soil C:N ratio [3]. Meanwhile, the increasing crop yields of land under sustainable agricultural management can enhance the input crop residues and roots, which can in turn, increase the SOM content and promote the accumulation of SOC and TN [19–21].

Generally, the dynamics of soil C:N ratios are determined by the relative changes in magnitude of SOC and TN contents [22,23]. Many case studies have found that forest converted to cropland results in a decrease in soil C:N ratio because of the high uptake and storage of N by trees [1,3,24,25]. In contrast, agricultural management can promote maintenance of crop residues at the soil surface, which may have beneficial impacts on soil fertility and increase the soil C:N ratio [1]. Plantation age also shows complex effects on the accumulation and decomposition of SOC and TN. For example, soil C:N ratios of cropland showed a significant increase over time at a sub-tropical site, although little change was observed at three temperate sites [11]. The above studies illustrate that the variation in soil C:N ratios during agricultural reclamation and management is complex and is not yet fully understood.

The dynamics of SOC, N, and the C:N ratio, which are affected by land use coupled with other physical factors, present significant spatial and temporal variations. Agricultural reclamation activities mean exploiting other land uses for croplands. Outcomes of reclamation are primarily the result of changes in plant types associated with the changes in the litter input, detritus and litter decomposition, root secretion and soil mineralization [14,26,27]. These changes are related to the prior land use type before the reclamation, which directly or indirectly influence the soil C, N, and their balance after the reclamation. Meanwhile, knowledge of the agricultural duration is important [28], because it is related to the cultivation intensity and the soil mellowing process. Thus, land reclamation history, including the different prior land use types and the duration age of the reclamation, will influence the fate of soil C and N. However, the effects of reclamation history remain uncertain [29], and this leads to difficulties in decision making and agriculture management. This calls for a deeper understanding of the impact that past and future land use changes can have on soil C:N ratios.

Arid and semi-arid regions are especially susceptible to human disturbance and climate change [30], which easily lead to land degradation and soil nutrient element loss [31–33]. Oasis agriculture is a specific ecosystem surrounded by the desert in an arid area, where environmental conditions are milder than the desert, ensuring fertility and allowing desert farming [34]. Aridisols, called Yermosols in the FAO-UNESCO classification (Food and Agriculture Organization-United Nations Educational, Scientific and Cultural Organization) [35], are the major original soil types, undisturbed by humans in arid regions [36]. The expansion of oasis agriculture disturbs the soil environment and may cause the loss of soil C and N. Under long-term sustainable tillage, fertilizing and irrigation of the oasis agriculture [37,38], Aridisols are gradually developed to Anthrosols with changes in C and N storage [35]. Thus, this calls for a deep understanding of how oasis agriculture influences the soil C and N.

The Tarim Basin contains the Taklimakan Desert, the world's second largest shifting sand desert [39]. It has experienced a significant expansion in oasis agriculture from its native vegetation, and even the desert since 1949 [40]. The cropland area in this region has nearly doubled since the 1980s [41,42], which was much higher than the average increase rate in the world [43]. Due to the extreme dry and hot climate with annual limited precipitation of 50–100 mm, almost the entire oasis agriculture relies on irrigation for crop growth in harsh conditions and to resist drought [44]. For the last few decades, land reclamation and the development of high-efficiency irrigation techniques has played an important role in this region [45], making the Tarim Basin one of the main agricultural production regions in China and one of the largest areas of oasis agriculture in the world.

This dramatic expansion in cropland has significantly changed the nutrients in the cultivated soils, and has also caused the overuse of water sources and land desertification [46], which in turn, influence the terrestrial C and N cycles [47]. Proper management of such changes requires a deep understanding of the effects of the land reclamation history on the variation in SOC and TN and their balance. Thus, the Tarim Basin was selected as the study area to explore the relationship between the land reclamation history and changes in the soil C, N, and C:N ratio. The objectives of this study were to (1) design a soil sampling campaign from land use change histories to quantify the changes in the SOC and TN dynamics of the cultivated topsoil, and (2) investigate the effects of the land reclamation history on these changes and the coupled relationships between SOC and TN.

2. Methods

2.1. Study Area

The Tarim Basin, located in the south of Xinjiang Province, Northwest China, was selected as our study region. It is located between 38°–43° N latitude and 76°–87° E longitude. Five typical sub-basins, the Aksu River Basin, the Kaidu-Kongque River Basin, the mainstream of Tarim River Basin, the Weigan River Basin, and the Yarkant River Basin, were selected as the case study sites with a total area of 311,765 km^2 (Figure 1). They lie within an extremely dry climate zone, with high temperatures of 20–30 °C in the summer and low temperatures of −10 to −20 °C in winter, a mean annual precipitation of 50–100 mm, and an annual evaporation of 2000–3000 mm [48]. Landscapes in the study area majorly consist of the desert, grassland, cropland, and forest. Native forests with shrubs, mainly including the *Populus euphratica* Oliv. and *Tamarix ramosissima* Ledeb. [49], are located in the riparian zone with an adequate water supply for vegetation growth. These climatic conditions have resulted in a specific oasis agriculture, with widespread use of highly efficient water-saving irrigation techniques.

2.2. Conceptual Framework for Exploring Impacts of Land Reclamation History on Soil C and N

Figure 2 showed the framework for exploring how land reclamation history influences the soil properties. Using the land use maps at different time nodes and spatial overlaying techniques, the land change histories at each location were identified from ca. 1978 to 2015. Also, the specific land reclamation history, including the prior land use before the reclamation and the duration age after the reclamation were extracted. Meanwhile, the simultaneous changes of SOC, TN, and C:N ratio were collected by purposive survey sampling in 2015 and the soil dataset in ca. 1978. Based on the statistical test, the impact of land reclamation history on SOC, TN, and the C:N ratio can be analyzed.

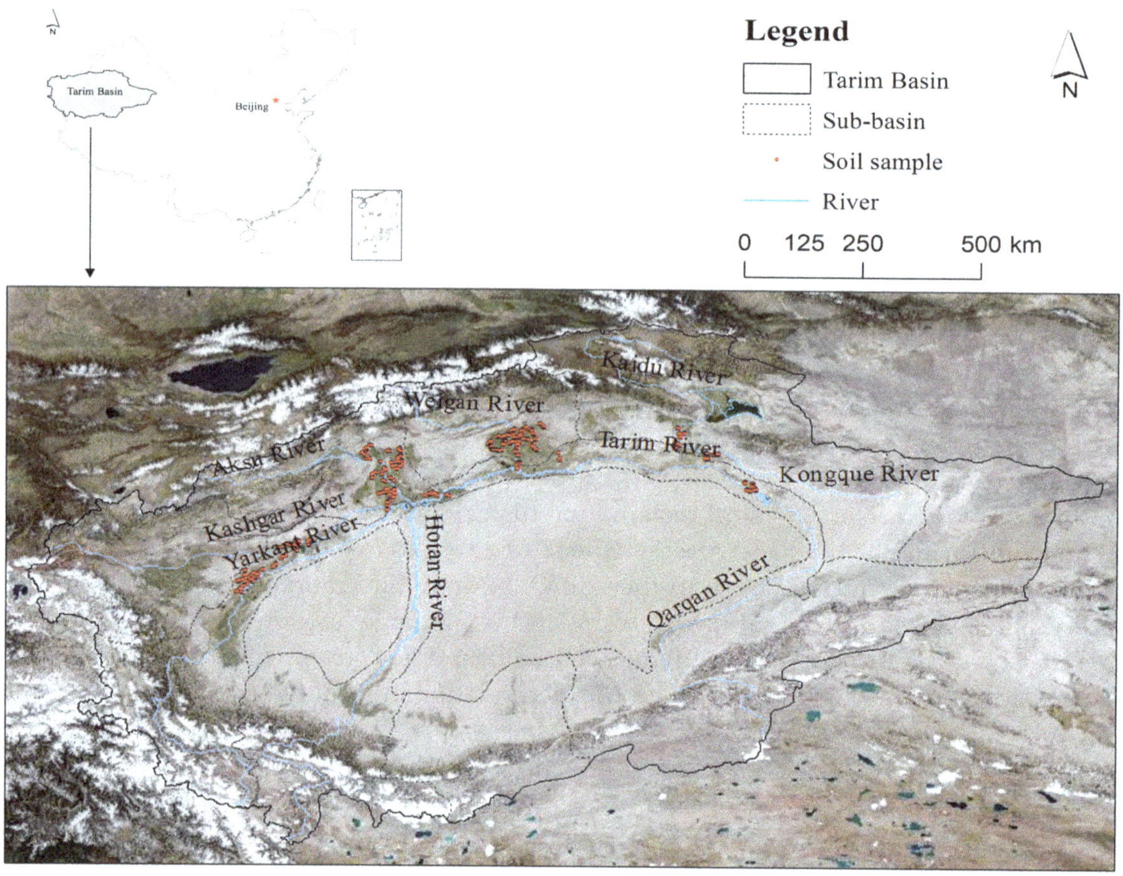

Figure 1. Location and soil sampling sites in the area of the Tarim Basin.

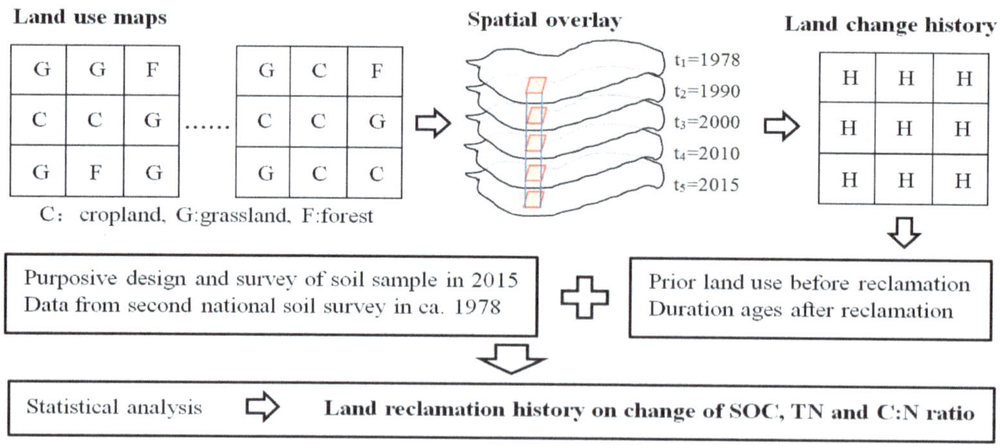

Figure 2. Conceptual framework for exploring impacts of land reclamation history on soil C and N.

2.3. *Land Use Change History Detection*

The soil sampling strategy was designed on the basis of the land reclamation history, which included land use prior to reclamation indicating the original soil nutrient status, while the cropland age was used to characterize the duration of management practices. To examine the land use change history on the SOC and TN dynamics from ca. 1980 (national soil surveys in 1978 to 1982) to 2015 (soil survey in this study), five time nodes, i.e., 1978, 1990, 2000, 2010, and 2015, of remote sensing images from Landsat MSS/TM and Landsat-8 Operational Land Imager (OLI) images were collected to map the land use via visual interpretation with a local expert tutorial on land use interpretation by ourselves. The paths and rows covering the study area are shown in Table S1. Preprocessing, including the application of geometric and atmospheric corrections, was conducted using ENVI 5.1 (ITT Visual

Information Solutions Inc.). Images within ±2 y of the time node were used as a proxy when there were no effective cloud-free images from a particular node.

The land use maps were divided into seven classes: cropland, forest, grassland, water body, built-up land, bare land, and glacier. As the majority of the croplands were derived from grassland, with a minority from forest and other land use types [47], this study focused on the effects of grassland and forest conversion to croplands on the SOC and TN dynamics. Using spatial overlay techniques in ArcGIS 10.1 (ESRI Inc., Redlands, CA, USA) for the land use vector maps with seven classes at the five time nodes, the overlaying vector polygon representing the land use change trajectory from ca. 1978 to 2015 was obtained. With the overlaying land use maps, the time nodes where the reclamations occurred were obtained, and the land use prior to the reclamation was identified. Grassland, forest, and cropland were labeled as G, F, and C, respectively, and the land use types for each of the five time nodes were identified. For example, "G→G→C→C→C" indicates the reclamation of grassland to cropland in the period of 1990 to 2000. The label "C→C→C→C→C", indicates that the cropland was assarted before 1978, and land with this history of use was used as a reference for this study. Following labeling for land use, the assarted croplands were grouped according to age as follows: assarted between 2010 and 2015 (new, 1–5 years), 2000 and 2010 (young, 6–15 years), 1990 and 2000 (medium, 16–25 years), 1978 and 1990 (old, 26–37 years). In addition, in some areas land use changes occurred more than once, and cases where cropland transferred to other land use types because of abandonment or build-up and occupation were excluded in this study. Thus, cropland, grassland, and forest were the three prior land use types in this study. Finally, three land use change types, i.e., unchanged cropland, cropland from grassland, and cropland from forest over nearly four decades, with five discrete reclamation ages were identified for the design of the soil sampling campaign.

2.4. Soil Sampling Design, Collection, and Laboratory Methods

Since we were particularly interested in the effect of land reclamation history, soil sampling conducted in 2015 was only carried out for cropland with different reclamation histories. First, the range of cropland in 2015 was extracted and 280 soil plots were identified in the croplands of five sub-basins by a random sampling strategy. A minimum distance of 1000 m between plots of the same category of land use change history was set. Next, three land use change types and five discrete reclamation ages from 1978 to 2015 in the designed soil plots were identified based on the spatial overlay techniques mentioned in Section 2.1. If the area of the vector polygon for overlaying land use maps was smaller than 1 km^2, the corresponding soil plots were excluded to minimize geolocalization errors of land use interpretation and make the resolution of land use maps consistent with soil C and N data. Finally, we collected samples from a total of 270 soil sample plots, including 126, 103, and 41 plots of unchanged cropland, cropland from grassland, and cropland from forest between 1978 and 2015, respectively. At each sampling plot of 10 m × 10 m, five sites, four at the corners and one in the center of the plot, were selected. According to the Genetic Soil Classification of China [38], the main soil type of the soil samples is irrigation-silted soil (Aric Anthrols in the FAO-UNESCO classification). Irrigation-silted soil is a type of Anthrosol and formed the new anthraquic epipedon in the soil irragric process and mellowing process under cultivation, which means soil materials are silted with irrigation water [38].

The soil C and N distribution within soil profile is highly influenced by management practices, especially tillage systems [50]. As soil disturbance by agricultural tillage is a primary cause of the historical change of soil nutrients, the tillage zone or management zone, mainly including the 0–20 cm soil layer, was sampled using a soil auger of 5 cm diameter. Soil samples were transported to the laboratory and preprocessing was conducted. Soil clods were crushed, and litter and living roots, stones (>2 mm), and visible plant remains were removed. Next, the soil samples were air-dried in the shade and passed through a 2 mm sieve. SOC and TN were measured by the H_2SO_4-$K_2Cr_2O_7$ oxidation method [51] and Kjeldahl procedure [52], respectively. Soil C:N fractions were calculated as the mass ratio of SOC to TN.

To compare simultaneous changes in soil properties and land use, the SOM and TN from ca. 1978 were obtained from the soil characteristics database of China, which were measured by the same chemical analysis methods as those used in 2015. This data has a resolution of 30 × 30 arc-seconds, and was generated from soil data collected in the 1979–1982 national soil survey. As the designed soil plots were located in the overlaying land use vector polygon with an area of larger than 1 km^2 (close to 30 × 30 arc-seconds), the spatial resolution of land reclamation histories and soil properties in the plot were consistent and could be compared and analyzed.

The data was provided by the Cold and Arid Regions Science Data Center at Lanzhou [53] and downloaded from their website (http: //westdc.westgis.ac.cn/data). The SOC was then calculated from SOM values, using the Bemmelen index of 0.58, which was widely used in previous studies [54,55]. Since the soil data set consists of eight vertical layers with different depths, the SOC and TN at the study depth of 0–20 cm were estimated by a weighted depth method provided by Yan, et al. [19].

2.5. Statistical Analysis

Statistical analysis was carried out using the SPSS18.0 statistical package (SPSS Inc., New York, NY, USA). Differences in variation of SOC, TN, and C:N ratio for the 270 soil plots between ca. 1978 and 2015 were calculated using paired-samples *t*-tests, using 270 pairs in the same spatial coordinates. The significance of differences in the levels and variations of SOC, TN, and C:N ratio in plots with different land reclamation histories were tested with one-way analysis of variance (ANOVA) and least-significant-difference (LSD) methods, followed by Tamhane post hoc tests ($p < 0.05$).

3. Results

3.1. Land Reclamation from 1978 to 2015 in Typical Areas of the Tarim Basin

The classification accuracy of land use maps was assessed by a total of 2000 ground-reference data-points in the five sub-basins, including the field validation investigations with GPS in 2015, and points from high resolution images from Google Earth in 2010 and 2015. The ground-reference data-points were randomly collated by the stratified land use types. This assessment found a high classification accuracy, with kappa coefficients of 0.913 and 0.906 and an overall accuracy of 92.30% and 91.35% for the 2010 and 2015 data, respectively. Since the visual interpretation methods used for other year were similar, it was assumed that the classification accuracies of land use maps for 1990 and 2000 were similar to those for the 2010 and 2015 maps. Similar to previous studies using the multiple Landsat MSS/TM images for long time series of land use change detection [56,57], the land use map in 1978 interpreted by the same executor with the same satellite and similar images could be relied upon. Figure 3 shows the land use maps for 1978 and 2015.

Croplands accounted for 10.37% of the total area of the five sub-basins of the Tarim Basin in 2015. Areas of croplands were 1.72×10^4, 1.89×10^4, 2.11×10^4, 2.72×10^4, and 3.23×10^4 km^2 for the five time nodes of 1978, 1990, 2000, 2010, and 2015, respectively. The cropland area nearly doubled with a significantly increased rate of 88.37% from 1978 to 2015. Most of them mainly expanded in the vicinity of their existing areas. The assarted croplands were converted from grassland, forest, bare land, and water body, with the proportions of 70.35%, 16.89%, 9.56%, and 3.20%, respectively.

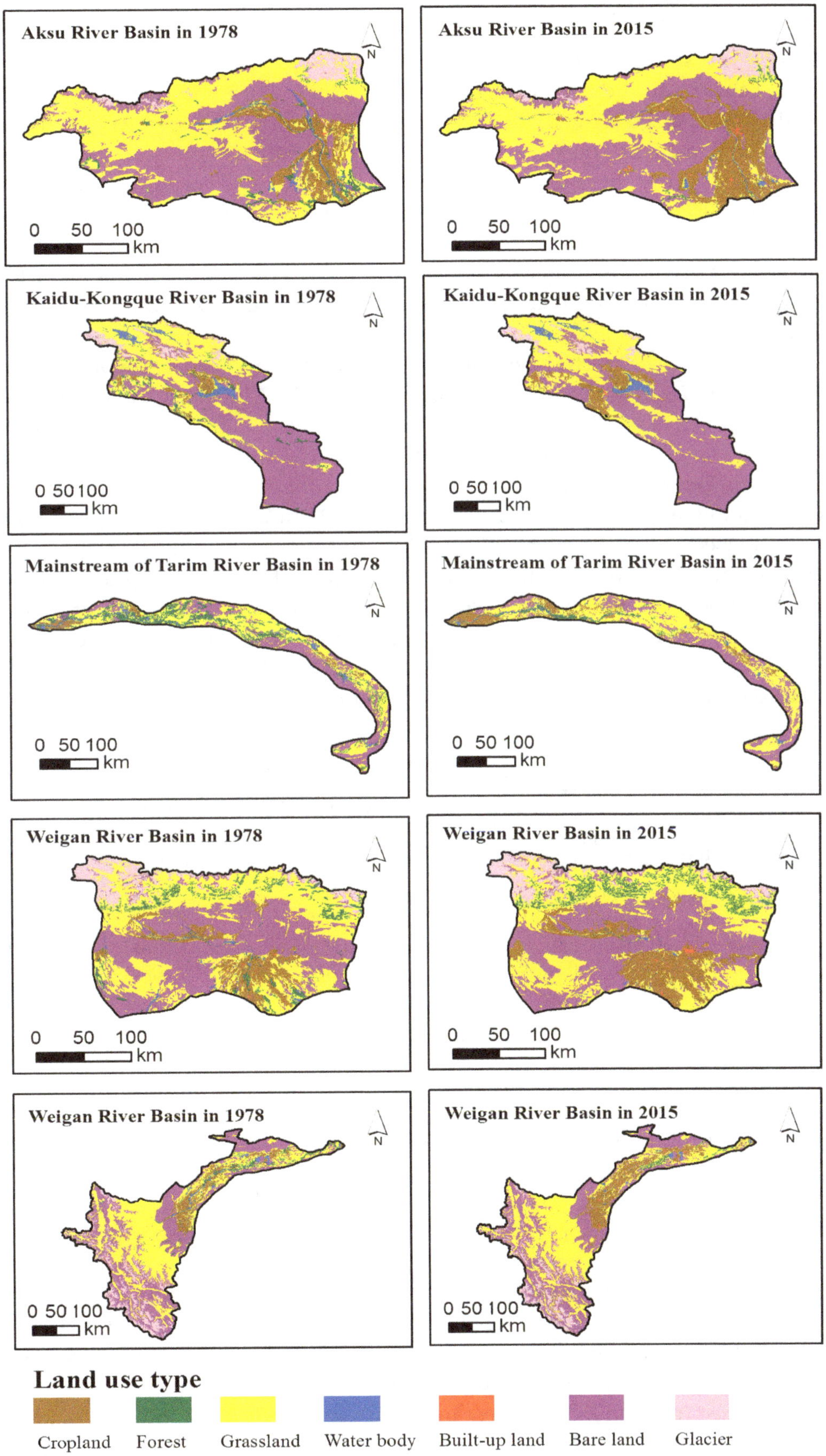

Figure 3. Land use maps of 1978 and 2015 in typical areas of the Tarim Basin.

3.2. *Status and Change in Soil C, N and C:N*

Table 1 shows the statistical characteristics of SOC, TN, and C:N ratios in the Tarim Basin. The mean values of these parameters increased from ca.1978 to 2015 with varying magnitudes, but do not show a significant change ($p < 0.05$). The mean SOC content is 6.61 g/kg, with an increase of 0.17 g/kg from ca.1978 to 2015. The TN contents, meanwhile, increased from 0.63 g/kg to 0.65 g/kg. It was found that the TN content is highly correlated with the SOC content in 270 soil sample plots, showing high Pearson's correlation coefficients of 0.938 and 0.925 in ca. 1978 and 2015, respectively. Although the range of contents of SOC and TN increased from ca. 1978 to 2015 (Table 1), their standard deviations decreased. With cultivation practices, the standard deviation of SOC and TN in 270 plots show increases from 1.96 g/kg and 0.22 g/kg, respectively, in ca. 1978, to 2.90 g/kg and 0.28 g/kg, respectively, in 2015.

The change in statistical characteristics for the C:N ratio is similar to those of SOC and TN content. The mean C:N ratio is 10.22 in 2015, ranging from 6.11 to 17.25. Most of these values fall in the range from 9.0 (10% quantile) to 11.5 (90% quantile). A larger increase in SOC over TN resulted in a slight increase of 0.11 in the C:N ratio from circa (ca.) 1978 to 2015, with no significant change at the 95% confidence level. Long-term cultivation has tended to homogenize the soil C:N ratio, resulting in a decrease in standard deviation, from 1.73 in ca. 1978 to 1.21 in 2015, separating the soil plots as two groups, i.e., the increased values from ca. 1978 to 2015 and decreased values at this period. Using the paired-samples *t*-test, two-third of the total observations showed an increasing trend in SOC and TN at a significant level of 99%, and other one-third showed a significant decrease. In contrast, about 59.26% of the total observations showed an increasing trend in C:N ratio at a non-significant level of 95%, with the remainder showing a decrease at a significant level of 95%. Thus, a deep exploration of change in soil C and N based on the different land reclamation histories is called for.

Table 1. Statistical characteristics of soil properties.

		Mean	Standard Deviation	Maximum	Minimum
Soil bulk density (g cm^{-3})	ca.1978	1.30	0.18	1.75	0.93
	2015	1.29	0.16	1.61	1.05
SOC content (g/kg)	ca.1978	6.44	2.90	16.85	3.42
	2015	6.61	1.96	17.65	2.26
TN content (g/kg)	ca.1978	0.63	0.28	1.67	0.29
	2015	0.65	0.22	1.76	0.19
C:N ratio	ca.1978	10.11	1.73	14.90	7.33
	2015	10.22	1.21	17.25	6.11

3.3. *Effects of Land Reclamation History on SOC and TN*

The selected sites presented significant differences ($p < 0.05$) in SOC and TN contents in 2015, depending on their prior land uses (Figure 4a,b). The SOC and TN contents of cropland soils for lands assarted before 1978 were 7.29 g/kg and 0.74 g/kg, respectively. These values were significantly higher than those for soils from land with prior use as grassland and forest assarted in the study period. The SOC content in previously forested land (6.21 g/kg) is slightly higher than that for land that was previously grassland (5.93 g/kg), and the TN content of soil from forest (0.59 g/kg) is close to that from grassland (0.58 g/kg). Comparing the SOC and TN contents in ca. 1978 and 2015, differences in change values according to prior land use were also significant ($p < 0.05$) (Figure 4c,d). In 2015, SOC and TN content in those sites previously used as cropland had significantly increased, as indicated by the paired-samples *t*-test ($p < 0.05$), the largest increases being 1.59 g/kg and 0.15 g/kg, respectively. Sites where prior land use was grassland showed an increase in SOC content of 0.17 g/kg and a slight decrease in TN content. In contrast, the SOC and TN content in previously forested land showed a significant decrease of 4.18 g/kg and 0.30 g/kg, respectively.

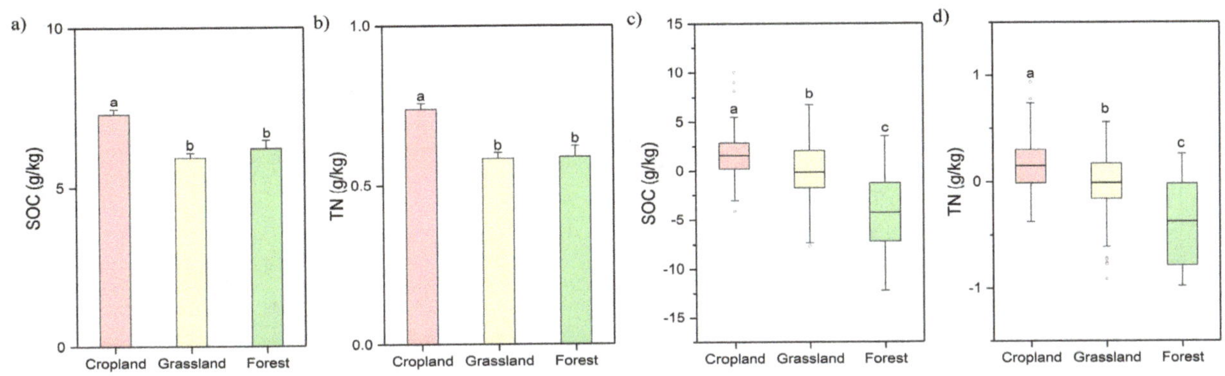

Figure 4. Soil C and N for different prior land uses: (**a**) Soil organic carbon content in 2015, (**b**) Total nitrogen content in 2015, (**c**) Change of soil organic carbon content, (**d**) Change of total nitrogen content. Note: 1. Group differences after one-way ANOVA ($p < 0.05$) was indicated by different lowercase letters. 2. Bar length gives the mean value, vertical whisker of bar for each column indicate standard errors of the mean. 3. Line within the boxes gives the median value, box means the 25th and 75th percentile, whisker of box represents 1.5 times the length of the box from either end of the box (1.5 times the interquartile range), circle represents outliers and extremes. The same as below.

With the increasing reclamation age, the accumulation of SOC and TN in soil from land derived from grassland and forest increased (Figure 5a–d). The mean SOC and TN content in the newly asserted soils were the lowest, and were significantly lower than those in the old asserted soils ($p < 0.05$). In addition, SOC and TN content for old asserted soils were lower than the reference cropland asserted before 1978, but were not significantly different.

The SOC content in land that was previously grassland and forest increased significantly from 5.17 and 5.23 g/kg in the newly asserted soils to 6.60 and 6.95 g/kg in the old asserted soils, respectively (Figure 5a,c). Similarly, the TN content in land that was previously grassland and forest increased significantly from 0.51 and 0.47 g/kg in the newly asserted soils to 0.66 and 0.66 g/kg in the old asserted soils, respectively (Figure 5b,d).

An increase in the reclamation age resulted in an accumulation in SOC and TN at a significance level of 95% (Figure 5e–h). Compared to the initial SOC and TN levels prior to reclamation, changes in content for new, young, and medium asserted soils were negative. The decreases in the newly asserted soils were the largest, and increasing reclamation age resulted in increases in SOC and TN stocks. A slight decrease in SOC and TN content in the sites which were previously grassland was observed, with a larger decrease for previously forested sites.

For example, the changes in SOC content for new, young, and medium croplands derived from grassland are −0.48, −0.32, and −0.07 g/kg, respectively; and those for croplands derived from forest are −6.03, −3.90, and −4.19 g/kg, respectively. With a longer cultivation age, the change in SOC and TN in old croplands derived from grassland was positive, with mean values of 0.89 and 0.09 g/kg, respectively, which were significantly higher than those of newly asserted croplands with mean changes of −0.48 and −0.06 ($p < 0.05$) (Figure 5e,f). In contrast, changes in SOC and TN in soils from old croplands derived from forested lands were still negative, with mean values of −2.91 and −0.25 g/kg, respectively (Figure 5g,h).

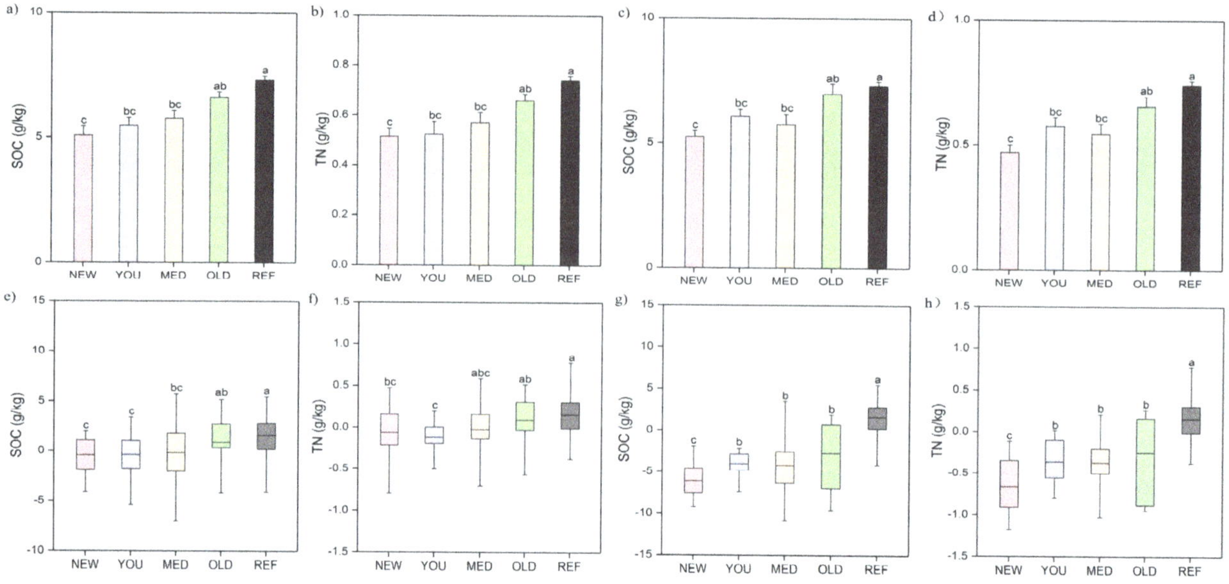

Figure 5. Impact of reclamation ages on soil C and N: (**a**) Soil organic carbon content in 2015 assarted from grassland, (**b**) Total nitrogen content in 2015 assarted from grassland, (**c**) Soil organic carbon content in 2015 assarted from forest, (**d**) Total nitrogen content in 2015 assarted from forest, (**e**) Change of soil organic carbon content assarted from grassland, (**f**) Change of total nitrogen content assarted from grassland, (**g**) Change of soil organic carbon content assarted from forest, (**h**) Change of total nitrogen content assarted from forest. NEW = newly assarted (1–5 years), YOU = young assarted (6–15 years), MED = medium assarted (16–25 years), OLD = old assarted (26–37 years), REF = referenced cropland assarted before 1978.

3.4. Effects of Land Reclamation History on Soil C:N Ratio

Compared to the difference in SOC and TN levels for different prior land uses, the C:N ratios showed opposite characteristics (Figure 6a). The C:N ratio in previously forested land (10.85) was significantly higher than where the prior land use was cropland or grassland (9.96 and 10.29, respectively). Comparing the changes with the three prior land uses in the past four decades, the C:N ratios showed non-significant variations ($p < 0.05$) (Figure 6b), of 0.05, 0.32, and −0.25 for sites with prior land use as cropland, grassland, and forest, respectively. The C:N ratio for sites with prior land use as cropland increased slightly, from 9.91 to 9.96 in two periods of ca. 1978 and 2015, with a non-significant change as indicated by the paired-samples t-test ($p < 0.05$). In contrast, the C:N ratio for soils from grassland sites increased from 9.97 to 10.29, and the ratio for forest sites decreased from 11.10 to 10.85.

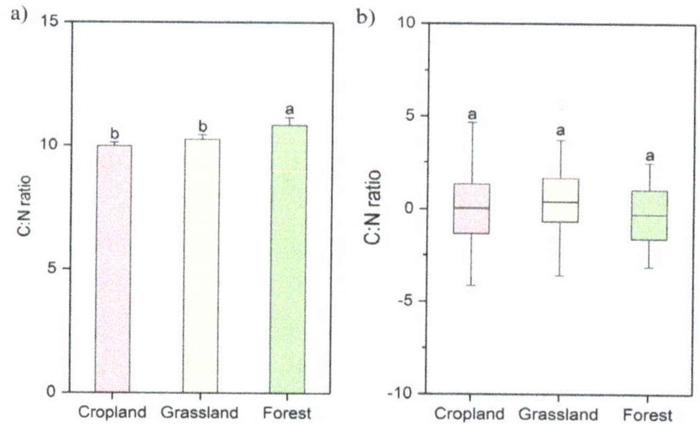

Figure 6. Soil C:N ratio for different prior land uses: (**a**) Values in 2015, (**b**) Changes from ca.1978 to 2015.

Although the SOC and TN changed significantly with increasing reclamation age, the change in soil C:N ratios shows different trends (Figure 7a,b). With an increase in the reclamation age, C:N ratios with prior land use as grassland significantly decreased from 10.79 in the newly assarted soils to 9.92 in the old assarted soils at a 95% confidence level (Figure 7a). The C:N ratio of old assarted soils was close to the C:N ratio of 9.96 in the reference cropland soils assarted prior to 1978. In contrast, relatively little variation, with a non-significant difference in C:N ratios, was found for soils from previously forested land (Figure 7b). C:N ratios from the previously forested land are 10.98, 11.01, 10.83, and 10.87 for the new, young, medium, and old assarted soils, respectively.

Changes in C:N ratios in two periods of ca. 1978 and 2015 were found to have relatively less variation (Figure 7c,d). The changes in C:N ratios for grassland tended to be positive, but that increased magnitude tended to diminish with increasing reclamation age (Figure 7c). The mean increase in C:N ratio was 0.89 for newly assarted soils, which is significantly higher than those in medium and old assarted soils, with mean values of 0.28 and 0.06 at a 95% confidence level, respectively. Moreover, for previously forested land, the change in the magnitude of the C:N ratios did not tend to vary significantly with an increase in the reclamation age, being −0.43, −0.21, −0.23, −0.15 for new, young, medium, and old assarted soils, respectively (Figure 7d).

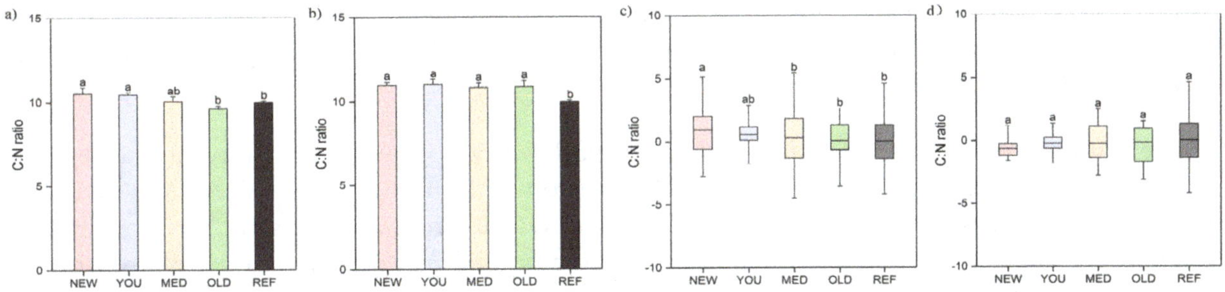

Figure 7. Impact of reclamation ages on soil C:N ratio: (**a**) Values in 2015 assarted from grassland, (**b**) values in 2015 assarted from forest, (**c**) Changes assarted from grassland, (**d**) Changes assarted from forest. NEW = newly assarted (1–5 years), YOU = young assarted (6–15 years), MED = medium assarted (16–25 years), OLD = old assarted (26–37 years), REF = referenced cropland assarted before 1978. NEW = newly assarted (1–5 years), YOU = young assarted (6–15 years), MED = medium assarted (16–25 years), OLD = old assarted (26–37 years), REF = referenced cropland assarted before 1978.

4. Discussion

Land use change history is closely related to soil C and N dynamics [1,29]. Numerous studies have explored the relationship between soil C and N, however, the balance between them and their coupling relationship remains uncertain [55]. This study investigated how large-scale cultivation has affected the SOC, TN, and C:N ratio. Using remote sensing and GIS techniques, this research identified the land use change history, which indicated that cropland has nearly doubled in typical regions of the Tarim Basin over the last four decades. According to different land reclamation histories, large-scale soil sampling was conducted to quantify the spatial variability of soil C and N. The multiple temporal changes of land use and soil proprieties using a repeated soil sampling strategy allowed the investigation of the time effects of reclamation on the soil C:N ratio [22,58]. Results found that the changes in soil C:N ratios are significantly different from the changes in SOC and TN under different land reclamation histories. This research can serve as a better reference for the soil C and N balance in ecological interactions and processes [59].

Land reclamation from grassland and forest in the study area resulted in SOC and TN loss during the initial reclamation stages, but these recovered with increasing reclamation age (Figure 5), which is in accordance with previous studies [18,60,61]. Anthropogenic disturbances frequently destroy the initial soil structure and accelerate the mineralization and decomposition process of SOM, which can intensify losses of C and N [62–64]. In the study area, the scarce water resources and low rate of bioaccumulation limited the growth of uncultivated grassland, primarily desert steppe [65], which resulted in a low level of initial SOC and TN [47,66]. After reclamation, the conditions of soil moisture and biomass growth improved in the arid regions; processes which can increase the above- and below-ground biomass, and result in soil C and N accumulation with increasing reclamation age [67]. The high-efficiency agricultural production in the Tarin Basin increases the crop yield with enhanced soil C and N input through crop residues and also enhances the anthropogenic mellowing process of soils to improve the soil structure and function, which further promotes the accumulation of soil C and N [68]. Improved agricultural practices, including applying organic manure, and conservative tillage measures are used in oasis agriculture, which enhances the soil C and N storage by providing water, organic amendments, and increased crop residue is returned to the soil [19,69,70].

C and N cycling are tightly coupled in ecosystems [71] and TN change in soils is generally assumed to follow variation in SOC, since both elements are bound into organic compounds [72,73]. The high correlation between these parameters was confirmed in this study. Land use change and management practices can alter C and N bio-geochemical cycling, and the change in soil C:N ratio can be used to determine the relative input and output of SOC and TN [23]. This study found that land reclamation history change had an opposite effect on C:N ratio, compared to the trends observed for SOC and TN. The magnitude of changes observed for C:N ratios were also smaller than those observed for SOC and TN (Figures 4–7). This implied that the change in C:N ratio is complex and does not show consistency with either C or N variations.

Different land use regulates C and N source availability, soil microbial activities, and therefore litter decomposition rates [74–76]. In general, forested soils had a higher C:N ratio, which is to be expected since trees take up more N and store it within their biomass rather than in the soil, and also, microbial decomposers of forested soils have higher C:N ratios than cultivated soils [3,25]. This resulted in a significant decrease in C:N ratios when forests were converted to croplands (Figure 6). In contrast, the conversion from grassland to cropland resulted in an increase in C:N ratios but a decrease in SOC and TN (Figures 4 and 6) in the initial reclamation process, indicating a greater loss of N compared to that of SOC. This may be partly explained by baseline effects [58]. In the Tarim Basin, the uncultivated grassland, primarily desert steppe, was limited by water resources and a lower rate of bioaccumulation [65], resulting in a lower level of initial SOC than TN [47,66]. Cultivation with adequate water supplies and fertilizer application would significantly improve soil physicochemical properties [77]. Land reclamation enhanced the C input of the litter returned to the soil and sped up nutrient cycling, resulting in an increase in the soil C:N ratio [23,78]. This, in turn, can result in relatively lower SOC losses, and more SOC accumulation compared to TN to increase the C:N ratio at the initial reclamation stage.

Moreover, as the age of cultivated soils increased, the different accumulation rates of C and N in the soils resulted in a change of C:N ratio [79]. With increasing reclamation age, it was found that the increases in C:N ratios for sites previously used as grassland tended to be less (Figure 7c). The increased N accumulation in the soil may be partly due to the overuse of N fertilizer, an effective way to increase crop biomass and yield, enhancing the N deposition in soils [80]. An experimental site receiving synthetic N fertilization over a 40–50 year period indicated a net decline in soil C [81]. Thus, the C:N ratio decreased according to the smaller change rate of SOC accumulation than N deposition. In addition, C:N ratios from previously forested land stayed relatively stable, indicating similar accumulation rates and a relative balance between SOC and TN. The variations in soil C:N ratio are not only caused by C and N inputs, but also by decomposition in soils [11]. Soil micro-organisms in the previously forested soils may enhance nitrogen-induced increases in the C uptake in the soil [82]. The underlying mechanisms of these variations and the longer time monitoring of C:N ratios require further study.

The global mean value of soil C:N ratios (0–30 cm) in cropland is 12.5 [2], which is much higher than the mean level for our study area (10.22), and the highest value for sites derived from forest (10.85). In addition, the C:N ratios for sites derived from grassland with low cropland ages tended to decrease with increasing reclamation age. A reduction in the soil C:N ratio could disrupt the balance between soil C and N cycling and cause loss of N through leaching and denitrification processes [1,22,83]. These results highlight the importance of effective agricultural management for the soils in the study area. Although the change in C:N ratios for cropland assarted before 1978 has remained relatively stable over the last four decades, it was found that over 10% of these C:N ratios were larger than 12.50, indicating that it is possible to increase the C:N ratio with the use of sustainable agricultural management. For example, no-tillage can increase soil C:N ratios as compared to the use of traditional pillow tillage [1,84,85]. In addition, using N fertilizer more efficiently and decreasing the overall amounts used would effectively decrease the input of N to soils [80,86]. Management efforts in the region should focus on how to enhance the accumulation of soil C and N, and also to better understand their coupling characteristics in order that N deposition might be matched with C sequestration [87].

5. Conclusions

To reveal the effect of land use changes on the balance of soil C and N, a purpose designed sampling strategy, along with the gathering of information on different land use change histories, was conducted in representative areas within the Tarim Basin. Based on the visual interpretation of data from remote sensing, cropland area nearly doubled in four decades in the study area, where over 70% of new cropland was assarted mainly from grasslands, a small fraction from forest, and few from other land use types. Using GIS spatial analysis techniques, three prior land uses and five types of cropland age were identified. The status and change in soil C, N, and their fractions were investigated under the different land reclamation histories. Results found that the prior land use significantly influenced the soil C and N. Meanwhile, the relative magnitudes of C:N ratios for the various land uses were,: forest > grassland > cropland. For soils assarted before 1978, SOC and TN significantly increased, but the C:N ratio increased slightly by 0.05 without a significant difference. An increase was found in the SOC, TN, and C:N ratio where prior land use was grassland, but these decreased in soils derived from forest between 1978 and 2015. The change in magnitudes of SOC and TN with a significant difference between three prior land uses is larger than that C:N ratio with a non-significant difference. SOC and TN decreased in the initial stage of the reclamation process and did not recover in the short term. After 30 years of cultivation, the decrease in initial SOC and TN values for land recovered from grassland had recovered. SOC and TN loss from land derived from forest, however, was not completely recovered even after 40 years of agricultural management. With increasing reclamation age, the increase in C:N ratios of soils from land derived from grassland was less, but remained relatively stable for land derived from forest. The findings in this study indicated near opposite effects of land reclamation history on the soil C and N and C:N ratios. Differences in the variation of the soil C:N ratio are determined by the relative input and output of SOC and TN under the different land reclamation processes and associated cultivation practices. Since the C:N ratios are much lower than the global average and show a decreasing trend in soils derived from grasslands, the application of sustainable agricultural management is suggested to increase not only SOC and TN, but also the C:N ratio, matching N deposition with the currently larger amount of carbon sequestration.

Author Contributions: E.X., H.Z. and Y.X. conceived and designed the experiments. E.X. and Y.X. analyzed the data. E.X. and H.Z. wrote and revised the paper.

Acknowledgments: We are thankful to the National Natural Science Foundations of China for their financial support of the present study and the comments and suggestions of the editor and reviewers.

References

1. Liu, M.; Ussiri, D.A.; Lal, R. Soil organic carbon and nitrogen fractions under different land uses and tillage practices. *Commun. Soil Sci. Plant Anal.* **2016**, *47*, 1528–1541. [CrossRef]
2. Xu, X.; Thornton, P.E.; Post, W.M. A global analysis of soil microbial biomass carbon, nitrogen and phosphorus in terrestrial ecosystems. *Glob. Ecol. Biogeogr.* **2013**, *22*, 737–749. [CrossRef]
3. Murty, D.; Tarinm, M.U.; Mcmurtrie, R.E.; Mcgilvray, H. Does conversion of forest to agricultural land change soil carbon and nitrogen? A review of the literature. *Glob. Chang. Biol.* **2002**, *8*, 105–123. [CrossRef]
4. Stiglitz, R.; Mikhailova, E.; Sharp, J.; Post, C.; Schlautman, M.; Gerard, P.; Cope, M. Predicting Soil Organic Carbon and Total Nitrogen at the Farm Scale Using Quantitative Color Sensor Measurements. *Agronomy* **2018**, *8*, 212. [CrossRef]
5. Huang, B.; Sun, W.; Zhao, Y.; Zhu, J.; Yang, R.; Zou, Z.; Ding, F.; Su, J. Temporal and spatial variability of soil organic matter and total nitrogen in an agricultural ecosystem as affected by farming practices. *Geoderma* **2007**, *139*, 336–345. [CrossRef]
6. Wiesmeier, M.; Hübner, R.; Barthold, F.; Spörlein, P.; Geuß, U.; Hangen, E.; Reischl, A.; Schilling, B.; von Lützow, M.; Kögel-Knabner, I. Amount, distribution and driving factors of soil organic carbon and nitrogen in cropland and grassland soils of southeast Germany (Bavaria). *Agric., Ecosyst. Environ.* **2013**, *176*, 39–52. [CrossRef]
7. Kindler, R.; Siemens, J.; Kaiser, K.; Walmsley, D.C.; Bernhofer, C.; Buchmann, N.; Cellier, P.; Eugster, W.; Gleixner, G.; Grũnwald, T. Dissolved carbon leaching from soil is a crucial component of the net ecosystem carbon balance. *Glob. Change Biol.* **2011**, *17*, 1167–1185. [CrossRef]
8. Lovett, G.M.; Weathers, K.C.; Arthur, M.A. Control of Nitrogen Loss from Forested Watersheds by Soil Carbon: Nitrogen Ratio andTree Species Composition. *Ecosystems* **2002**, *5*, 712–718. [CrossRef]
9. Cools, N.; Vesterdal, L.; De Vos, B.; Vanguelova, E.; Hansen, K. Tree species is the major factor explaining C:N ratios in European forest soils. *For. Ecol. Manag.* **2014**, *311*, 3–16. [CrossRef]
10. Klemedtsson, L.; Von Arnold, K.; Weslien, P.; Gundersen, P. Soil CN ratio as a scalar parameter to predict nitrous oxide emissions. *Glob. Chang. Biol.* **2005**, *11*, 1142–1147. [CrossRef]
11. Cong, R.; Wang, X.; Xu, M.; Zhang, W.; Xie, L.; Yang, X.; Huang, S.; Wang, B. Dynamics of soil carbon to nitrogen ratio changes under long-term fertilizer addition in wheat-corn double cropping systems of China. *Eur. J. Soil Sci.* **2012**, *63*, 341–350. [CrossRef]
12. Obu, J.; Lantuit, H.; Myers-Smith, I.; Heim, B.; Wolter, J.; Fritz, M. Effect of terrain characteristics on soil organic carbon and total nitrogen stocks in soils of Herschel Island, Western Canadian Arctic. *Permaf. Periglac. Processes* **2017**, *28*, 92–107. [CrossRef]
13. Lozano-García, B.; Muñoz-Rojas, M.; Parras-Alcántara, L. Climate and land use changes effects on soil organic carbon stocks in a Mediterranean semi-natural area. *Sci. Total Environ.* **2017**, *579*, 1249–1259. [CrossRef] [PubMed]
14. Yimer, F.; Ledin, S.; Abdelkadir, A. Changes in soil organic carbon and total nitrogen contents in three adjacent land use types in the Bale Mountains, south-eastern highlands of Ethiopia. *For. Ecol. Manag.* **2007**, *242*, 337–342. [CrossRef]
15. Soosaar, K.; Mander, Ü.; Maddison, M.; Kanal, A.; Kull, A.; Lõhmus, K.; Truu, J.; Augustin, J. Dynamics of gaseous nitrogen and carbon fluxes in riparian alder forests. *Ecol. Eng.* **2011**, *37*, 40–53. [CrossRef]
16. Edmondson, J.L.; Davies, Z.G.; McCormack, S.A.; Gaston, K.J.; Leake, J.R. Land-cover effects on soil organic carbon stocks in a European city. *Sci. Total Environ.* **2014**, *472*, 444–453. [CrossRef] [PubMed]
17. Mishra, U.; Ussiri, D.A.; Lal, R. Tillage effects on soil organic carbon storage and dynamics in Corn Belt of Ohio USA. *Soil Till. Res.* **2010**, *107*, 88–96. [CrossRef]
18. Kopittke, P.M.; Dalal, R.C.; Finn, D.; Menzies, N.W. Global changes in soil stocks of carbon, nitrogen, phosphorus, and sulphur as influenced by long-term agricultural production. *Glob. Chang. Boil.* **2017**, *23*, 2509–2519. [CrossRef]

19. Yan, X.; Cai, Z.; Wang, S.; Smith, P. Direct measurement of soil organic carbon content change in the croplands of China. *Glob. Chang. Biol.* **2011**, *17*, 1487–1496. [CrossRef]
20. Maaroufi, N.I.; Nordin, A.; Hasselquist, N.J.; Bach, L.H.; Palmqvist, K.; Gundale, M.J. Anthropogenic nitrogen deposition enhances carbon sequestration in boreal soils. *Glob. Chang. Boil.* **2015**, *21*, 3169–3180. [CrossRef]
21. Alvarez, R. A review of nitrogen fertilizer and conservation tillage effects on soil organic carbon storage. *Soil Use Manag.* **2005**, *21*, 38–52. [CrossRef]
22. Gao, Y.; He, N.; Yu, G.; Chen, W.; Wang, Q. Long-term effects of different land use types on C, N, and P stoichiometry and storage in subtropical ecosystems: A case study in China. *Eco. Eng.* **2014**, *67*, 171–181. [CrossRef]
23. Brady, N.; Weil, R. *The Nature and Properties of Soils*; Pearson Prentice Hall: Upper Saddle River, NJ, USA, 2008.
24. Liu, X.; Ma, J.; Ma, Z.-W.; Li, L.-H. Soil nutrient contents and stoichiometry as affected by land-use in an agro-pastoral region of Northwest China. *Catena* **2017**, *150*, 146–153. [CrossRef]
25. Livesley, S.J.; Ossola, A.; Threlfall, C.G.; Hahs, A.K.; Williams, N.S.G. Soil Carbon and Carbon/Nitrogen Ratio Change under Tree Canopy, Tall Grass, and Turf Grass Areas of Urban Green Space. *J. Environ. Qual.* **2016**, *45*, 215–223. [CrossRef] [PubMed]
26. Hobbie, S.E.; Reich, P.B.; Oleksyn, J.; Ogdahl, M.; Zytkowiak, R.; Hale, C.; Karolewski, P. Tree species effects on decomposition and forest floor dynamics in a common garden. *Ecology* **2006**, *87*, 2288–2297. [CrossRef]
27. Li, Z.; Liu, C.; Dong, Y.; Chang, X.; Nie, X.; Liu, L.; Xiao, H.; Lu, Y.; Zeng, G. Response of soil organic carbon and nitrogen stocks to soil erosion and land use types in the Loess hilly–gully region of China. *Soil Till. Res.* **2017**, *166*, 1–9. [CrossRef]
28. Post, W.M.; Kwon, K.C. Soil carbon sequestration and land-use change: Processes and potential. *Glob. Chang. Boil.* **2000**, *6*, 317–327. [CrossRef]
29. Sartori, F.; Lal, R.; Ebinger, M.H.; Parrish, D.J. Potential soil carbon sequestration and CO2 offset by dedicated energy crops in the USA. *Crit. Rev. Plant Sci.* **2006**, *25*, 441–472. [CrossRef]
30. Hulme, M. Recent climatic change in the world's drylands. *Geophys. Res. Lett.* **1996**, *23*, 61–64. [CrossRef]
31. Reynolds, J.F.; Smith, D.M.S.; Lambin, E.F.; Turner, B.; Mortimore, M.; Batterbury, S.P.; Downing, T.E.; Dowlatabadi, H.; Fernández, R.J.; Herrick, J.E. Global desertification: Building a science for dryland development. *Science* **2007**, *316*, 847–851. [CrossRef]
32. Qiu, L.; Wei, X.; Zhang, X.; Cheng, J.; Gale, W.; Guo, C.; Long, T. Soil organic carbon losses due to land use change in a semiarid grassland. *Plant Soil* **2012**, *355*, 299–309. [CrossRef]
33. Lal, R. Carbon sequestration in dryland ecosystems. *Environ. Manag.* **2004**, *33*, 528–544. [CrossRef] [PubMed]
34. Mekki, I.; Jacob, F.; Marlet, S.; Ghazouani, W. Management of groundwater resources in relation to oasis sustainability: The case of the Nefzawa region in Tunisia. *J. Environ. Manag.* **2013**, *121*, 142–151. [CrossRef] [PubMed]
35. FAO-UNESCO. *Soil Map of the World: Revised Legend (with Corrections and Updates), World Soil Resources Report 60*; FAO: Rome, Italy, 1988.
36. Eswaran, H.; Van Den Berg, E.; Reich, P. Organic carbon in soils of the world. *Soil Sci. Soc. Am. J.* **1993**, *57*, 192–194. [CrossRef]
37. Su, Y.-Z.; Wang, F.; Zhang, Z.-H.; Du, M.-W. Soil Properties and Characteristics of Soil Aggregate in Marginal Farmlands of Oasis in the Middle of Hexi Corridor Region, Northwest China. *Agric. Sci. China* **2007**, *6*, 706–714. [CrossRef]
38. Gong, Z.; Zhang, G.; Luo, G. Diversity of Anthrosols in China. *Pedosphere* **1999**, *9*, 193–204.
39. Liu, F.; Zhang, H.; Qin, Y.; Dong, J.; Xu, E.; Yang, Y.; Zhang, G.; Xiao, X. Semi-natural areas of Tarim Basin in northwest China: Linkage to desertification. *Sci. Total Environ.* **2016**, *573*, 178–188. [CrossRef] [PubMed]
40. Chen, Y. *Water Resources Research in Northwest China*; Springer Science & Business Media: Dordrecht, The Netherlands, 2014.
41. Liu, J.; Kuang, W.; Zhang, Z.; Xu, X.; Qin, Y.; Ning, J.; Zhou, W.; Zhang, S.; Li, R.; Yan, C. Spatiotemporal characteristics, patterns, and causes of land-use changes in China since the late 1980s. *J. Geogr. Sci.* **2014**, *24*, 195–210. [CrossRef]
42. Bai, L.Y.; Feng, J.Z.; Ma, Y.X.; Ran, Q.Y.; Wang, K.; Zhao, Y. Analysis of spatial pattern Change of LU/LC over the upper Tarim River region since 1990 using remote sensing data. *IOP Conference Series: Earth Environ. Sci.* **2017**, *57*, 012038. [CrossRef]

43. Song, X.-P.; Hansen, M.C.; Stehman, S.V.; Potapov, P.V.; Tyukavina, A.; Vermote, E.F.; Townshend, J.R. Global land change from 1982 to 2016. *Nature* **2018**, *560*, 639. [CrossRef]
44. Wichern, F.; Luedeling, E.; Müller, T.; Joergensen, R.G.; Buerkert, A. Field measurements of the CO_2 evolution rate under different crops during an irrigation cycle in a mountain oasis of Oman. *Appl. Soil Ecol.* **2004**, *25*, 85–91. [CrossRef]
45. Liu, Y.; Fang, F.; Li, Y. Key issues of land use in China and implications for policy making. *Land Use Policy* **2014**, *40*, 6–12. [CrossRef]
46. Wang, T.; Yan, C.; Song, X.; Xie, J. Monitoring recent trends in the area of aeolian desertified land using Landsat images in China's Xinjiang region. *ISPRS J. Photogramm. Remote Sens.* **2012**, *68*, 184–190. [CrossRef]
47. Wang, Y.; Luo, G.; Zhao, S.; Han, Q.; Li, C.; Fan, B.; Chen, Y. Effects of arable land change on regional carbon balance in Xinjiang. *Acta Geogr. Sin.* **2014**, *69*, 110–120. (In Chinese) [CrossRef]
48. Yaning, C.; Changchun, X.; Xingming, H.; Weihong, L.; Yapeng, C.; Chenggang, Z.; Zhaoxia, Y. Fifty-year climate change and its effect on annual runoff in the Tarim River Basin, China. *Quat. Int.* **2009**, *208*, 53–61. [CrossRef]
49. Zhang, Y.; Chen, Y.; Pan, B. Distribution and floristics of desert plant communities in the lower reaches of Tarim River, southern Xinjiang, People's Republic of China. *J. Arid Environ.* **2005**, *63*, 772–784. [CrossRef]
50. Olson, K.R.; Al-Kaisi, M.M. The importance of soil sampling depth for accurate account of soil organic carbon sequestration, storage, retention and loss. *CATENA* **2015**, *125*, 33–37. [CrossRef]
51. Nelson, D.W.; Sommers, L.E. Total carbon, organic carbon, and organic matter. *Methods Soil Anal. Part 3—Chem. Methods* **1996**, 961–1010.
52. Gallaher, R.; Weldon, C.; Boswell, F. A Semiautomated Procedure for Total Nitrogen in Plant and Soil Samples 1. *Soil Sci. Soc. Am. J.* **1976**, *40*, 887–889. [CrossRef]
53. Shangguan, W.; Dai, Y.; Liu, B.; Zhu, A.; Duan, Q.; Wu, L.; Ji, D.; Ye, A.; Yuan, H.; Zhang, Q.; et al. A China data set of soil properties for land surface modeling. *J. Adv. Model. Earth Syst.* **2013**, *5*, 212–224. [CrossRef]
54. Yang, Y.; Fang, J.; Smith, P.; Tang, Y.; Chen, A.; Ji, C.; Hu, H.; Rao, S.; Tan, K.; HE, J.S. Changes in topsoil carbon stock in the Tibetan grasslands between the 1980s and 2004. *Glob. Chang. Biol.* **2009**, *15*, 2723–2729. [CrossRef]
55. Zhou, Z.; Wang, C. Reviews and syntheses: Soil resources and climate jointly drive variations in microbial biomass carbon and nitrogen in China's forest ecosystems. *Biogeosciences* **2015**, *12*, 6751–6760. [CrossRef]
56. Renó, V.F.; Novo, E.M.; Suemitsu, C.; Rennó, C.D.; Silva, T.S. Assessment of deforestation in the Lower Amazon floodplain using historical Landsat MSS/TM imagery. *Remote Sens. Environ.* **2011**, *115*, 3446–3456. [CrossRef]
57. Vittek, M.; Brink, A.; Donnay, F.; Simonetti, D.; Desclée, B. Land cover change monitoring using Landsat MSS/TM satellite image data over West Africa between 1975 and 1990. *Remote Sens.* **2014**, *6*, 658–676. [CrossRef]
58. Chen, L.; Smith, P.; Yang, Y. How has soil carbon stock changed over recent decades? *Glob. Change Boil.* **2015**, *21*, 3197–3199. [CrossRef] [PubMed]
59. Sterner, R.W.; Elser, J.J. *Ecological Stoichiometry: The Biology of Elements from Molecules to the Biosphere*; Princeton University Press: Princeton, NJ, USA, 2002.
60. Houghton, R.; Skole, D.; Nobre, C.A.; Hackler, J.; Lawrence, K.; Chomentowski, W.H. Annual fluxes of carbon from deforestation and regrowth in the Brazilian Amazon. *Nature* **2000**, *403*, 301–304. [CrossRef] [PubMed]
61. Lal, R. Soil carbon dynamics in cropland and rangeland. *Environ. Pollut.* **2002**, *116*, 353–362. [CrossRef]
62. Gelaw, A.M.; Singh, B.; Lal, R. Soil organic carbon and total nitrogen stocks under different land uses in a semi-arid watershed in Tigray, Northern Ethiopia. *Agric., Ecosyst. Environ.* **2014**, *188*, 256–263. [CrossRef]
63. Berihu, T.; Girmay, G.; Sebhatleab, M.; Berhane, E.; Zenebe, A.; Sigua, G.C. Soil carbon and nitrogen losses following deforestation in Ethiopia. *Agron. Sustain. Dev.* **2017**, *37*, 1. [CrossRef]
64. Lozano-García, B.; Parras-Alcántara, L.; Cantudo-Pérez, M. Land use change effects on stratification and storage of soil carbon and nitrogen: Application to a Mediterranean nature reserve. *Agric., Ecosyst. Environ.* **2016**, *231*, 105–113. [CrossRef]
65. Shan, L.; Zhang, X.; Wang, Y.; Wang, H.; Yan, H.; Wei, J.; Xu, H. Influence of moisture on the growth and biomass allocation in Haloxylon ammodendron and Tamarix ramosissima seedlings in the shelterbelt along the Tarim Desert Highway, Xinjiang, China. *Chin. Sci. Bull.* **2008**, *53*, 93–101. [CrossRef]

66. Huang, Y.; Sun, W. Changes in topsoil organic carbon of croplands in mainland China over the last two decades. *Chin. Sci. Bull.* **2006**, *51*, 1785–1803. [CrossRef]
67. Li, X.-G.; Li, F.-M.; Rengel, Z.; Wang, Z.-F. Cultivation effects on temporal changes of organic carbon and aggregate stability in desert soils of Hexi Corridor region in China. *Soil Till. Res.* **2006**, *91*, 22–29. [CrossRef]
68. Shen, W.; Lin, X.; Gao, N.; Zhang, H.; Yin, R.; Shi, W.; Duan, Z. Land use intensification affects soil microbial populations, functional diversity and related suppressiveness of cucumber Fusarium wilt in China's Yangtze River Delta. *Plant Soil* **2008**, *306*, 117–127. [CrossRef]
69. Congreves, K.; Hooker, D.; Hayes, A.; Verhallen, E.; Van Eerd, L. Interaction of long-term nitrogen fertilizer application, crop rotation, and tillage system on soil carbon and nitrogen dynamics. *Plant Soil* **2017**, *410*, 113–127. [CrossRef]
70. Smith, P. Land use change and soil organic carbon dynamics. *Nutr. Cycl. Agroecosyst.* **2008**, *81*, 169–178. [CrossRef]
71. Luo, Y.; Hui, D.; Zhang, D. Elevated CO_2 stimulates net accumulations of carbon and nitrogen in land ecosystems: A meta-analysis. *Ecology* **2006**, *87*, 53–63. [CrossRef] [PubMed]
72. Manzoni, S.; Porporato, A. Soil carbon and nitrogen mineralization: Theory and models across scales. *Soil Biol. Biochem.* **2009**, *41*, 1355–1379. [CrossRef]
73. Yuan, J.; Ouyang, Z.; Zheng, H.; Xu, W. Effects of different grassland restoration approaches on soil properties in the southeastern Horqin sandy land, northern China. *Appl. Soil Ecol.* **2012**, *61*, 34–39. [CrossRef]
74. Fierer, N.; Craine, J.M.; McLauchlan, K.; Schimel, J.P. Litter quality and the temperature sensitivity of decomposition. *Ecology* **2005**, *86*, 320–326. [CrossRef]
75. Kemmitt, S.J.; Wright, D.; Goulding, K.W.; Jones, D.L. pH regulation of carbon and nitrogen dynamics in two agricultural soils. *Soil Biol. Biochem.* **2006**, *38*, 898–911. [CrossRef]
76. Pietri, J.A.; Brookes, P. Nitrogen mineralisation along a pH gradient of a silty loam UK soil. *Soil Biol. Biochem.* **2008**, *40*, 797–802. [CrossRef]
77. Singh Brar, B.; Singh, J.; Singh, G.; Kaur, G. Effects of Long Term Application of Inorganic and Organic Fertilizers on Soil Organic Carbon and Physical Properties in Maize–Wheat Rotation. *Agronomy* **2015**, *5*, 220–238. [CrossRef]
78. Su, Y.Z.; lin Zhao, H. Soil properties and plant species in an age sequence of Caragana microphylla plantations in the Horqin Sandy Land, north China. *Eco. Eng.* **2003**, *20*, 223–235. [CrossRef]
79. O'Brien, S.L.; Jastrow, J.D.; Grimley, D.A.; Gonzalez-Meler, M.A. Moisture and vegetation controls on decadal-scale accrual of soil organic carbon and total nitrogen in restored grasslands. *Glob. Chang. Biol.* **2010**, *16*, 2573–2588. [CrossRef]
80. Liu, X.; Zhang, Y.; Han, W.; Tang, A.; Shen, J.; Cui, Z.; Vitousek, P.; Erisman, J.W.; Goulding, K.; Christie, P. Enhanced nitrogen deposition over China. *Nature* **2013**, *494*, 459. [CrossRef] [PubMed]
81. Khan, S.A.; Mulvaney, R.L.; Ellsworth, T.R.; Boast, C.W. The Myth of Nitrogen Fertilization for Soil Carbon Sequestration. *J. Environ. Qual.* **2007**, *36*, 1821–1832. [CrossRef] [PubMed]
82. Reay, D.S.; Dentener, F.; Smith, P.; Grace, J.; Feely, R.A. Global nitrogen deposition and carbon sinks. *Nat. Geosci.* **2008**, *1*, 430. [CrossRef]
83. Li, Z.-P.; Zhang, T.-L.; Chen, B.-Y. Changes in organic carbon and nutrient contents of highly productive paddy soils in Yujiang county of Jiangxi province, China and their environmental application. *Agric. Sci. China* **2006**, *5*, 522–529. [CrossRef]
84. Mazzoncini, M.; Antichi, D.; Di Bene, C.; Risaliti, R.; Petri, M.; Bonari, E. Soil carbon and nitrogen changes after 28 years of no-tillage management under Mediterranean conditions. *Eur. J. Agron.* **2016**, *77*, 156–165. [CrossRef]
85. Mazzoncini, M.; Sapkota, T.B.; Barberi, P.; Antichi, D.; Risaliti, R. Long-term effect of tillage, nitrogen fertilization and cover crops on soil organic carbon and total nitrogen content. *Soil Till. Res.* **2011**, *114*, 165–174. [CrossRef]
86. Yang, Y.-C.; Zhang, M.; Zheng, L.; Cheng, D.-D.; Liu, M.; Geng, Y.-Q. Controlled release urea improved nitrogen use efficiency, yield, and quality of wheat. *Agron. J.* **2011**, *103*, 479–485. [CrossRef]
87. Gu, B.; Ju, X.; Wu, Y.; Erisman, J.W.; Bleeker, A.; Reis, S.; Sutton, M.A.; Lam, S.K.; Smith, P.; Oenema, O. Cleaning up nitrogen pollution may reduce future carbon sinks. *Glob. Environ. Chang.* **2018**, *48*, 56–66. [CrossRef]

12

Discrimination of Tomato Plants (*Solanum lycopersicum*) Grown under Anaerobic Baffled Reactor Effluent, Nitrified Urine Concentrates and Commercial Hydroponic Fertilizer Regimes using Simulated Sensor Spectral Settings

Mbulisi Sibanda [1,*], Onisimo Mutanga [1], Lembe S. Magwaza [2,3], Timothy Dube [4], Shirly T. Magwaza [2], Alfred O. Odindo [3], Asanda Mditshwa [3] and Paramu L. Mafongoya [1]

1 Discipline of Geography, School of Agricultural, Earth and Environmental Sciences, University of KwaZulu-Natal, P/Bag X01, Scottsville, Pietermaritzburg 3209, South Africa
2 Discipline of Horticulture, School of Agricultural, Earth and Environmental Sciences, University of KwaZulu-Natal, P/Bag X01, Scottsville, Pietermaritzburg 3209, South Africa
3 Discipline of Crop Science, School of Agricultural, Earth and Environmental Sciences, University of KwaZulu-Natal, P/Bag X01, Scottsville, Pietermaritzburg 3209, South Africa
4 Institute for Water Studies, Department of Earth Sciences, University of the Western Cape, Private Bag X17, Bellville 7535, South Africa
* Correspondence: sibandam3@ukzn.ac.za

Abstract: We assess the discriminative strength of three different satellite spectral settings (HyspIRI, the forthcoming Landsat 9 and Sentinel 2-MSI), in mapping tomato (*Solanum lycopersicum* Linnaeus) plants grown under hydroponic system, using human-excreta derived materials (HEDM), namely, anaerobic baffled reactor (ABR) effluent and nitrified urine concentrate (NUC) and commercial hydroponic fertilizer mix (CHFM) as main sources of nutrients. Simulated spectral settings of HyspIRI, Landsat 9 and Sentinel 2-MSI were resampled from spectrometric proximally sensed data. Discriminant analysis (DA) was applied in discriminating tomatoes grown under these different nutrient sources. Results showed that the simulated spectral settings of HyspIRI sensor better discriminate tomatoes grown under different fertilizer regimes when compared to Landsat 9 OLI and Sentinel-2 MSI spectral configurations. Using the DA algorithm, HyspIRI exhibited high overall accuracy (OA) of 0.99 and a kappa statistic of 0.99 whereas Landsat OLI and Sentinel-2 MSI exhibited OA of 0.94 and 0.95 and 0.79 and 0.85 kappa statistics, respectively. Simulated HyspIRI wavebands 710, 720, 690, 840, 1370 and 2110 nm, Sentinel 2-MSI bands 7 (783 nm), 6 (740 nm), 5 (705 nm) and 8a (865 nm) as well as Landsat bands 5 (865 nm), 6 (1610 nm), 7 (2200 nm) and 8 (590 nm), in order of importance, were selected as the most suitable bands for discriminating tomatoes grown under different fertilizer regimes. Overall, the performance of simulated HyspIRI, Landsat 9 OLI-2 and Sentinel-2 MSI spectral bands seem to bring new opportunities for crop monitoring.

Keywords: hydroponic; vegetable monitoring; crop production; spectral simulation; hyperspectral data

1. Introduction

Food insecurity is a large and growing challenge in sub-Saharan Africa [1,2]. It is estimated that at least one out of four people are hungry and undernourished in sub-Saharan Africa. The World Bank estimates that in 2030, nearly 9 in 10 extremely poor people will be living in Sub-Saharan Africa [3]. This is exacerbated by droughts and soil nutrients deficiencies resulting from limited fertilizer applications [1,2]. This is in turn associated with high fertilizer and food prices, amongst other factors. According to FAO [4] the annual food inflation increased from 5% in 2014 to 6% in 2018 whereas in Europe it remained stable and declined in Latin America, Asia, and Oceania. Subsequently, the improvement of crop production which leads to food security has been amongst the principal priorities required to fulfil the goals of sustainable human development as well as the African Union's Agenda 2063 [5]. Furthermore, food demand is anticipated to triple in sub-Saharan Africa after the projected 2.5-fold increase in population increase [6]. Specifically, a 60% increase in agricultural and horticultural production will be required by the increasing population in the light of diminishing water and soil nutrient resources [6]. The major concern is the current dietary transition, which is in favour of vegetables such as tomatoes amongst other crops, which is projected to increase, especially in urban areas, while water and soil nutrients are in decline [6–8].

Tomato fruits have a critical dietary role in providing folate, vitamins A, C and E; as well as antioxidants (lycopene, beta-carotene, gamma-carotene); trace elements of flavonoids; phytosterols and water-soluble vitamins important for human health [7]. To circumvent the challenge of decreasing soil nutrients and increase the production of vegetables (tomatoes) within a small land area, efforts have been exerted towards improving soil fertility and reducing expenses associated with commercial hydroponic fertilizer mix (CHFM) through the use of anaerobic baffled reactor (ABR) effluents and nitrified urine concentrate (NUC) both as a source of soil nutrients and water [8,9]. Smith and Smith [9], for instance, noted that nitrogen recovered from wastewater supported a high increase in tomato (*Solanum lycopersicum*) plant canopy volume, flower and fruit production when compared to plants treated with commercial hydroponic fertilizer mix (CHFM) which contained N, P, K, Ca, Mg and Si. Al-Hamdan, Cruise et al. (2014) in Jordan noted that treatment of tomato crop using waste water facilitated and increased their fruit size by up to 2 cm in diameter, and weight up to 78.7 g in relation to those administered with potable water in their field experiment. However, the challenge that has been lurking in the agricultural sector is the lack of comprehensive spatial explicit frameworks as well as objective criteria for crop growth and productivity monitoring. Spatially explicit data is important in effectively and precisely managing production both in the field and greenhouses to meet the increasing demand of high quality and safe agricultural products such as tomatoes. Information on vegetable crop type, growth, productivity or health status was previously measured in situ or done through routine field surveys which are often time consuming and date lagged [10,11]. Despite the fact that these in situ methods obtained plausible levels of accuracy in characterizing crops, they lacked spatial representativeness. Consequently, there is need for spatial explicit techniques that can be operationally used not only to characterize the crop's areal extent, but also their physiognomies due to lack or excess of soil nutrients. This information can help deduce and understand crop quality, growth and productivity patterns, which are critical in ensuring food security and coming up with well-informed intervention mechanisms or management strategies where necessary.

Meanwhile, earth observation technologies offer spatially explicit non-destructive synoptic views, innovative and economically feasible timely spatial scale means of generating farm scale crop monitoring. Literature shows that crop/plant physiology and structure such as leaf area index, water content plant pigment content, canopy architecture and canopy density are associated with specific key spectral wavebands [12–15]. Subsequently the variations in these biochemical and physical characteristics of crop plants caused by various crop management practices (i.e., different fertilizer regimes) facilitates unique variations in their spectral finger prints. This makes remote sensing and earth observation facilities to be critical reservoirs of spatially and crop explicit information required in ensuring good quality of crop produce. Remotely sensed data is robust and very sensitive to subtle vegetation traits such as those induced by different water and nutrient regimes. Rajah et al. showed that hyperspectral remotely sensed data could discriminate common dry beans that were rain-fed from those that were irrigated. Lu, et al. [16] discriminated tomato crops that were infected with multi-diseases at different phenological stages using hyperspectral data. Their results exhibited a high overall classification accuracy of 100% in discriminating multi-diseases during the early, asymptomatic and late stages of leaves growth. Above all, the advent and advancement of earth observation facilities has unveiled opportunities for assessing previously unresolved crop quality, growth and productivity related questions linked to plant physiognomies such as that induced by different fertilizer regimes on plant spectral characteristics [11]. Despite hyperspectral data's trade-off between cost and accuracy, it remains the most accurate spatial data for monitoring crop growth and productivity. Unlike broadband satellite data, hyperspectral data has numerous contiguous spectral channels with a potential ability to detect and characterize subtle differences in plant traits such as those induced by different fertilizer application regimes in relation. For example, Česonienė, et al. [17] demonstrated that hyperspectral data could discriminate between conventionally and organically grown carrots in Lithuania to Jeffries-Matusita distances ranging between 1.98 and 2.00. They attributed their results to the ability of hyperspectral data to discern on the variations in the cell structure conditions and canopy structure of the carrots grown under different farming methods. Meanwhile, multispectral sensors like Landsat, Satellite Pour l'Observation de la Terre (SPOT), MODerate Resolution Imaging Spectroradiometer (MODIS) are characterized with broad bands making it difficult to discern subtle plant traits as they tend to mask out critical plant information.

Although hyperspectral sensors provide accurate datasets, a number of sensors have been or are being developed with improved sensing capabilities because of the exorbitant acquisition costs associated with it [18]. For example, the earth observation community recently witnessed the launching of Sentinel 2 multispectral imager (MSI) and Landsat 8 OLI etc. Sentinel 2 MSI has been the first freely available sensor with a set of spectral wavebands covering the red-edge section of the electromagnetic spectrum (B5 (705 nm), 6 (740 nm), and 7 (783 nm)) at a relatively fine spatial resolution of 20m. The sensor has a wide swath-width of 290 km, coupled with a high spatial resolution of 10 m as well as a five-day temporal resolution making it a better facility for crop mapping and monitoring. Both sensors (i.e., Sentinel 2 MSI and Landsat 8 OLI) have been tested in various environmental application areas with plausible findings and conclusions [19–22]. However, in some instances they have been reported to experience challenges, especially when applied at farm level monitoring. This has been attributed to the presence of broad wavebands which are perceived to be concealing most important information. As a result, now new sensors such as the proposed Landsat 9 OLI-2 (with improved noise-to-signal ratio), Environmental Mapping and Analysis Program (EnMAP) and Hyperspectral Infrared Imager (HyspIRI) are being developed. The National Aeronautics and Space Administration agency (NASA) is looking forward to launching the state-of-the-art HyspIRI instruments covering the visible and near-infrared section (Vis/NIR) as well as the thermal infrared (TIR). There is a need to evaluate their performance in discriminating subtle plant properties resulting from crop management practices in relation to the available broadband multispectral sensors. These sensors could help by providing a spatially explicit non-destructive method of characterizing crop quality, growth and productivity patterns which are required in the food industry for pricing as well as in agricultural production for ensuring food security.

The upcoming hyperspectral instruments have a potential to supply the much-needed spatially explicit, accurate, consistent information on vegetable crops. Both of these instruments will be spectrometric covering the spectral ranges of 420–2450 nm and 380–2510 nm at different sampling distances of 6.5 nm for EnMAP's VNIR and 10 nm for EnMAP's SWIR section as well as HyspIRI's VSWIR [23,24]. The swath width of HyspIRI will be 185 km at 30 and 60 m spatial resolutions whereas EnMAP have 30-km-wide coverage across-track at a ground-sampling unit of 30m. The temporal resolution of EnMAP will be 4 days at the equator whereas that of HyspIRI will be 5 days. The fine spectral, spatial and temporal resolutions of these sensors make them relatively more suitable for agriculture applications. The major advantage with such instruments is that they will avail quality data at relatively low costs for data scarce regions such as the sub-Saharan Africa where resources are limited. In this regard, there is need to compare the performance of these hyperspectral sensors to the recently launched freely available and forthcoming multispectral sensors (i.e., Sentinel 2 MSI and Landsat 9 OLI-2 with improve spectral settings) so as to ascertain their full potential.

Considering the fact that literature states that hyperspectral data is generally characterized by high collinearity and that there is no specific algorithm that is suitable for discriminating vegetation characteristics at different places and times, this study also examined the performance of discriminant analysis (DA) and partial least squares discriminant analysis (PLS-DA) in discriminating tomato plants grown under different fertilizer regimes. PLS-DA and DA have been widely used in discriminating plants with different characteristics using both hyperspectral and broadband sensors data [25–27]. These Algorithms have been widely used because they offer an opportunity to evaluate and interpret the minute spectral pattern variations in plants, especially those grown under different management regimes. PLS-DA and DA classification ensembles construct a distinctive spectrum that represents the spectral signatures of the plants samples while simplifying the discrimination process when compared with other methods such as K-nearest neighbours [28–30]. Despite the optimal performance of these methods in vegetation discrimination, the robustness and accuracy of these two is yet to be established [31] particularly in discriminating fine spectral variations of vegetable crops (i.e., tomatoes) induced by different fertilizer application regimes. Furthermore, hyperspectral data could be widely renowned for its exceptional performance in literature in relation to broad bands sensors but the performance of these forthcoming sensors remains undocumented. This study therefore, sought to assess the discriminative performance of HyspIRI, Landsat 9 OLI-2 and Sentinel 2 MSI in characterizing tomato (*Solanum lycopersicum*) crops grown under commercial hydroponic fertilizer mix, anaerobic baffled reactor effluent and nitrified urine concentrate as nutrient sources. HyspIRI was selected as a representative of hyperspectral sensors in this study due to the fact that both EnMAP and HyspIRI will have similar spectral and spatial characteristics and that HyspIRI is going to cover a larger spectral portion of the electromagnetic spectrum particularly in the VSWIR sensor in relation to that of EnMAP. In addition, this study also assessed the performance of DA and PLS-DA in discriminating tomato crops grown under different fertilizer regimes using HyspIRI, Landsat 9 OLI-2 and Sentinel 2 MSI spectral data simulated from hyperspectral data.

2. Materials and Methods

2.1. Experimental Set-Up

A pot experiment was conducted in a hydroponic system that was set up in a polyethylene tunnel located at Newlands-Mashu Research Station under eThekwini Municipality, Durban, South Africa (29°46′25.648″ E 30°58′28.329″ S). The hydroponic system was designed to run three nutrients streams namely, anaerobic baffle reactor (ABR) effluents, nitrified urine concentrate (NUC) and commercial hydroponic fertilizer mix (CHFM) as a control. Each hydroponic system consisted of 150 L tank and the nutrient solution for each system was enclosed in a 100 L container stacked on the ground at the foot of each system.

Six-weeks-old, seedlings of 'Monica', a determinate tomato cultivar purchased from a local nursery (Sunshine seedlings, Pietermaritzburg, South Africa) were transplanted to 30 cm polyethylene pots filled with pine sawdust as a growing medium. The nutrient solution for each nutrient source was supplied to the plants using a pressure pump (DAB Model K30/70M, DAB Pumps, MarcoPolo, Mestrino, Italy) via a 20 m irrigation line. A 20 cm drip irrigation emitters (2 L) were placed and irrigation was performed at six intervals of 5 min/duration daily using a timer. The study was arranged using a complete randomised design with three replications of five plants each, giving fifteen experimental units per nutrient source.

Tomato plants of the control treatment were irrigated with a commercial hydroponic fertilizer mix (Hygroponic® and Solu-cal®) at the rate of 800 g + 620 g/1000 L of water as recommended for hydroponic tomato production; NUC, commercial fertilizer application rate was used as a standard as recommended by Jonsson et al. (2004) and ABR effluents with no specified application rate. For the CHFM and NUC treatment, the fertilizer was mixed using municipal tap water whereas for the ABR treatment only effluent from the anaerobic baffle reactor component was used as nutrient and irrigation source. The experiment was allowed to run for 12 weeks before the crop was harvested.

Remotely sensed proximal spectra data was collected at vegetative stage (i.e., four weeks after transplanting) and two weeks after flowering. In this study all requirements for the photosynthetic activity of the tomato plants were the same except for the fertilisation treatments. This was done so as to make sure that any spectral differences between the CHFM, ABR, NUC treated tomato plants were mainly due to the fertiliser applications since all conditions were the same.

2.2. Remotely Sensed Data

The FieldSpec-3 (ASD Inc., Boulder, CO, USA) Analytic Spectral Device (ASD) FieldSpec instrument was used to acquire the spectral reflectance of tomatoes plants receiving ABR effluents, NUC and CHFM. The ASD measured the radiation at 1.4 nm intervals for the 350–1000 nm and 2 nm intervals for the 1000–2500 nm spectral regions. The reflectance measurements were conducted using bare fibre-optic held at nadir position ~0.5m above the tomato canopies resulting in a field of view with a diameter of ~0.225m. This diameter was found to be adequate to capture the reflectance of the tomato canopies. The normalization of spectral measurements was conducted after every 5 to 10 spectral measurements, using a standard white spectralon. A standard spectralon is optically flat to ±4% over the range of 250–2500 nm and ±1% over the photopic region of the spectrum [32,33]. Its diffuse reflectance standards are highly Lambertian [33]. This was done to circumvent the possible changes in weather conditions as well as irradiance from the sun [34]. The spectral measurements were conducted under clear skies during the day between 10:00 and 14:00 since this is the time with maximum net radiation. A total of 900 spectral samples were measured on canopies of tomatoes treated with CHFM (n = 300), ABR (n = 300) and NUC (n = 300). Noise regions between 350 and 399, 1355 and 1420 nm, 1810 and 1940 nm and 2470–2500 nm were cleaned prior to any analysis of the data [35]. The spectrometrically sensed proximal spectra were then resampled into the spectral settings of HyspIRI, Sentinel 2_MSI and Landsat 9 OLI-2 (Table 1). To simulate the HyspIRI VSWIR sampling rate, 21 successive ASD bands were averaged to create one proxy HyspIRI band using a Gaussian averaging window as detailed in Prasad et al. (2009) [36] and Samiappan et al. (2010) [37]. To simulate the ASD data into the spectral settings of broad spectral sensors, the PRISM in IDL-ENVI was used.

Table 1. Properties of HyspIRI, Sentinel 2-MSI and Landsat 9 OLI-2 sensors.

Sensor	Orbital Altitude	Revisit Time (days)	Swath Width (Km)	Spectral Bands	Band Centre	Spatial Resolution
HyspIRI (VSWIR)	626 km	16	185 km	Contiguous (10 nm) 380–2500 nm		30
Landsat 9 OLI-2	705 km	16	185 km	1Coastal/Aerosol	443	30
				2Blue	482	30
				3Green	562	30
				4Red	655	30
				5NIR	865	30
				6SWIR 1	1610	30
				7SWIR 2	2200	30
				8Panchromatic	590	15
				9Cirrus	1375	30
				10Thermal	10,800	100
				11Thermal	12,000	100
Sentinel 2-MSI	786 km	5	280km	1Coastal aerosols	443	60
				2Blue	490	10
				3Green	560	10
				4Yellow	665	10
				5Red edge	705	20
				6Red edge	740	20
				7Red edge	783	20
				8NIR	842	10
				8aNIR	865	20
				9NIR	945	60
				10SWIR	1375	60
				11SWIR	1610	20
				12SWIR	2190	20

2.3. *Discriminating Tomato Plants Grown under Different Fertilizer Regimes*

Exploratory data analysis was conducted to determine if the data followed a normal distribution curve. Normality test was performed using the Kolmogorov-Smirnov test. Further, we assessed spectral separability as well as administered a pre-filter [38] after hyperspectral data resampling it to HyspIRI, Sentinel-2 MSI and Landsat 8/9 OLI spectral configurations. Resampling was based on the Analysis of Variance test (ANOVA). We then conducted post hoc test to establish the channels that exhibited significant differences between the spectral data of the tomato crops receiving nutrients from ABR effluent, NUC and CHFM.

The other objective of this study was to assess the accuracies of partial least squares discriminant analysis and discriminant analysis algorithms in characterizing tomato crops grown under the three fertilizer regimes. In that regard, we used the PLSDA and the DA to classify the spectral reflectance of tomato crops growing under UNC, UNF and ABR fertilizer treatments. Details about DA and PLS DA are provided in Zhang, et al. [39] and Boulesteix [29]. Prior to conducting PLS-DA and DA, the spectral samples were partitioned into training (70%) and testing (30%) data.

2.4. *Classification Accuracy Assessment*

The 30 percent of the samples were used for classification accuracy assessment (90 per treatment). Confusion matrices were computed and used to evaluate the classification accuracies of the DA and PLS DA models. We further computed the overall, producer and user accuracies, as well as the kappa statistics for each set of spectral settings as classified by the two algorithms based on the confusion matrices. To compare the performance of the two algorithms, a McNemar's test was conducted as detailed by Manandhar, et al. [40] and de Leeuw, et al. [41].

3. Results

Analysis of variance tests results showed significant differences ($\alpha = 0.05$) between tomato plants treated with different fertilizer combinations based on the spectral settings of HyspIRI, Sentinel-2MSI and Landsat 9 OLI. Figure 1 illustrates spectral signatures (mean spectral reflectance) of tomato plants grown under ABR, CHFM and NUC treatments. Specifically, the spectral signatures of tomatoes grown under NUC exhibited higher reflectance curves when compared to those growing under CHFM and ABR in all the sections of the electromagnetic spectrum across all the sensors. The CHFM spectral signature was the lowest in comparison to the other to fertilizer treatments across all the spectral signatures and sensors. For HyspIRI resampled data, significant differences were observed in the visible, NIR as well as the SWIR portions of the electromagnetic spectrum (Figure 1).

Meanwhile the most glaring differences in the reflectance of tomato plants grown under ABR, CHFM and NUC treatments were observed in the NIR portion of the electromagnetic spectrum based on the simulated Sentinel 2 MSI and Landsat 9 OLI-2 data (Figure 1). HyspIRI spectral settings exhibited more potential spectral windows of separability between tomato plants grown under ABR, CHFM and NUC treatments when compared with the broadband sensors. Potential spectral windows of separability exhibited by Sentinel 2 MSI and Landsat 9 OLI-2 were only in the near infrared regions (Figure 1).

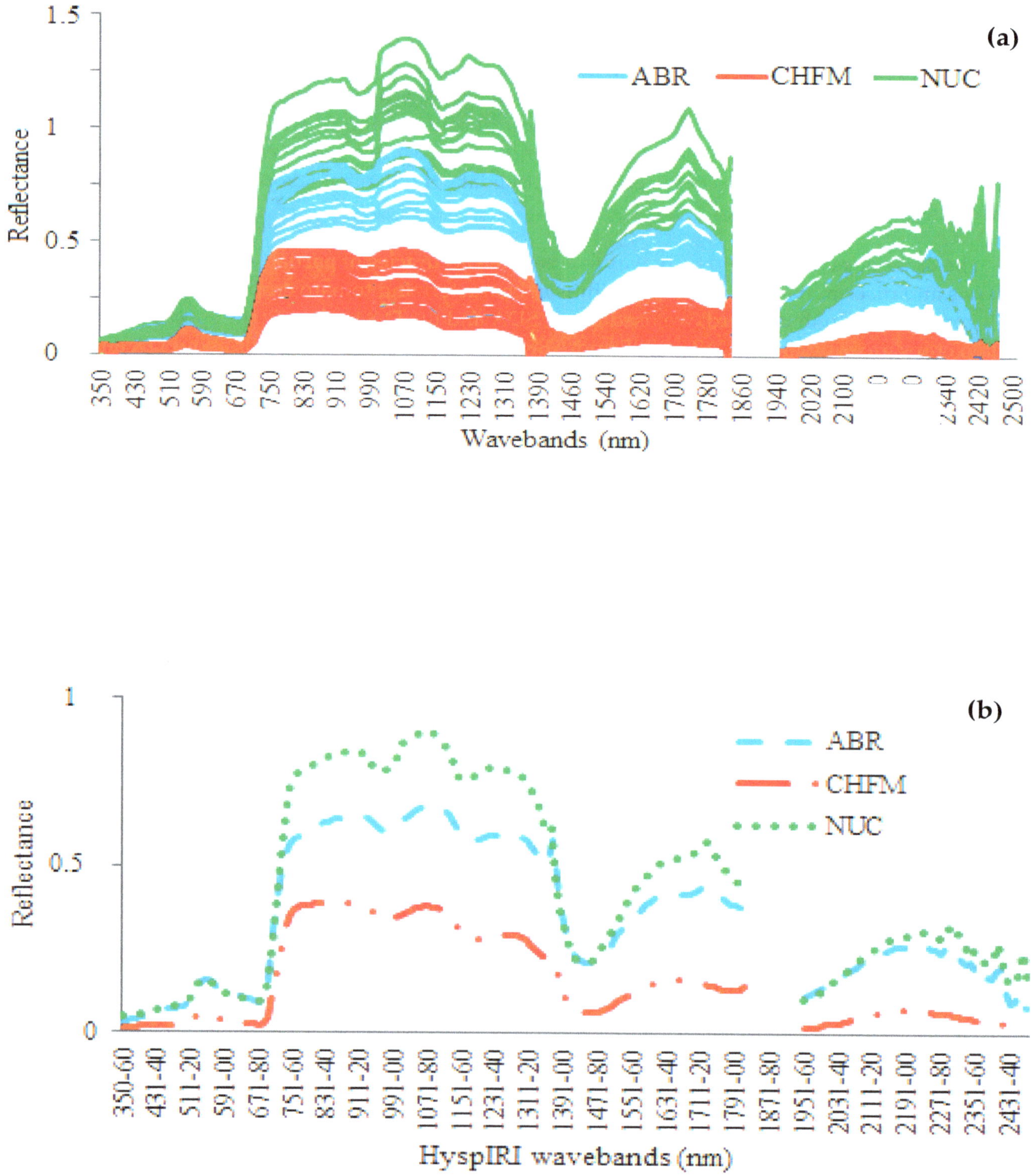

Figure 1. *Cont.*

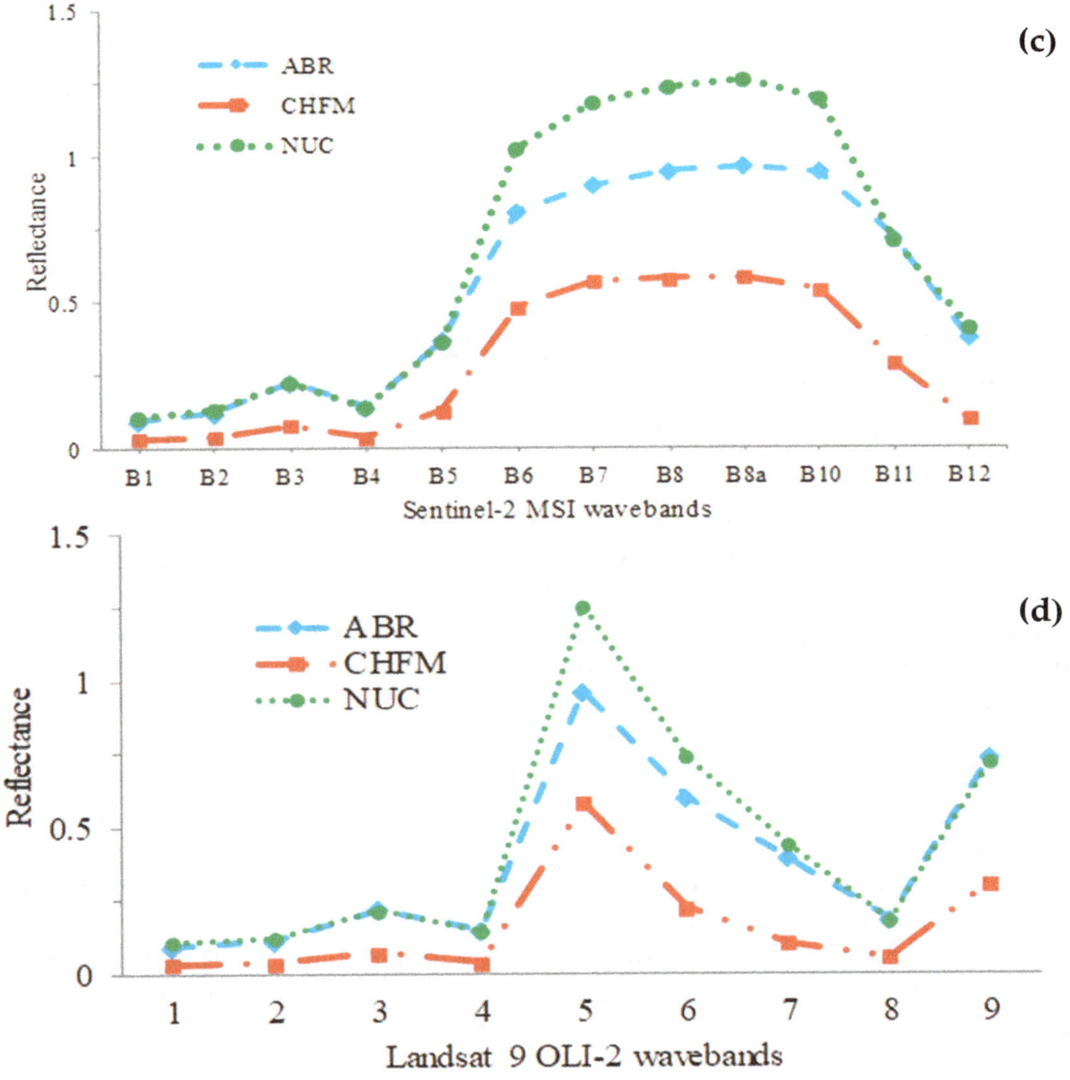

Figure 1. Collected spectra (**a**) and Mean spectral signatures of tomato crops grown under ABR effluent, NUC and CHFM treatments based on (**b**) HyspIRI (**c**) Sentinel -2 MSI and (**d**) Landsat 9 OLI-2 spectral settings.

3.1. Discriminating Tomatoes under ABR Effluent, NUC and CHFM

All sensors better characterized tomato crops administered with ABR when compared to those administered with CHFM and NUC. Specifically, high producer and user accuracies ranging between 91% and 100% were observed in characterizing tomatoes treated with ABR (Table 2). Meanwhile tomatoes administered with ABR and CHFM were characterized with slightly lower accuracies (Table 2). Moreover, HyspIRI, produced high accuracies characterized by kappa statistics of 0.99, whereas the spectral settings of Sentinel-2 MSI and Landsat 9 OLI-2's spectral settings exhibited kappa statistics of 0.85 and 0.79, respectively. HyspIRI spectral settings exhibited high producer accuracies of 100%, 92% and 100% for ABR, CHFM and NUC respectively. Meanwhile Sentinel-2 MSI exhibited slightly lower producer accuracies (ABR = 91%, CHFM = 86% and NUC = 100%). Landsat OLI-2 exhibited producer accuracies that were comparable to those of Sentinel 2 MSI which were 91% for ABR, 83% for CHFM and 100% for NUC. The same trend could be observed on the user accuracies (Table 2). In general, the CHFM treatments had slightly lower-class accuracies across all sensors in relation to the HEDM treatments.

Table 2. Classification accuracies derived using HyspIRI, Sentinel 2 MSI and Lands 9 OLI-2 spectral settings.

Sensor		PLS-DA		DA		PLS-DA		DA	
		PA	UA	PA	UA	OA	Kappa	OA	Kappa
	ABR	100	95	100	100	0.97	0.90	0.99	0.99
HyspIRI	CHFM	92	100	100	100				
	NUC	100	94	100	100				
Sentinel 2 MSI	ABR	91	100	100	100	0.90	0.69	0.95	0.85
	CHFM	86	100	89	100				
	NUC	100	65	100	82				
Landsat 9 OLI-2	ABR	91	95	95	100	0.89	0.63	0.94	0.79
	CHFM	83	100	89	100				
	NUC	100	65	100	76				

3.2. Performance of DA and PLS-DA Algorithms in Discriminating Tomatoes under ABR Effluent, NUC and CHFM

When comparing the performance of algorithms, DA exhibited very high accuracies. For instance, DA's producer accuracies derived using DA across all fertilizer treatments and sensor simulations ranged from a minimum of 89% whereas PLS-DA had a slightly lower minimum of 83% (Table 2). The user accuracies derived using DA ranged from a minimum of 76% whereas those derived using PLS-DA ranged from 65% to 100%. The overall accuracies derived using DA were higher (i.e., 0.94–0.99) when compared to those derived using PLS-DA (i.e., 0.3–0.97). Uniformly, the kappa statistics derived using DA were higher, ranging between 0.79 and 0.99 whereas those derived using PLS-DA were lower ranging between 0.63 and 0.90 (Table 2). Table 3 summarizes potential spectral variables for discriminating tomato plants grown under ABR, CHFM and NUC treatments using HyspIRI, Sentinel 2 MSI and Landsat 9 OLI-2. HyspIRI wavebands 710, 720, 690, 840, 1370 and 2110 nm, Sentinel 2-MSI bands 7, 6, 5 and 8a as well as Landsat bands 5, 6, 7 and 8, in order of importance, were identified as the most optimal spectral variables for discriminating tomato plants grown under different fertilizer regimes. NIR was the most prominent optimal section with a potential of discriminating tomatoes plants grown under different fertilizer regimes across the three sensors, although it did not outperform the red-edge section of the electromagnetic spectrum which was the most influential region.

Table 3. Influential bands in discriminating tomato plants grown under different fertilizer regimes.

Spectrum	Sensor					
	HyspIRI		Sentinel		Landsat	
	DA	PLS-DA	DA	PLS-DA	DA	PLS-DA
Visible	5	4	2	1	2	1
Red edge	12	10	3	3		
NIR	10	10	2	2	1	1
SWIR	8	7	1	1	2	2
M/FWIR	7	7				
Total	42	38	8	7	5	4

NB M/FWIR means Mid/Far wavelength infrared.

4. Discussion

We sought to compare the strength of HyspIRI's spectral configuration in relation to Landsat 9 OLI-2 and Sentinel 2 MSI spectral settings in characterizing tomato (*Solanum lycopersicum*) crops grown under CHFM, ABR and NUC treatment regimes. Results of this study showed that tomatoes that were fertilized using ABR could be optimally discriminated (i.e., Kappa statics ranging 0.79 to 0.99) from those that were administered with CHFM and NUC. The spectral reflectance of HDME fertilized crops was higher than that of those treated with CHFM. This indicates the potential of HEDM fertilizer for sustaining the health of vegetables in a manner comparable to that of conventional chemical fertilizers. This could be attributed to the fact that ABR effluents have nutrient properties favorable for tomato plants, which facilitate excessive vegetative growth, high biomass accumulation, delayed or uneven maturity [42–44]. Tomato crops with excessive vegetative growth, high biomass accumulation, delayed or uneven maturity tend to be easily detected and discriminated by satellite sensors compared to those which are not. Subsequently, tomatoes growing under the ABR treatments have a different spectral signature from those that are fertilized using CHFM and NUC. Literature illustrates that ABR tends to facilitate high biomass accumulation (i.e., increased leaf area index) hence the high classification accuracies exhibited by all remotely sensed data in this study in characterizing the HEDM fertiliser treatments [11,28,29]. For example, Al-Lahham, El Assi and Fayyad [45] illustrated that tomato crops that were administered with high quantities of waste water had big fruit sizes hence high biomass accumulation in relation to those that were administered with potable water. In a related study, Zavadil (2009) noted that primary treated waste water which contained an average of 14-fold nitrogen amounts (70.6 mg/L, which was 89% ammonia form) and also a 3-fold of the total phosphorus, resulted in high biomass accumulation and yields of lettuce salad, radishes and carrots in their study assessing the influence of sewage in relation to potable water on vegetable growth. Subsequently, the increases in biomass accumulation associated with wastewater treated vegetables could explain the discrimination of ABR treated tomato crops in comparison to those administered with CHFM and NUC, in this study. Furthermore, the most influential wavebands that facilitated high classification accuracies in discriminating tomato plants grown under human-excreta derived materials (HEDM) from those growing under CHFM conditions were from the red-edge section of the electromagnetic spectrum which is generally associated with healthy plants. The red-edge section of the electromagnetic spectrum is very sensitive to high chlorophyll, leaf angle distribution, leaf area index levels associated with plants grown in more nutritious environments such as the those exhibited by ABR. As aforementioned, HEDM have a high turnover of N P K plant nutrients which is comparable to that of commercial soluble fertilizers [46–48]. In this regard, the high turnover of plant nutrients such as nitrogen facilitates an increase in biomass, LAD, LAI which in turn makes the canopy spectral signature of those plants treated with HEDM to be discriminable from other plants especially in the red-edge and the NIR sections of the electromagnetic spectrum [49,50]. Subsequently, the red-edge and the NIR regions of the electromagnetic spectrum illustrate that the spectral signatures and the health of plants grown under HEDM are comparable to those grown under the CHFM.

When assessing the performance of sensors, HyspIRI outperformed the two multispectral sensors namely, Sentinel-2 MSI and Landsat OLI spectral settings in discriminating tomato crops grown under different fertilizer regimes. This could be explained by the fact that HyspIRI is a hyperspectral sensor characterized by narrow spectral wavebands that are more sensitive to the spectral reflectance of tomato crops grown under different fertilizer regimes than broadband sensor settings such as those of Landsat which could be masking out those minute tomato crops spectral variations. There is a consistently growing body of literature that supports the claim that hyperspectral sensors are more sensitive to minute vegetation spectral variabilities compared to broadband sensors due to the narrowed bandwidths configuration [51–56]. Specifically, Thenkabail, Smith and De Pauw [55,57] illustrated that narrow bands characterized different crop traits such as yield as well as spectral variations when compared to broadband spectral data. Also, they were able to better characterize wheat from barley using hyperspectral data in relation to the data from broadband sensors. They attributed this to the variation in spectral settings (bandwidths) of the sensors they used. These variations in spectral settings affected the detail that determined the accuracy of their models for plant trait characterization. Meanwhile, Clark [57] noted that there was no significant variation in the performance of HyspIRI and Sentinel-2 MSI as well as Landsat OLI in landcover classification of the San Francisco Bay Area in northern California, USA. However, their results confirmed that HyspIRI exhibited higher classification accuracies in their study.

Results of this study also illustrated that Sentinel 2 MSI and Landsat 9 performed satisfactorily in discriminating tomatoes grown under different fertilizer regimes, although Sentinel-2 MSI outperformed

Landsat OLI. This could be explained by the fact that Sentinel-2 MSI spectral settings cover the red-edge portion of the electromagnetic spectrum which is critical in mapping and detecting various vegetation traits. Moreover, there is a large and growing body of literature illustrating that Sentinel-2 MSI performs better than Landsat OLI in vegetation mapping [22,57,58]. The study by Colkesen and Kavzoglu [58] illustrated that Sentinel-2 MSI outperformed Landsat OLI in discriminating alfalfa, sugar beet and bean in the agricultural lands of the Ferizli district, Turkey. Shoko and Mutanga [22] also illustrated the robustness of Sentinel-2 MSI remotely sensed data in better discriminating C3 *Festuca costata* Nees from the C4 *Themeda triandra* Forssk grasses in a mountainous area in South Africa. They also attributed the optimal performance of Sentinel-2 MSI to the presence of red-edge bands in discriminating *Festuca costata* from *Themeda triandra* grasses.

Even though this study successfully illustrated the robustness of the spectral settings of these sensors in discriminating tomato crops grown under different fertilizer regimes using simulated data, there is still need to assess their satellite remotely sensed data. This study only sought to evaluate the spectral configuration of these sensors, hence there is still need to evaluate their radiometric and spatial resolutions in a similar application setup. Generally, the reflectance of remotely sensed satellite data tends to be affected by numerous factors associated with the sensor's platform, atmospheric related influences etc. [59–61]. Factors that affect satellite remotely sensed data are different from those that affect spectrometric proximally sensed data. Spectrometric proximally sensed data tends to be less affected by atmospheric related influences; hence, it offers a robust dataset suitable for testing the spectral settings of sensors, particularly the forth coming ones. The performance of the same sensors' spectral settings derived using satellite platforms, therefore, still needs to be evaluated.

Although this was not the major objective of the study, DA outperformed PLS-DA in discriminating tomato crops grown under different fertilizer regimes. In this study PLS-DA failed to derive unnecessary variables for characterizing tomato crops grown under different fertilizer regimes. On the other hand, there are numerous studies that have illustrated the optimal performance of DA in dimension reduction as well as feature extraction [62–64]. Our results also indicated that DA selected more spectral bands in relation to PLS-DA as optimal spectral variables for discriminating tomatoes grown under different fertilizer treatment regimes. The optimal performance of DA in relation to PLS-DA could be attributed to the conservative nature of PLS-DA in classifying tomato plants [31].

Implications of the Study'S Findings for Horticultural Crop Production

The optimal performance of HyspIRI's spectral settings in characterizing tomato plants grown under different fertilizer treatments illustrates its great potential in providing additional invaluable spatially explicit information urgently required in horticulture management and decision-making processes particularly for successful site and crop specific management practices. For instance, this study shows that based on remotely sensed data HEDM fertilizers have a great potential of improving vegetable crops health in manner comparable to that of conventional fertilisers. The fact that this study showed that HyspIRI, Sentinel 2 MSI and Landsat 9 OLI-2 could detect the differences between HEDM fertilized plants and those that are grown under conventional fertilizers underscores the potential of these sensors in providing information that could also be used in other applications such as crop inventory, condition, production forecast, assessment of nutrient deficiencies as well as growth and health of tomato plants and other horticultural crops. Furthermore, these sensors offer cheap, fast and reliable spatial explicit information on crop conditions required in the administration of precise and proper fertilizer applications while optimizing resources and increasing net returns especially in sub-Saharan countries with limited data and financial resources for facilitating high production of horticultural crops. However, more and extensive research efforts still need to be exerted in fully ascertaining and exploiting the potential of remotely sensed data in other horticultural crop-management applications.

5. Conclusions

The prime objective of this study was to compare the strength of the forthcoming hyperspectral sensor HyspIRI's spectral settings in the context of characterizing the effects of different fertilizer-treatment regimes on tomato crops. Furthermore, the study assessed the performance of PLS-DA in relation to DA in discriminating tomatoes treated with ABR, NUC and CHFM. Based on the results exhibited by this study we conclude that:

- The forthcoming HyspIRI sensor has the potential to accurately map tomato crops under various fertilizer regimes. Landsat and Sentinel, performed comparably to HyspIRI spectral settings.
- Overall, all the sensors were able to characterise the comparable impact of HEDM in relation to that of CHFM fertilizers on the spectral characteristics of tomato plants as a proxy of their health.
- DA offers optimal accuracies in characterizing tomatoes grown under different fertilizer regimes when compared to PLS-DA.

These findings are a substantial foundation upon which comprehensive precision agricultural assessments initiatives could be formed. These initiatives are required in order to attain sustainable agriculture as well as food security in regions such as sub-Saharan Africa where agricultural crop monitoring is currently hindered by the limited access to robust spatial data sets.

Author Contributions: Conceptualization, (M.S., O.M. and L.S.M.); Data curation, (L.S.M., S.T.M., A.O.O. and A.M.); Formal analysis, (M.S.); Investigation, (M.S., S.T.M. and A.M.); Project administration, (O.M. and P.L.M.); Resources, (O.M. and P.L.M.); Writing–original draft, (M.S.); Writing–review & editing, (M.S. and T.D.).

Acknowledgments: The Pollution Research Group (PRG) is acknowledged for technical support and for providing human-excreta derived materials as well as space for experimentation for this study. We also extend our gratitude to the anonymous reviewers for their constructive criticism.

References

1. Mabhaudhi, T.; Chibarabada, T.; Modi, A. Water-Food-Nutrition-Health Nexus: Linking Water to Improving Food, Nutrition and Health in Sub-Saharan Africa. *Int. J. Environ. Res. Public Health* **2016**, *13*, 107. [CrossRef]
2. Van Ittersum, M.K.; Van Bussel, L.G.; Wolf, J.; Grassini, P.; Van Wart, J.; Guilpart, N.; Claessens, L.; de Groot, H.; Wiebe, K.; Mason-D'Croz, D.; et al. Can Sub-Saharan Africa Feed Itself? *Proc. Natl. Acad. Sci. USA* **2016**, *113*, 14964–14969. [CrossRef] [PubMed]
3. The-World-Bank. *Poverty & Equity Data Portal*; The World Bank: Wastington, DC, USA, 2019.
4. FAO. *Inflation in Consumer Price Index for Food*; FAO: Rome, Italy, 2019.
5. Conceição, P.; Levine, S.; Lipton, M.; Warren-Rodríguez, A. Toward a Food Secure Future: Ensuring Food Security for Sustainable Human Development in Sub-Saharan Africa. *Food Policy* **2016**, *60*, 1–9. [CrossRef]
6. Nordey, T.; Basset-Mens, C.; De Bon, H.; Martin, T.; Déletré, E.; Simon, S.; Parrot, L.; Despretz, H.; Huat, J.; Biard, Y.; et al. Protected Cultivation of Vegetable Crops in Sub-Saharan Africa: Limits and Prospects for Smallholders. A Review. *Agron. Sustain. Dev.* **2017**, *37*, 53. [CrossRef]
7. Beecher, G.R. Nutrient Content of Tomatoes and Tomato Products. *Proc. Soc. Exp. Biol. Med.* **1998**, *218*, 98–100. [CrossRef]
8. Busari, T.I.; Senzanje, A.; Odindo, A.O.; Buckley, C.A. Evaluating the Effect of Irrigation Water Management Techniques on (Taro) Madumbe (Colocasia Esculenta (L.) Schott) Grown with Anaerobic Filter (Af) Effluent at Newlands, South Africa. *J. Water Reuse Desalin.* **2019**, *9*, 203–212. [CrossRef]
9. Smith, D.P.; Smith, N.T. Recovery of Wastewater Nitrogen for Solanum Lycopersicum Propagation. *Waste Biomass Valorization* **2019**, *10*, 1192–1202. [CrossRef]
10. Al-Hamdan, M.; Cruise, J.; Rickman, D.; Quattrochi, D. Forest Stand Size-Species Models Using Spatial Analyses of Remotely Sensed Data. *Remote Sens.* **2014**, *6*, 9802–9828. [CrossRef]

11. Petropoulos, G.; Srivastava, P.; Piles, M.; Pearson, S. Earth Observation-Based Operational Estimation of Soil Moisture and Evapotranspiration for Agricultural Crops in Support of Sustainable Water Management. *Sustainability* **2018**, *10*, 181. [CrossRef]
12. Peng, Y.; Nguy-Robertson, A.; Arkebauer, T.; Gitelson, A. Assessment of Canopy Chlorophyll Content Retrieval in Maize and Soybean: Implications of Hysteresis on the Development of Generic Algorithms. *Remote Sens.* **2017**, *9*, 226. [CrossRef]
13. Nguy-Robertson, A.L.; Peng, Y.; Gitelson, A.A.; Arkebauer, T.J.; Pimstein, A.; Herrmann, I.; Karnieli, A.; Rundquist, D.C.; Bonfil, D.J. Estimating Green Lai in Four Crops: Potential of Determining Optimal Spectral Bands for a Universal Algorithm. *Agric. For. Meteorol.* **2014**, *192*, 140–148. [CrossRef]
14. Luo, S.; Wang, C.; Xi, X.; Nie, S.; Fan, X.; Chen, H.; Yang, X.; Peng, D.; Lin, Y.; Zhou, G. Combining Hyperspectral Imagery and Lidar Pseudo-Waveform for Predicting Crop Lai, Canopy Height and above-Ground Biomass. *Ecol. Indic.* **2019**, *102*, 801–812. [CrossRef]
15. Zhao, T.; Koumis, A.; Niu, H.; Wang, D.; Chen, Y. Onion Irrigation Treatment Inference Using a Low-Cost Hyperspectral Scanner. In Proceedings of the Paper Presented at the Multispectral, Hyperspectral, and Ultraspectral Remote Sensing Technology, Techniques and Applications VII, Honolulu, HI, USA, 24–26 September 2018.
16. Lu, J.; Ehsani, R.; Shi, Y.; de Castro, A.I.; Wang, S. Detection of Multi-Tomato Leaf Diseases (Late Blight, Target and Bacterial Spots) in Different Stages by Using a Spectral-Based Sensor. *Sci. Rep.* **2018**, *8*, 2793. [CrossRef] [PubMed]
17. Česonienė, L.; Masaitis, G.; Mozgeris, G.; Gadal, S.; Šileikienė, D.; Karklelienė, R. Visible and near-Infrared Hyperspectral Imaging to Describe Properties of Conventionally and Organically Grown Carrots. *J. Elem.* **2019**, *24*, 421–435.
18. Transon, J.; d'Andrimont, R.; Maugnard, A.; Defourny, P. Survey of Hyperspectral Earth Observation Applications from Space in the Sentinel-2 Context. *Remote Sens.* **2018**, *10*, 157. [CrossRef]
19. Korhonen, L.; Packalen, P.; Rautiainen, M. Comparison of Sentinel-2 and Landsat 8 in the Estimation of Boreal Forest Canopy Cover and Leaf Area Index. *Remote Sens. Environ.* **2017**, *195*, 259–274. [CrossRef]
20. Dube, T.; Mutanga, O. Evaluating the Utility of the Medium-Spatial Resolution Landsat 8 Multispectral Sensor in Quantifying Aboveground Biomass in Umgeni Catchment, South Africa. *ISPRS J. Photogramm. Remote Sens.* **2015**, *101*, 36–46. [CrossRef]
21. Ahmadian, N.; Ghasemi, S.; Wigneron, J.P.; Zölitz, R. Comprehensive Study of the Biophysical Parameters of Agricultural Crops Based on Assessing Landsat 8 Oli and Landsat 7 Etm+ Vegetation Indices. *GISci. Remote Sens.* **2016**, *53*, 337–359. [CrossRef]
22. Shoko, C.; Mutanga, O. Examining the Strength of the Newly-Launched Sentinel 2 Msi Sensor in Detecting and Discriminating Subtle Differences between C3 and C4 Grass Species. *ISPRS J. Photogramm. Remote Sens.* **2017**, *129*, 32–40. [CrossRef]
23. Guanter, L.; Kaufmann, H.; Segl, K.; Foerster, S.; Rogass, C.; Chabrillat, S.; Kuester, T.; Hollstein, A.; Rossner, G.; Chlebek, C.; et al. The Enmap Spaceborne Imaging Spectroscopy Mission for Earth Observation. *Remote Sens.* **2015**, *7*, 8830–8857. [CrossRef]
24. Lee, C.M.; Cable, M.L.; Hook, S.J.; Green, R.O.; Ustin, S.L.; Mandl, D.J.; Middleton, E.M. An Introduction to the Nasa Hyperspectral Infrared Imager (Hyspiri) Mission and Preparatory Activities. *Remote Sens. Environ.* **2015**, *167*, 6–19. [CrossRef]
25. Sheik, O.M.; Kabir, P.; Ilaria, G.; Onisimo, M.; Zakariyyaa, O. Detecting Canopy Damage Caused by Uromycladium Acaciae on South African Black Wattle Forest Compartments Using Moderate Resolution Satellite Imagery. *S. Afr. J. Geomat.* **2019**, *8*, 69–83.
26. Huang, H.C.; Liu, S.C.; Wang, C.; Xia, K.B.; Zhang, D.J.; Wang, H.Z.; Zhan, S.Y.; Huang, H.; He, S.Y.; Liu, C.C. On-Site Visualized Classification of Transparent Hazards and Noxious Substances on a Water Surface by Multispectral Techniques. *Appl. Opt.* **2019**, *58*, 4458–4466. [CrossRef] [PubMed]
27. Li, X.Y.; Zhang, F.L.; Jane, Y. Locally Weighted Discriminant Analysis for Hyperspectral Image Classification. *Remote Sens.* **2019**, *11*, 109. [CrossRef]
28. Corbane, C.; Alleaume, S.; Deshayes, M. Mapping Natural Habitats Using Remote Sensing and Sparse Partial Least Square Discriminant Analysis. *Int. J. Remote Sens.* **2013**, *34*, 7625–7647. [CrossRef]
29. Boulesteix, A.L. Pls Dimension Reduction for Classification with Microarray Data. *Stat. Appl. Genet. Mol. Biol.* **2004**, *3*, 1–32. [CrossRef] [PubMed]

30. Bagheri, N.; Mohamadi-Monavar, H.; Azizi, A.; Ghasemi, A. Detection of Fire Blight Disease in Pear Trees by Hyperspectral Data. *Eur. J. Remote Sens.* **2018**, *50*, 1–10. [CrossRef]
31. Prats-Montalbán, J.M.; Ferrer, A.; Malo, J.L.; Gorbena, J. A Comparison of Different Discriminant Analysis Techniques in a Steel Industry Welding Process. *Chemom. Intell. Lab. Syst.* **2006**, *80*, 109–119. [CrossRef]
32. Julia, A. Colour Correction of Underwater Images Using Spectral Data. Ph.D. Thesis, Acta Universitatis Upsaliensis, Uppsala, Sweden, 2005.
33. Springsteen, A. Standards for Reflectance Measurements. *Appl. Spectrosc. A Compact. Ref. Pract.* **1998**, *15*, 247.
34. Abdel-Rahman, E.M.; Mutanga, O.; Odindi, J.; Adam, E.; Odindo, A.; Ismail, R. A Comparison of Partial Least Squares (Pls) and Sparse Pls Regressions for Predicting Yield of Swiss Chard Grown under Different Irrigation Water Sources Using Hyperspectral Data. *Comput. Electron. Agric.* **2014**, *106*, 11–19. [CrossRef]
35. Curran, P.J. Imaging Spectrqmetry-Its Present and Future Rôle in Environmental Research. In *Imaging Spectrometry—A Tool for Environmental Observations*; Springer: Berlin, Germany, 1994; pp. 1–23.
36. Prasad, S.; Lori, M.B.; Hemanth, K. Data Exploitation of Hyspiri Observations for Precision Vegetation Mapping. In Proceedings of the 2009 IEEE International Geoscience and Remote Sensing Symposium, Cape Town, South Africa, 12–17 July 2009.
37. Sathishkumar, S.; Prasad, S.; Bruce, M.L.; Robles, W. Nasa's Upcoming Hyspiri Mission—Precision Vegetation Mapping with Limited Ground Truth. In Proceedings of the 2010 IEEE International Geoscience and Remote Sensing Symposium, Honolulu, HI, USA, 25–30 July 2010.
38. Adelabu, S.; Mutanga, O.; Adam, E.; Sebego, R. Spectral Discrimination of Insect Defoliation Levels in Mopane Woodland Using Hyperspectral Data. Selected Topics in Applied Earth Observations and Remote Sensing. *IEEE J. Sel. Top. Appl. Earth Obs. Remote Sens.* **2014**, *7*, 177–186. [CrossRef]
39. Zhang, H.; Lan, Y.; Suh, C.P.; Westbrook, J.K.; Lacey, R.; Hoffmann, W.C. Differentiation of Cotton from Other Crops at Different Growth Stages Using Spectral Properties and Discriminant Analysis. *Trans. Am. Soc. Agric. Biol. Eng.* **2012**, *55*, 1623–1630. [CrossRef]
40. Manandhar, R.; Odeh, I.; Ancev, T. Improving the Accuracy of Land Use and Land Cover Classification of Landsat Data Using Post-Classification Enhancement. *Remote Sens.* **2009**, *1*, 330–344. [CrossRef]
41. de Leeuw, J.; Jia, H.; Yang, L.; Liu, X.; Schmidt, K.; Skidmore, A.K. Comparing Accuracy Assessments to Infer Superiority of Image Classification Methods. *Int. J. Remote Sens.* **2006**, *27*, 223–232. [CrossRef]
42. Pedrero, F.; Kalavrouziotis, I.; Alarcón, J.J.; Koukoulakis, P.; Asano, T. Use of Treated Municipal Wastewater in Irrigated Agriculture—Review of Some Practices in Spain and Greece. *Agric. Water Manag.* **2010**, *97*, 1233–1241. [CrossRef]
43. Maurer, M.A.; Davies, F.S.; Graetz, D.A. Reclaimed Wastewater Irrigation and Fertilization of Matureredblush'grapefruit Trees on Spodosols in Florida. *J. Am. Soc. Hortic. Sci.* **1995**, *120*, 394–402. [CrossRef]
44. Zavadil, J. The Effect of Municipal Wastewater Irrigation on the Yield and Quality of Vegetables and Crops. *Soil Water Res.* **2009**, *4*, 91–103. [CrossRef]
45. Al-Lahham, O.; El Assi, N.M.; Fayyad, M. Impact of Treated Wastewater Irrigation on Quality Attributes and Contamination of Tomato Fruit. *Agric. Water Manag.* **2003**, *61*, 51–62. [CrossRef]
46. Viskari, E.L.; Grobler, G.; Karimäki, K.; Gorbatova, A.; Vilpas, R.; Lehtoranta, S. Nitrogen Recovery with Source Separation of Human Urine–Preliminary Results of Its Fertiliser Potential and Use in Agriculture. *Front. Sustain. Food Syst.* **2018**, *2*, 32. [CrossRef]
47. Moya, B.; Parker, A.; Sakrabani, R. Challenges to the Use of Fertilisers Derived from Human Excreta: The Case of Vegetable Exports from Kenya to Europe and Influence of Certification Systems. *Food Policy* **2019**, *85*, 72–78. [CrossRef]
48. Moya, B.; Parker, R.S.; Mesa, B. Evaluating the Efficacy of Fertilisers Derived from Human Excreta in Agriculture and Their Perception in Antananarivo, Madagascar. *Waste Biomass Valorization* **2019**, *10*, 941–952. [CrossRef]
49. Bassegio, D.; Santos, R.F.; de Oliveira, E.; Wernecke, I.; Secco, D.; de Souza, S.N.M. Effect of Nitrogen Fertilization and Cutting Age on Yield of Tropical Forage Plants. *Afr. J. Agric. Res.* **2013**, *8*, 1427–1432.
50. Clevers, J.G.; Gitelson, A.A. Remote Estimation of Crop and Grass Chlorophyll and Nitrogen Content Using Red-Edge Bands on Sentinel-2 and-3. *Int. J. Appl. Earth Obs. Geoinf.* **2013**, *23*, 344–351. [CrossRef]
51. Adam, E.; Mutanga, O.; Rugege, D. Multispectral and Hyperspectral Remote Sensing for Identification and Mapping of Wetland Vegetation: A Review. *Wetl. Ecol. Manag.* **2010**, *18*, 281–296. [CrossRef]

52. Mansour, K.; Mutanga, O.; Everson, T. Remote Sensing Based Indicators of Vegetation Species for Assessing Rangeland Degradation: Opportunities and Challenges. *Afr. J. Agric. Res.* **2012**, *7*, 3261–3270.
53. Thenkabail, P.S.; Lyon, J.G. *Hyperspectral Remote Sensing of Vegetation: Knowledge Gain and Knowledge Gap after 50 Years of Research (Conference Presentation)*; Paper Presented at the Hyperspectral Imaging Sensors: Innovative Applications and Sensor Standards 2017; CRC Press: Boca Raton, FL, USA, 2017.
54. Thenkabail, P.S.; Enclona, E.A.; Ashton, M.S.; Legg, C.; De Dieu, M.J. Hyperion, Ikonos, Ali, and Etm+ Sensors in the Study of African Rainforests. *Remote Sens. Environ.* **2004**, *90*, 23–43. [CrossRef]
55. Thenkabail, P.S.; Smith, R.B.; De Pauw, E. Evaluation of Narrowband and Broadband Vegetation Indices for Determining Optimal Hyperspectral Wavebands for Agricultural Crop Characterization. *Photogramm. Eng. Remote Sens.* **2002**, *68*, 607–622.
56. Thenkabail, P.S.; Lyon, J.G. *Hyperspectral Remote Sensing of Vegetation*; CRC Press: Boca Raton, FL, USA, 2016.
57. Clark, M.L. Comparison of Simulated Hyperspectral Hyspiri and Multispectral Landsat 8 and Sentinel-2 Imagery for Multi-Seasonal, Regional Land-Cover Mapping. *Remote Sens. Environ.* **2017**, *300*, 311–325. [CrossRef]
58. Colkesen, I.; Kavzoglu, T. Ensemble-Based Canonical Correlation Forest (Ccf) for Land Use and Land Cover Classification Using Sentinel-2 and Landsat Oli Imagery. *Remote Sens. Lett.* **2017**, *8*, 1082–1091. [CrossRef]
59. Lillesand, T.; Ralph, W.K.; Jonathan, C. *Remote Sensing and Image Interpretation*; John Wiley & Sons: Hoboken, NJ, USA, 2015.
60. Thenkabail, P.S.; Mariotto, I.; Gumma, M.K.; Middleton, M.E.; Landis, R.D.; Huemmrich, F.K. Selection of Hyperspectral Narrowbands (Hnbs) and Composition of Hyperspectral Twoband Vegetation Indices (Hvis) for Biophysical Characterization and Discrimination of Crop Types Using Field Reflectance and Hyperion/Eo-1 Data. *Sel. Top. Appl. Earth Obs. Remote. Sens. IEEE J.* **2013**, *6*, 427–439. [CrossRef]
61. Mariotto, I.; Prasad, S.; Thenkabail, A.H.; Terrence Slonecker, E.; Platonov, A. Hyperspectral Versus Multispectral Crop-Productivity Modeling and Type Discrimination for the Hyspiri Mission. *Remote Sens. Environ.* **2013**, *139*, 291–305. [CrossRef]
62. Pu, R.; Liu, D. Segmented Canonical Discriminant Analysis of in Situ Hyperspectral Data for Identifying 13 Urban Tree Species. *Int. J. Remote. Sens.* **2011**, *32*, 2207–2226. [CrossRef]
63. Karimi Prasher, Y.S.O.; McNairn, H.; Bonnell, R.B.; Dutilleul, P.; Goel, P.K. Classification Accuracy of Discriminant Analysis, Artificial Neural Networks, and Decision Trees for Weed and Nitrogen Stress Detection in Corn. *Trans. ASAE* **2005**, *35*, 1261–1268. [CrossRef]
64. Filella, I.; Serrano, L.; Serra, J.; Penuelas, J. Evaluating Wheat Nitrogen Status with Canopy Reflectance Indices and Discriminant Analysis. *Crop Sci.* **1995**, *35*, 1400–1405. [CrossRef]

13

Prediction of Sugarcane Yield based on NDVI and Concentration of Leaf-Tissue Nutrients in Fields Managed with Straw Removal

Izaias Pinheiro Lisboa, Júnior Melo Damian *, Maurício Roberto Cherubin, Pedro Paulo Silva Barros, Peterson Ricardo Fiorio, Carlos Clemente Cerri and Carlos Eduardo Pellegrino Cerri

"Luiz de Queiroz" College of Agriculture, University of São Paulo, 11 Pádua Dias Avenue, Piracicaba, SP 13418-900, Brazil; iplisboaa@gmail.com (I.P.L.); cherubin@usp.br (M.R.C.); pedropaulo@usp.br (P.P.S.B.); fiorio@usp.br (P.R.F.); cerri@cena.usp.br (C.C.C.); cepcerri@usp.br (C.E.P.C.)

* Correspondence: damianjrm@usp.br

Abstract: The total or partial removal of sugarcane (*Saccharum* spp. L.) straw for bioenergy production may deplete soil quality and consequently affect negatively crop yield. Plants with lower yield potential may present lower concentration of leaf-tissue nutrients, which in turn changes light reflectance of canopy in different wavelengths. Therefore, vegetation indexes, such as the normalized difference vegetation index (*NDVI*) associated with concentration of leaf-tissue nutrients could be a useful tool for monitoring potential sugarcane yield changes under straw management. Two sites in São Paulo state, Brazil were utilized to evaluate the potential of *NDVI* for monitoring sugarcane yield changes imposed by different straw removal rates. The treatments were established with 0%, 25%, 50%, and 100% straw removal. The data used for the *NDVI* calculation was obtained using satellite images (CBERS-4) and hyperspectral sensor (FieldSpec Spectroradiometer, Malvern Panalytical, Almelo, Netherlands). Besides sugarcane yield, the concentration of the leaf-tissue nutrients (N, P, K, Ca, and S) were also determined. The *NDVI* efficiently predicted sugarcane yield under different rates of straw removal, with the highest performance achieved with *NDVI* derived from satellite images than hyperspectral sensor. In addition, leaf-tissue N and P concentrations were also important parameters to compose the prediction models of sugarcane yield. A prediction model approach based on data of *NDVI* and leaf-tissue nutrient concentrations may help the Brazilian sugarcane sector to monitor crop yield changes in areas intensively managed for bioenergy production.

Keywords: crop residue management; remote sensing; satellite images; hyperspectral sensor; vegetation index; yield monitoring

---

1. Introduction

Brazil is the largest sugarcane producer in the world, representing 40% of global production [1]. Sugarcane area extends for approximately 9 million ha, with an ethanol production at around 27 billion liters and about 40 million tonnes of sugar [2]. Although Brazil is a major player in the global biofuel market, growing internal and external demands have been stimulated an increase in ethanol production. Considering the current global production of 98.6 billion liters [3], it will be necessary to double ethanol production to meet the estimated global demand of 200 billion liters in 2021 [4]. In addition, bioelectricity represents an important proportion of the revenues in the industry, and it is projected that about 17% of the domestic electric energy production in Brazil will be provided by sugarcane biomass by 2023 [5].

An alternative to increase bioenergy production, without expanding area, is to use lignocellulosic raw materials, such as bagasse and straw. In this process, crop residues, rich in carbohydrates, are converted into simple sugars using enzymatic hydrolysis, followed by yeast fermentation to produce ethanol, or crop residues are burnt to generate bioelectricity [6].

More recently, sugarcane straw has been considered as an alternative raw material to address the increasing demand for bioenergy production [7]. Since adoption of mechanized harvesting of sugarcane, about 10 to 20 Mg ha^{-1} (as dry mass) remains on the field annually [8]. However, integral or partially straw retention on the soil surface is essential to support several soil functions [9,10], such as thermal regulation [11], control of soil compaction [12], water retention [13], carbon storage [14], nutrient cycling [15], biological activities [16], and control of erosion [17]. Consequently, changes in soil quality can affect, directly or indirectly, crop production [14,18,19]. Therefore, monitoring the yield of sugarcane could be a strategy to evaluate the effects of varying the amount of straw in the soil–plant system.

Crop yield can be monitored by vegetation indexes. Indeed, Tucker et al. [20] reported that up to 64% of the grain yield variation was explained by spectral data. The applications of *NDVI* has been broadly increased, not only because *NDVI* can be derived from data of either hyperspectral sensors or satellite images [21], but also because it is closely correlated with yield of different crops [22], such as maize (*Zea mays* L.) [23], soybean (*Glycine max* L.) [24], and sugarcane [25]. Additionally, deficiency and excess of nutrients in the leaves can be diagnosed through spectral responses obtained by sensors and, consequently, by vegetation indexes (e.g., *NDVI*) [26]). The variation in the concentration of nutrients in the leaves indicates physiological stress, which can be caused by diverse limiting factors that occurring in the production environment [27].

Although *NDVI* has been used for different purposes, as previously stated, in our knowledge, there are no studies in the literature in which *NDVI* and concentration of leaf-tissue nutrients under sugarcane straw removal management were used to develop predictive models for stalk yield. Thus, the following hypotheses were established: (i) the *NDVI* and concentration of leaf-tissue nutrients are efficient parameters to predict sugarcane yield under different straw removal rates; (ii) the *NDVI* obtained using data from satellite images or hyperspectral sensors have similar performance for predicting sugarcane yield; and (iii) the nutritional status of the crop (evaluated by the leaf-tissue nutrient concentrations) is affected by the different straw removal rates. To test the hypotheses, a study was conducted in areas managed with straw removal in southeastern Brazil to predict sugarcane yield based on *NDVI* and plant nutritional status. In addition, the efficiency of the *NDVI* data for predicting sugarcane yield, obtained using remote sensing imagery from various platforms, was compared.

2. Materials and Methods

2.1. Study Sites

The field study was conducted at two sites in southeastern Brazil, which represents the largest sugarcane-producing region of the country. These areas are located in São Paulo state, near to Capivari, at the Bom Retiro mill (Area 1) (Lat. 22°59′42″ S; Long. 47°30′34″ W) and Valparaiso, at the Univalem mill (Area 2) (Lat. 21°14′48″ S; Long. 50°47′04″ W) (Figure 1).

Soil is classified as Rhodic Kandiudox [28] with a sandy clay loam texture in Area 1, whereas in Area 2, the soil is classified as Kanhaplic Haplustults [28] with a sandy loam texture.

The regional climate for the Area 1 is humid subtropical-Cwa type (Köppen classification) characterized by dry winters and hot summers, with a mean annual temperature of 21.8 °C and annual precipitation of 1289 mm (Figure 2A). In the Area 2, the climate is tropical-Aw type, characterized by dry winters, with a mean annual temperature of 23.4 °C and annual precipitation of 1241 mm (Figure 2B). Rainfall at both sites is concentrated in the spring and summer (October to April), while the dry season is in the autumn and winter (May to September).

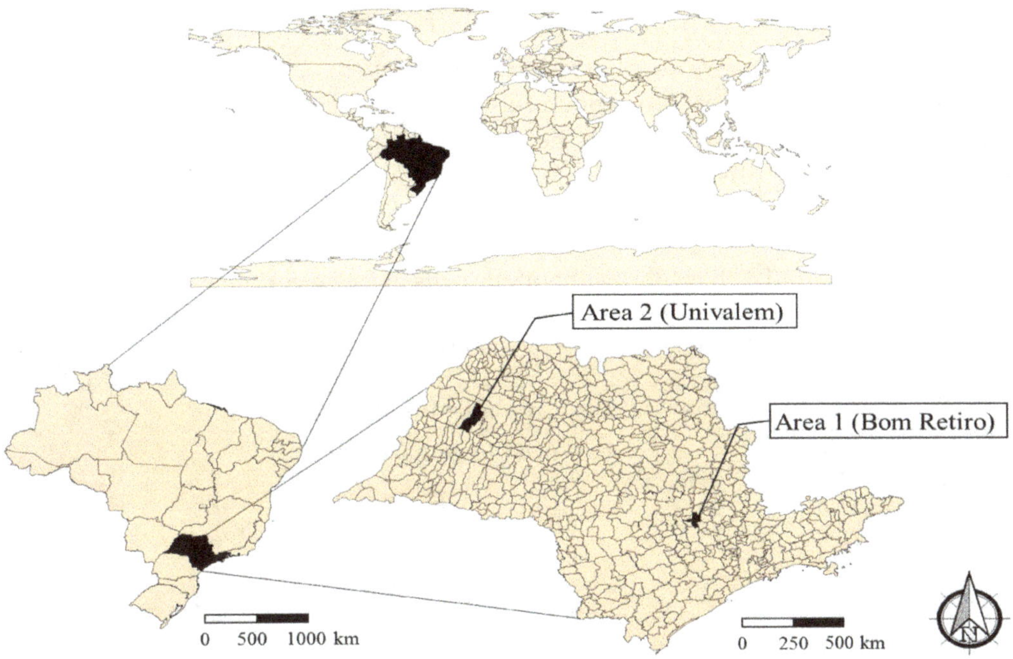

Figure 1. Geographical location of the experimental areas used in the study.

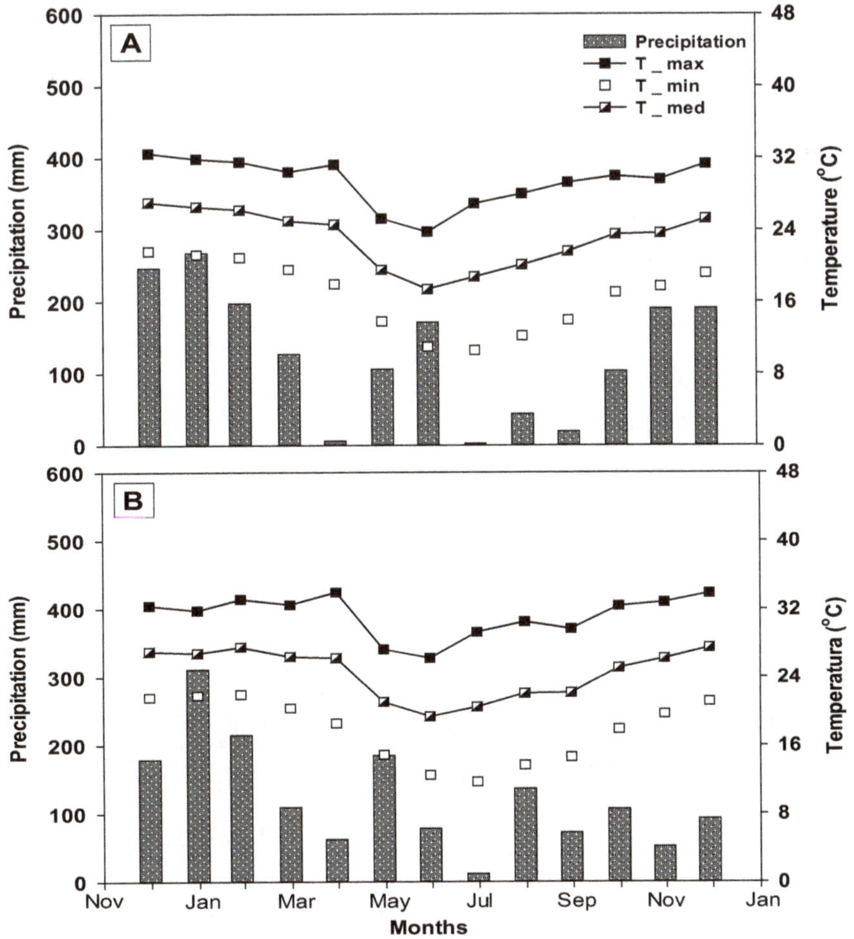

Figure 2. Mean monthly temperatures (maximum, mean, and minimum) (°C) and monthly precipitation (mm) in Area 1 (Capivari, SP) (**A**) and Area 2 (Valparaiso, SP) (**B**) from 1 December 2015 to 31 December 2016. Sources: CEPAGRI [29] and ESALQ [30].

2.2. Experimental Design

Within each site, sugarcane was planted in 15 February 2013, using the varieties CTC 14 and RB 867515 in Area 1 and Area 2, respectively. These varieties are broadly cultivated in this sugarcane-producing region of Brazil. Before the experiment was installed, the areas were sampled in order to obtain a chemical and physical characterization of the soils (Table 1), as previously presented by Satiro et al. [12].

Table 1. Initial characterization of the soils in the experimental areas.

Soil Layer (cm)	pH	C	P	K	Ca	Mg	Base Saturation	Al Saturation	Clay
	H_2O	g kg^{-1}	mg dm^{-3}	mmol$_c$ dm^{-3}			%		
				Area 1 (Bom Retiro)					
0–10	5.2	11.3	29.3	9.3	26.1	7.7	68.8	0.8	33
10–20	4.8	11.0	24.9	5.1	19.0	5.9	54.7	3.5	33
20–30	4.5	9.4	22.1	3.3	12.5	2.95	36.8	4.2	34
				Area 2 (Univalem)					
0–10	5.2	6.1	17.4	3.3	9.3	2.9	51.1	2.4	11
10–20	4.8	5.5	14.1	2.6	4.8	1.5	34.8	5.6	11
20–30	4.5	4.9	12.7	2.1	3.6	1.0	27.5	7.4	12

The cane plant (i.e., first cycle) was cultivated for 20 months, thereafter the harvested was performed in 15 October 2014 and the experiments with different rates of straw removal were established. After 14 months, in 17 December 2015, the areas were harvested again and similar experiments were conducted. Leaf sampling and sensor data collection were done during the second sugarcane ratoon season (Figure 3). The experiments were mechanically established with harvester, which was set up with different angular velocities on the primary extractor fan while the secondary extractor fan was switched on or off to remove different quantities of sugarcane straw. Details on how extractors were set up are described in Lisboa et al. [7].

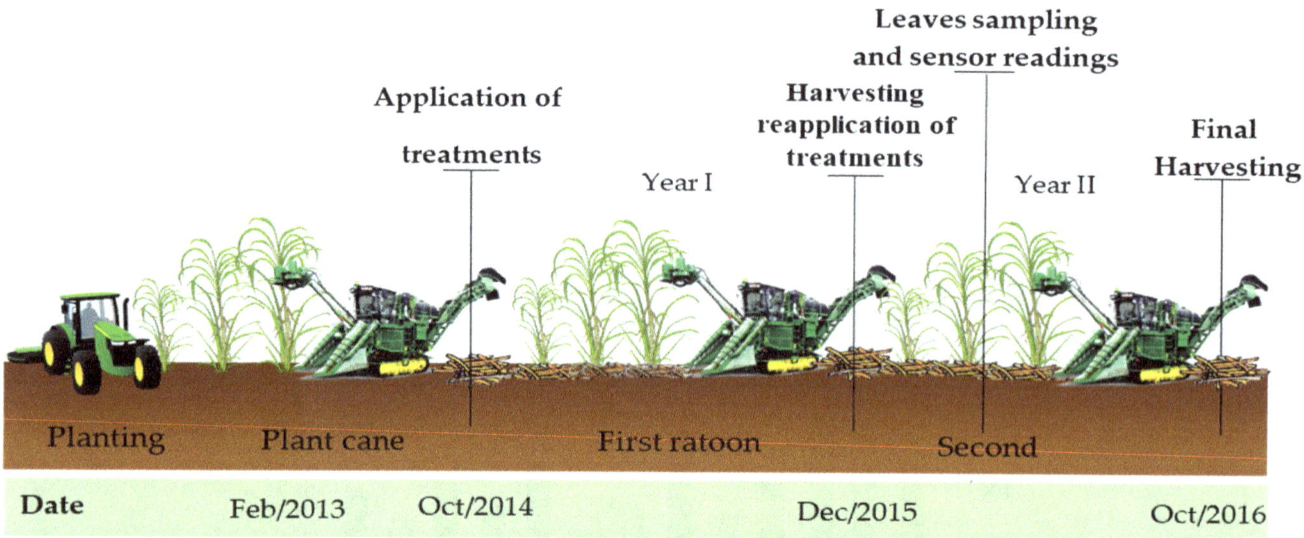

Figure 3. Timeline of the experiments establishment (at harvesting) and conduction over two years in southeast of Brazil. Source: adapted from Lisboa et al. [19].

The experimental design was randomized blocks with four treatments (i.e., straw removal rates), and four replications (four plots of ~50 × 25 m). Initially, the aim was to remove the amount of straw proportional to 0%, 25%, 50%, and 100% of the straw yield in each area. However, under field conditions, the exact proportion was hard to be achieved, but the removal rates were very close to those proposed (Table 2).

Table 2. Amount of sugarcane straw left on soil surface after each treatment in the first and second years in each experimental area.

Straw Removal Rate (%)	Area 1 (Bom Retiro)		Area 2 (Univalem)	
	Year I	Year II	Year I	Year II
	Amount of straw left on the soil surface (Mg ha^{-1}) #			
100	0.0 ± 0.0	0.0 ± 0.0	0.0 ± 0.0	0.0 ± 0.0
50	8.7 ± 0.9	5.5 ± 1.0	10.2 ± 0.9	6.8 ± 0.4
25	15.1 ± 1.5	10.5 ± 0.1	12.5 ± 0.9	11.6 ± 0.5
0	18.9 ± 1.6	13.6 ± 2.0	16.4 ± 1.4	13.7 ± 1.1

Dry mass; Source: adapted from Lisboa et al. [19].

At the start of each experiment period, the sugarcane straw was sampled to determine its elemental composition (Table 3).

Table 3. Carbon (C) and macronutrient (nitrogen—N, phosphorus—P, potassium—K, calcium—Ca, magnesium—Mg, sulphur—S) content of the sugarcane straw used in each experiment.

Sites	C	N	P	K	Ca	Mg	S	C:N Ratio
	g kg^{-1}							
Area 1 I (Bom Retiro)	479	2.58	0.39	1.66	1.96	1.38	0.45	177
Area 1 II (Bom Retiro)	470	3.10	0.34	0.56	2.44	1.24	0.35	152
Area 2 I (Univalem)	467	4.02	0.38	2.45	2.44	1.54	0.74	116
Area 2 II (Univalem)	422	6.04	0.58	1.30	8.55	2.55	0.95	73

I and II denote the first and second sugarcane ratoon, respectively. Source: adapted from Lisboa et al. [19].

Overall, this study frames the development of models based on *NDVI* and plant nutritional status to predict sugarcane yield in field managed with straw removal. Plant tillering, growth, and stalk yield under the same straw management were discussed in detail by Lisboa et al. [19]. Thus, part of the primary data (i.e., amount of straw, characterization and stalk yield) presented on this study (mainly in Material and Methods) were originally shown in Lisboa et al. [19].

2.3. Calculation of the NDVI

The data were obtained in the field using the FieldSpec Spectroradiometer (ASD–Analytical Spectral Devices Inc., Boulder, CO, USA) hyperspectral sensor. This equipment is classified as a passive hyperspectral sensor that uses the visible and infrared wavelengths (325 to 1075 nm), with a spectral resolution of 3 nm and a view angle of 25°. Before the canopy reading was initiated, the spectroradiometer was calibrated using the standard Lambertian plate that accompanies the apparatus. This calibration was repeated after reading approximately four plots or whenever a change of light intensity was observed in the field. The *NDVI* was calculated using Equation (1).

$$NDVI = \frac{(\mathrm{NIR} - \mathrm{R})}{(\mathrm{NIR} + \mathrm{R})} \quad (1)$$

where *NDVI* is the Normalized Difference Vegetation Index (unitless); NIR is the near infrared band and R is the red band; NIR represents the reflected light at the near infrared band and R represents the reflected light at the red band.

The field evaluations with the hyperspectral sensor were carried out on 28 April 2016 and 4 March 2016 for Areas 1 and 2, respectively, four months after harvesting. This period coincides with the recommended period for foliar diagnosis in sugarcane crop [31]. The evaluations were always carried out on sunny days, between 10:00 a.m. and 2:00 p.m., with the sensor positioned 1 m above the crop canopy and a field of view of 25°, which allowed the evaluation of a circular area of approximately 0.25 m^2 (Figure 4).

Figure 4. Representation of the hyperspectral sensor in the field (**a**) and Lambertini plate used for calibration (**b**).

The satellite images were obtained from the satellite CBERS-4 (China Brazil Earth Resources Satellite) on the 17 March 2016 and the 4 May 2016 for Areas 1 and 2, respectively. The images presented a spatial resolution of 5 m, radiometric resolution of 8 bits and wavelength of 0.50–0.840 μm. As the panchromatic band of CBERS-4 was supplied in digital numbers without atmospheric calibration, the images were submitted to atmospheric correction according to Vermote et al. [32]. To perform *NDVI* calculations, the image was converted to surface reflectance according to the method described by Carlotto [33].

2.4. Plant Parameters

After measurements to calculate the *NDVI*, 50 leaves (i.e., the diagnose leaf) were randomly collected from each plot for the determination of the leaf-tissue nutrient content. The third leaf from top to bottom of the stalk with clearly visible dewlap, is the most photosynthetically active and is used for monitoring the nutritional status of sugarcane crop [31].

At the end of the ratoon cycle, approximately one year after previous harvesting, the sugarcane yield was quantified. The fresh stalk mass was mechanically harvested from the five central rows (500 m long covering an area of 525 m^2) of each plot, weighed in the field using a wagon coupled to a balance and the results expressed in Mg ha^{-1}.

2.5. Data Analysis

The data were initially subjected to exploratory analysis (descriptive statistics), aiming to verify the position and dispersion of the data using the Statistical Analysis System (SAS) 9.4 Software Package for Windows 8 (SAS Inc, Cary, NC, USA).

In order to select the explaining variables of stalk yield, a stepwise multiple regression ($Y = a + b_1x^1 + b_2x^2 + ... + b_nx^n$) was applied. This functions by the systematic addition or removal of variables in the regression, based on a statistical test of significance for each variable ($p < 0.05$). The final model includes only those variables that have a decisive influence on the dependent variable. The models were evaluated according to the coefficient of determination, residual standard error and the Durbin–Watson test. The Durbin–Watson test ranges from 0 to 4, where values closer to 2 indicate optimal values [34], i.e., the absence of autocorrelation in the data. The means of variables selected by stepwise multiple regression were compared according to Tukey's test ($p < 0.05$). In addition, the overall model that pools data from both studied areas was tested through sensitivity analysis, where position measurements (maximum, minimum, and mean) of each selected variable were inserted independently in the model, aiming to evaluate the resulting variations in sugarcane yield.

Finally, a model validation process was carried out with 30% of the data that were not used to generate the models. The relationship between the observed and predicted values were evaluated by the root mean square error (RMSE). The Pearson correlation matrix was also used to evaluate the relationship between observed and predicted productivity of sugarcane with variables selected with the stepwise models.

3. Results

3.1. NDVI Values

The *NDVI* obtained using the data from the hyperspectral sensor and the satellite images presented differences with respect to the distribution and magnitude of the values between the experimental areas (Figure 5). In Area 1, the *NDVI* values using the sensor and the satellite images ranged between 0.60 to 0.80 and 0.40 to 0.60, respectively. However, in Area 2 the *NDVI* values were smaller, ranging between 0.40 to 0.70 and 0.20 to 0.50 for the sensor and the satellite images, respectively. On average, *NDVI* values (from both methodologies) were 34% higher in Area 1 than in Area 2.

In both experimental areas, only the *NDVI* values derived from hyperspectral sensor followed a normal distribution (Figure 5). The largest differences were found in the *NDVI* values derived from the satellite images, especially when the removal rates were 25% and 50% for Area 1 and Area 2, which did not follow a normal data distribution. Although there are variations between *NDVI* derived by these two methods, the results showed a quite similar pattern in the variations of the *NDVI* data due to straw removal rates.

3.2. Sugarcane Leaf-Tissue Nutrient Concentration

The nutrient concentration in the sugarcane leaves varied according to the straw removal rate at both sites (Figures 6 and 7). The greatest variation was observed in the elements K (5 to 15 g/kg^{-1}) and Ca (2 to 6 g/kg^{-1}), whereas that the variation was less intense for the other elements. Despite that, for the majority of cases, nutrient data did not followed a normal distribution (Figures 6 and 7).

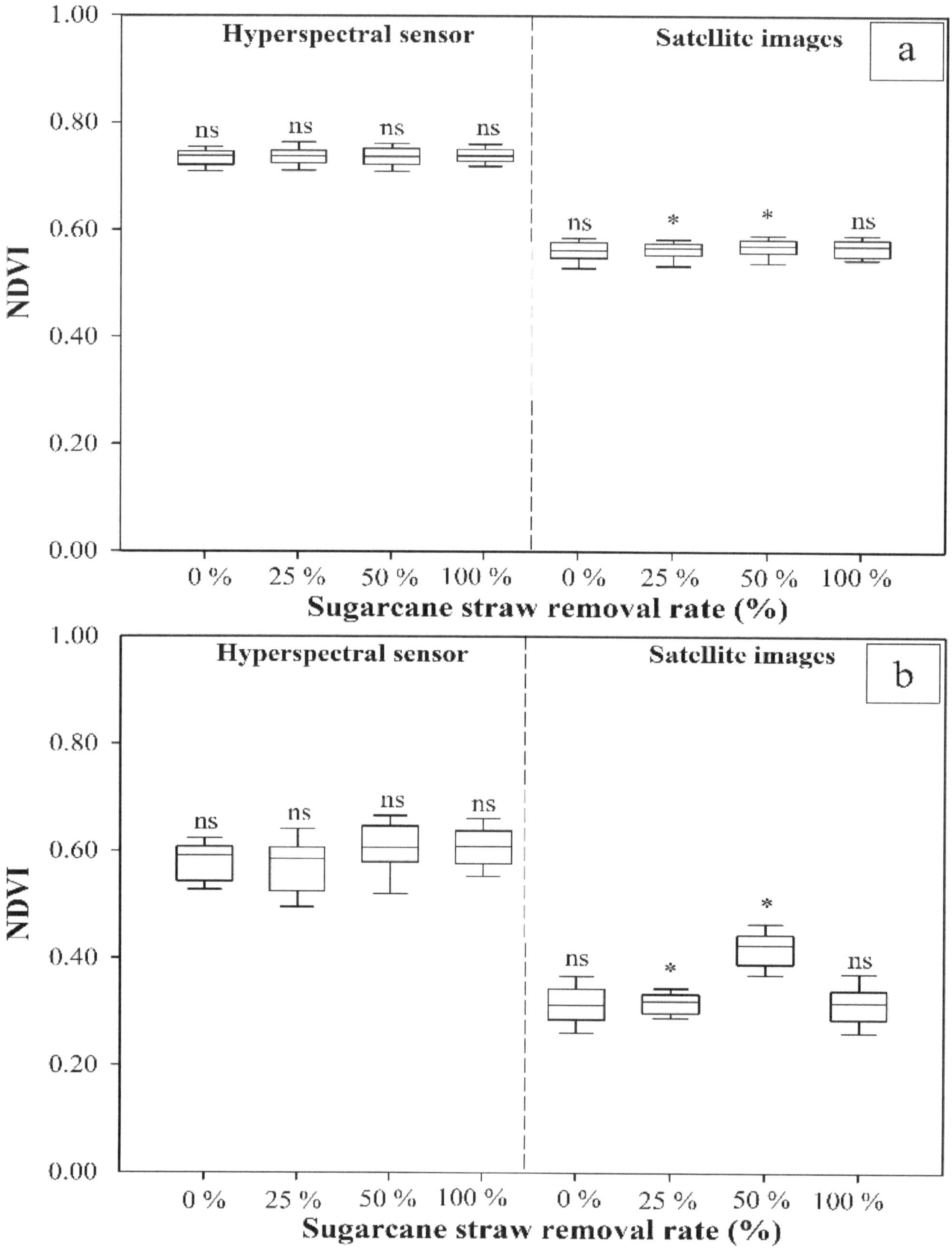

Figure 5. Values of *NDVI* for Areas 1 (**a**) and 2 (**b**) at the different rates of straw removal. Shapiro–Wilk Test for normal distribution, where: * significant ($p < 0.05$) and ns not significant. When significant, it indicates that the hypothesis for normal distribution is rejected.

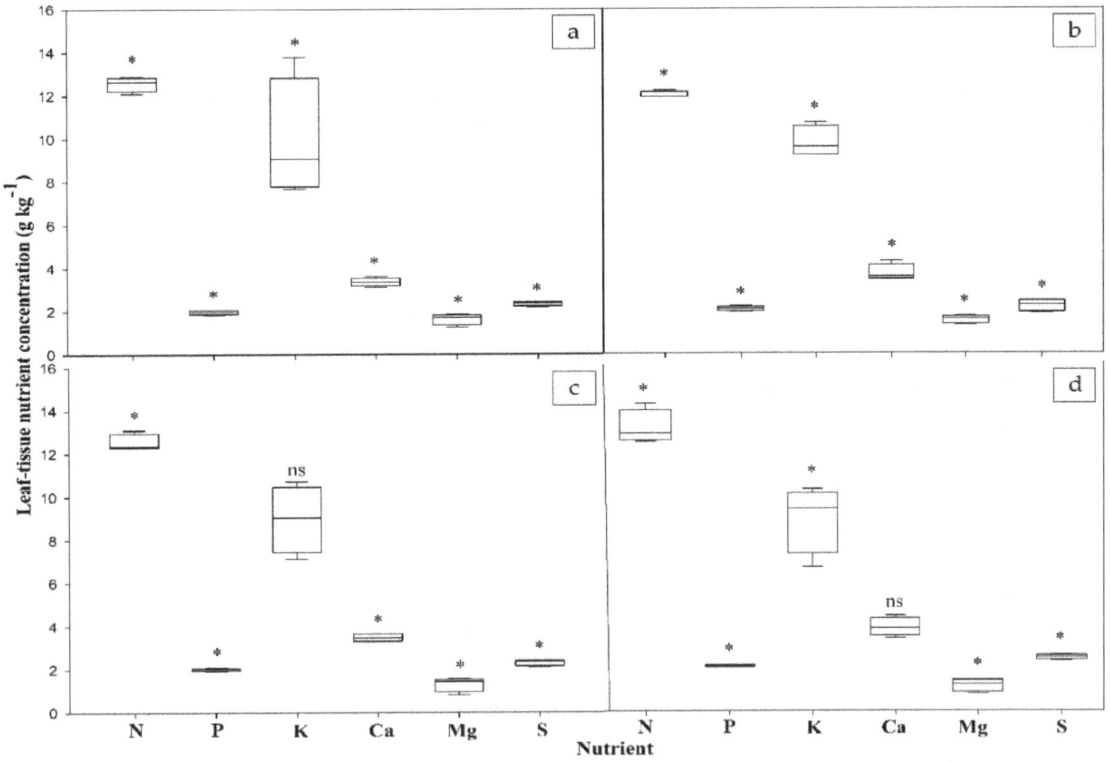

Figure 6. Leaf-tissue concentrations of N, P, K, Ca, Mg, and S in sugarcane grown under straw removal rates of 0 (**a**), 25 (**b**), 50 (**c**), and 100% (**d**) in Area 1. Shapiro–Wilk Test for normal distribution, where: * significant ($p < 0.05$) and ns not significant. When significant, it indicates that the hypothesis for normal distribution is rejected.

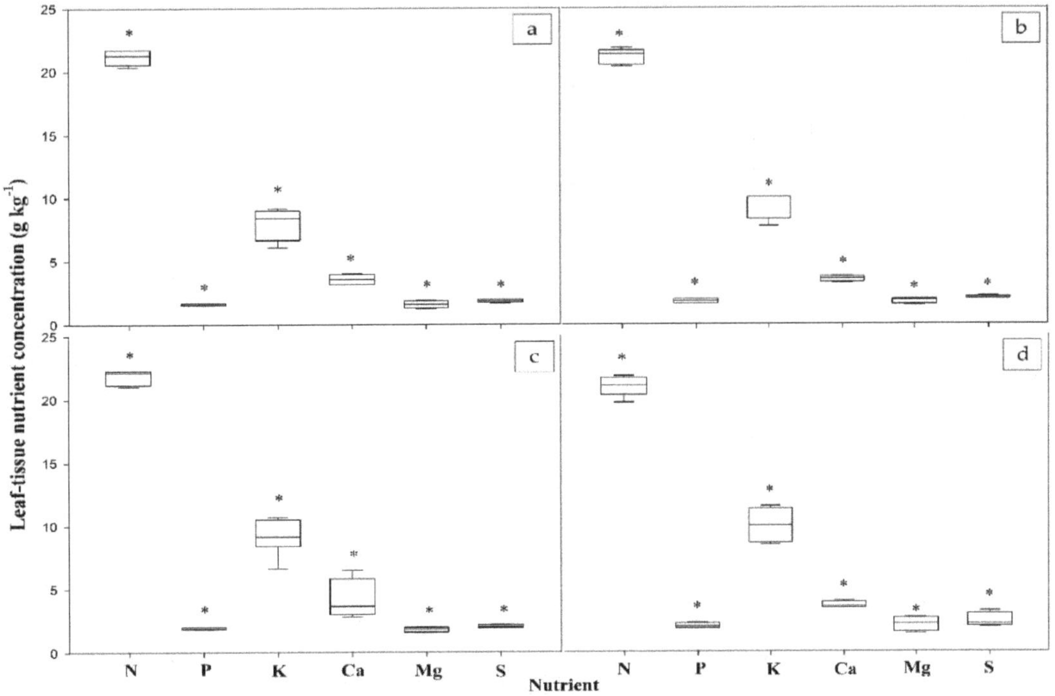

Figure 7. Leaf-tissue concentration of N, P, K, Ca, Mg, and S in sugarcane grown under straw removal rates of 0 (**a**), 25 (**b**), 50 (**c**), and 100% (**d**) in Area 2. Shapiro–Wilk Test for normal distribution, where: * significant ($p < 0.05$) and ns not significant. When significant, it indicates that the hypothesis for normal distribution is rejected.

3.3. Sugarcane Yield

Sugarcane yield varied from 60 to 85 Mg ha^{-1} and from 10 to 30 Mg ha^{-1} for the Areas 1 and 2, respectively. On average, sugarcane yield was three times higher in Area 1 than in Area 2 (Figure 8). Our findings showed that straw removal rates did not linearly affect sugarcane yield, but it depends on the site-specific characteristics related to soil, climate, and variety.

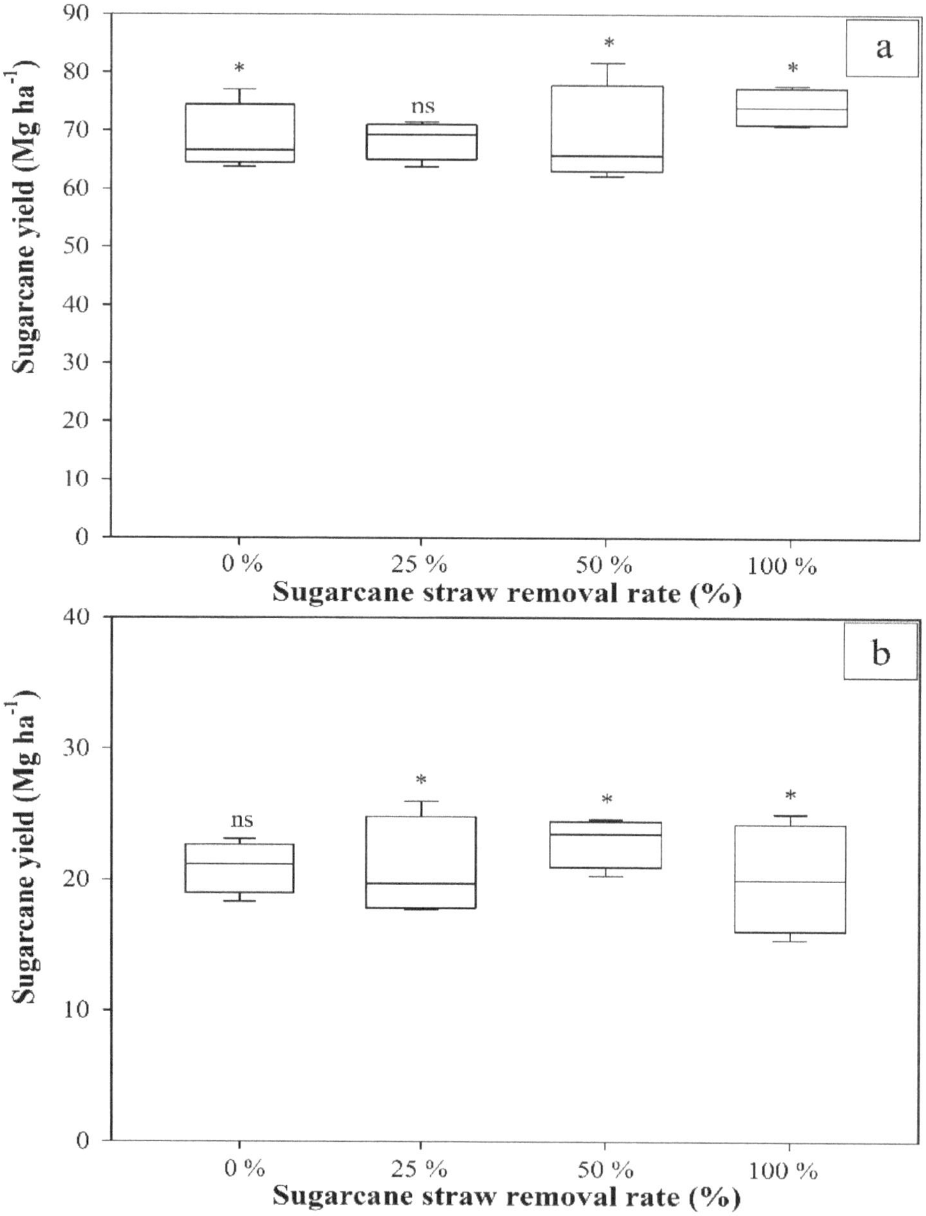

Figure 8. Sugarcane yield (Mg ha^{-1}) changes induced by straw removal rates in Areas 1 (**a**) and 2 (**b**). Shapiro–Wilk Test for normal distribution, where: * significant ($p < 0.05$) and ns not significant. When significant, it indicates that the hypothesis for normal distribution is rejected.

The data did not follow a normal distribution for the majority of the treatments (Figure 8), except for data from 25% and 0% straw removal plots in Area 1 and 2, respectively.

3.4. Modelling Sugarcane Yields by Stepwise Analysis

Explaining variable of sugarcane yield were selected for each straw removal rate using stepwise analysis (Table 4). The *NDVI* obtained from satellite images was more efficient to predict sugarcane yield compared to *NDVI* derived from hyperspectral sensor, and therefore, this parameter was included in all the models.

Table 4. Stepwise models for estimating the components of sugarcane yield under straw removal rates in Areas 1 and 2.

Straw Removal Rate (%)	Models	r^2 *	SRE	DW
	Area 1			
0	$y = 205.38 - 48.66\ \mathrm{P} - 0.83\ \mathrm{K} - 58.36\ NDVI_{SI}$	0.94	1.76	1.78
25	$y = 546.65 - 58.07\ \mathrm{P} - 30.31\ \mathrm{N} + 7.18\ NDVI_{SI}$	0.91	1.31	1.52
50	$y = 322.62 - 108.24\ \mathrm{P} - 6.33\ \mathrm{N} + 60.36\ NDVI_{SI}$	0.88	4.07	1.48
100	$y = 193.25 - 56.54\ \mathrm{P} + 4.25\ NDVI_{SI} + 0.27\ \mathrm{N}$	0.85	2.10	1.65
Overall (Area 1)	$y = 136.70 + 42.84\ NDVI_{SI} - 51.19\ \mathrm{P} + 1.61\ \mathrm{N}$	0.97	2.56	1.56
	Area 2			
0	$y = -13.20 + 1.75\ \mathrm{N} - 13.76\ NDVI_{SI}$	0.87	6.64	1.44
25	$y = -166.31 + 25.60\ \mathrm{P} + 6.51\ \mathrm{N} + 22.76\ NDVI_{SI}$	0.91	8.65	1.65
50	$y = 42.25 - 12.41\ \mathrm{P} + 22.23\ NDVI_{SI}$	0.76	4.50	1.59
100	$y = 134.25 - 20.90\ \mathrm{Ca} - 20.51\ \mathrm{P} + 31.89\ NDVI_{SI}$	0.94	4.29	1.56
Overall (Area 2)	$y = -127.21 + 3.80\ NDVI_{SI} - 0.26\ \mathrm{N} + 54.88\ \mathrm{P}$	0.88	3.86	1.61
Overall (Area 1 and Area 2)	$y = -150.028 + 79.11\ NDVI_{SI} - 3.65\ \mathrm{N} + 115.28\ \mathrm{P}$	0.89	1.79	1.72

SRE: standard residual error (%); DW: Durbin–Watson test. N: nitrogen; P: phosphorus; K: potassium; Ca: calcium; and $NDVI_{SI}$: *NDVI* obtained by satellite images. * correlation coefficient.

The nutrients that most influenced crop yield were P and N, being included in the model for at least one straw removal rate in both experimental areas; while the K and Ca were only included in the models for the removal rates of 0% and 25% for Area 1 and 100% for Area 2. Overall models for Area 1 and 2, separately and together, included the parameters such as *NDVI* obtained from the satellite images, leaf-tissue concentration of P and N.

The models presented efficient performance to predict the sugarcane yield (Table 4). In Area 1, the r^2 values varied between 0.85 and 0.94 with the highest SRE encountered under the 50% removal rate (4.07%). The r^2 values varied between 0.76 and 0.94 for Area 2, with the highest SRE observed under the 25% removal rate (8.65%). Although SRE of 8.65% was the highest value observed in the two areas it was considered acceptable by the validation of the model. In the overall models, the r^2 and SRE values for Areas 1 and 2 were 0.97% and 2.56%, and 0.88% and 3.86%, respectively, and for the general model representing both areas the r^2 and SRE values were 0.89% and 1.79%, respectively. For the models generated individually for each straw removal rate in each of the two areas, for the general models for each area and for the general model representing both areas, the Durbin–Watson test values were similar, with values ranging between 1.44 and 1.78. These values were close to 2 (ideal threshold), showing that the data do not present autocorrelation [34]. The sensitivity analysis of the overall model for Areas 1 and 2, showed that yield ranged 11% (i.e., 39.2 to 52.5 Mg ha^{-1}) when the selected parameter changed from the minimum to maximum value. Individually, variation on *NDVI*, P, and N impacts were 55.84%, 55.03%, and 53.10% in the sugarcane yield.

3.5. *Effect of Sugarcane Straw Removal on Selected Yield-Explaining Variables*

The comparison of the means of the selected variables by the stepwise model indicated that sugarcane yield was not altered by straw removal management within the study period (Figure 9a). In relation to the variables that explained the sugarcane yield, it was observed that 25% straw removal rate induced a reduction in the *NDVI*, using the satellite image data (Figure 9b), and the leaf-tissue P content (Figure 9d) in Area 1. However, 50% straw removal rate induced an increase in the *NDVI*, using the satellite image data (Figure 9b) and the leaf-tissue N (Figure 9c) and P (Figure 9d) contents in Area 2.

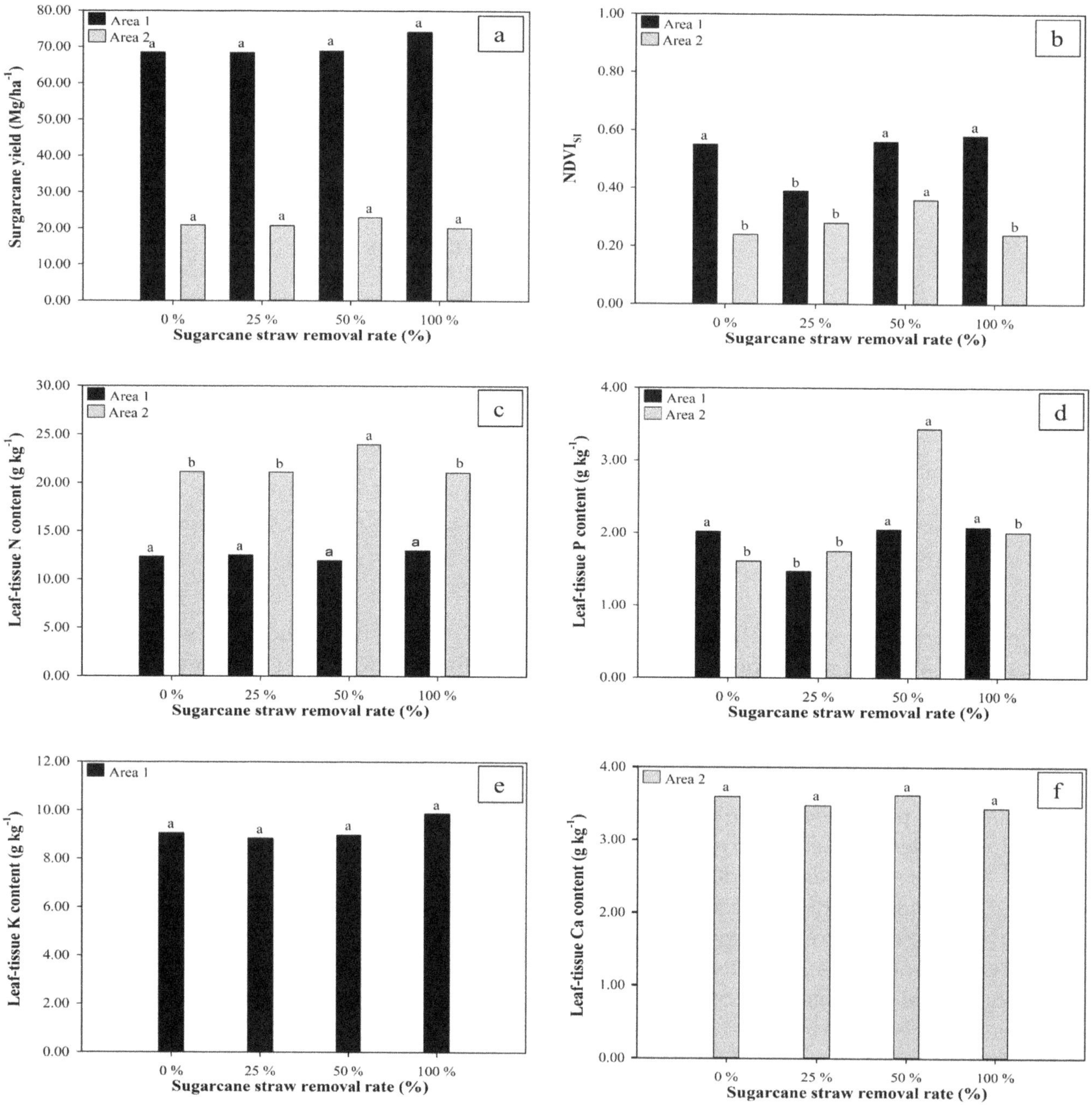

Figure 9. Comparison of the sugarcane straw removal rates with the variables selected by the stepwise model means. Sugarcane yield (**a**), the *NDVI* obtained from the satellite image data (**b**), leaf-tissue N (**c**), P (**d**), K (**e**), and Ca (**f**) contents in Areas 1 and 2. Means followed by the same letter are not statistically different by the Tukey test ($p \leq 0.05$).

3.6. Validation of Models

The models were validated based on relationship between observed and predicted values (Figure 10). In general, the results revealed increases of estimate errors (i.e., RMSE values) with the increase in the straw removal rate (0% to 100% straw removal rate). The highest values were observed for the rates of 50 (10.34 Mg ha^{-1}) and 100% (4.15 Mg ha^{-1}) for Areas 1 and 2 respectively. The overall model, including Areas 1 and 2, was efficient to predict the sugarcane yield, presenting RMSE of 0.77 Mg ha^{-1}.

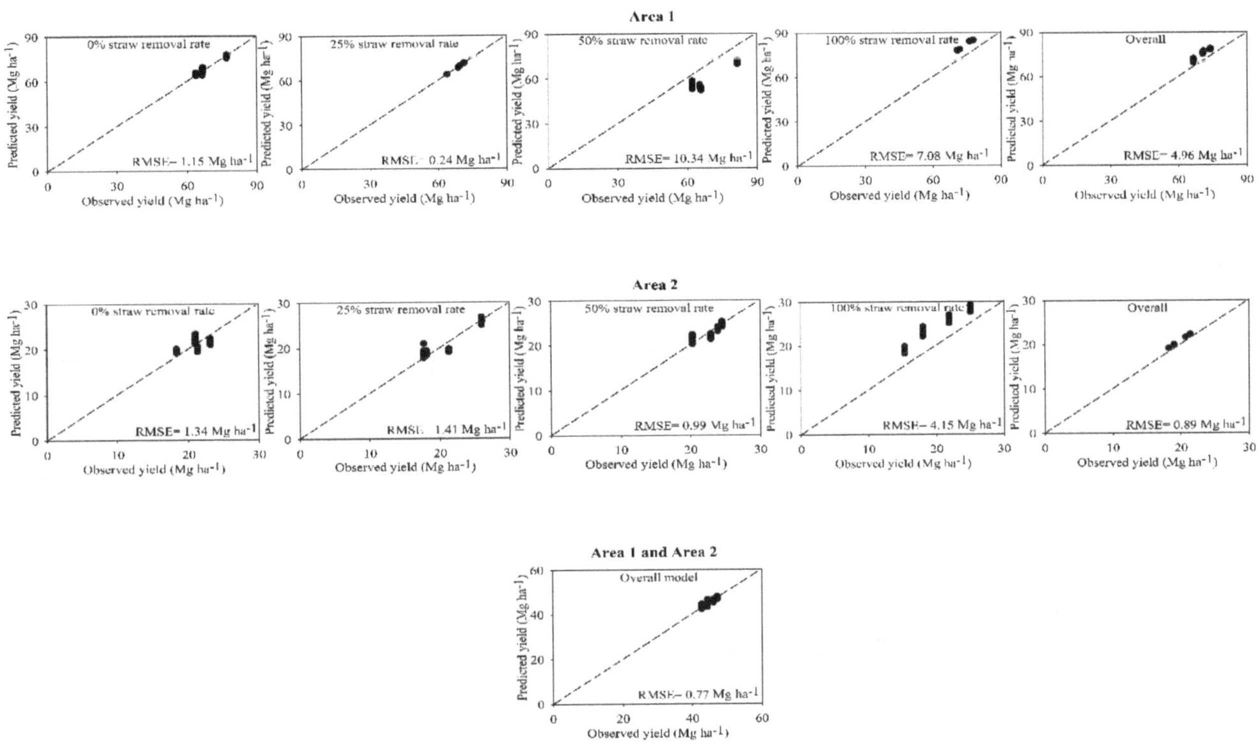

Figure 10. Comparison of sugarcane yield observed with predicted values with Stepwise models. The diagonal dashed line is the 1:1 line, whereas RMSE is the root mean square error.

Observed (OY) and predicted (PY) yield data presented satisfactory performance when compared to the variables selected through the stepwise models (Table 5). Leaf-tissue P concentration presented negative correlation with OY and PY yield data for the Area 1 subjected to all rates of straw removal (i.e., 0%, 25%, 50%, and 100%). Similar pattern was also observed to the general model for the same area. While for Area 2, P presented negative correlation with PY and OY data only under 50% of straw removal. In addition, N presented positive and significant correlations with OY and PY data from Area 1 under 50% and 100% of straw removal as well as for general model for this area, coefficients of correlation (r) ranged from 0.33 to 0.68. The correlations among $NDVI_{SI}$, OY, and PY data were positive for all rates of straw removal and overall model for the area 1. Same pattern was verified for all rates of straw removal in the Area 2, except for the rate of 25% of straw removal; coefficients of correlation ranged from -0.40 to 0.93 in Area 2.

It taking into account overall model established for both areas, leaf-tissue concentrations of P and N were significantly correlated with stalk yield, but in different order, as follow: P negatively correlated with OY ($r = -0.45$) and PY ($r = -0.50$) data, while N was positively correlated with OY ($r = 0.55$) and PY ($r = 0.58$) data. Similarly, $NDVI_{SI}$ correlated positively with OY ($r = 0.60$, $p < 0.05$) and PY ($r = 0.68$, $p < 0.05$) data.

Table 5. Correlation matrix between observed and predicted sugarcane yield values with selected variables with stepwise models.

Area 1							Area 2					
						0%						
	P	K	$NDVI_{SI}$	OY	PY			N	$NDVI_{SI}$	OY	PY	
P	-						N	-				
K	0.19	-					$NDVI_{SI}$	−0.14	-			
$NDVI_{SI}$	−0.23	0.22	-				OY	0.54 *	0.37 *	-		
OY	−0.84 **	0.60 **	0.71 **	-			PY	0.87 **	0.60 **	0.75 **	-	
PY	−0.86 **	0.62 **	0.74 **	0.97 **	-							
						25%						
	P	N	$NDVI_{SI}$	OY	PY			P	N	$NDVI_{SI}$	OY	PY
P	-						P	-				
N	−0.85 **	-					N	−0.49 *	-			
$NDVI_{SI}$	0.01	0.38	-				$NDVI_{SI}$	−0.56 *	0.02	-		
OY	−0.62 *	0.25	0.50 *	-			OY	0.62 **	0.23	−0.37 *	-	
PY	−0.71 **	0.23	0.45 *	0.98 **	-		PY	0.68 **	0.26	−0.40 *	0.90 **	-
						50%						
	P	N	$NDVI_{SI}$	OY	PY			P	$NDVI_{SI}$	OY	PY	
P	-						P	-				
N	0.15	-					$NDVI_{SI}$	−0.41 *	-			
$NDVI_{SI}$	−0.16	0.46 *	-				OY	−0.73 **	0.60 **	-		
OY	−0.90 **	0.33 *	0.84 **	-			PY	−0.88 **	0.76 **	0.89 **	-	
PY	−0.96 **	0.35 *	0.81 **	0.93 **	-							
						100%						
	P	N	$NDVI_{SI}$	OY	PY			P	Ca	$NDVI_{SI}$	OY	PY
P	-						P	-				
N	−0.62 **	-					Ca	−0.54 *	-			
$NDVI_{SI}$	−0.39 *	0.25	-				$NDVI_{SI}$	0.23	0.30	-		
OY	−0.96 **	0.56 **	0.73 **	-			OY	−0.22	0.64 **	0.30	-	
PY	−0.91 **	0.64 **	0.61 **	0.98 **	-		PY	−0.14	0.72 **	0.29	0.97 **	-
						Overall						
	P	N	$NDVI_{SI}$	OY	PY			P	N	$NDVI_{SI}$	OY	PY
P	-						P	-				
N	0.95 **	-					N	0.84 **	-			
$NDVI_{SI}$	−0.54 *	−0.28	-				$NDVI_{SI}$	−0.24	0.12	-		
OY	−0.84 **	0.68 **	0.86 **	-			OY	−0.28	−0.26	0.61 **	-	
PY	−0.81 **	0.61 **	0.93 **	0.96 **	-		PY	−0.24	−0.28	0.69 **	0.97 **	-

Table 5. *Cont.*

Area 1					**Area 2**
	Overall (Area 1 and Area 2)				
	P	N	$NDVI_{SI}$	OY	PY
P	-				
N	0.96 **	-			
$NDVI_{SI}$	−0.08	0.08	-		
OY	−0.45 *	0.55 *	0.60 **	-	
PY	−0.50 *	0.58 *	0.68 **	0.90 **	-

*, ** Significant at $p < 0.05$ and $p < 0.01$, respectively. OY: observed yield; PY: predicted yield; N: nitrogen; P: phosphorous; K: potassium; Ca: calcium and $NDVI_{SI}$: *NDVI* obtained by satellite images.

4. Discussion

4.1. Effect of Straw Removal on the Leaf-Tissue Nutrient Concentration and Sugarcane Yield

Sugarcane straw removal had little or even no impacts on leaf-tissue nutrient content (Figures 6 and 7). Nutrient cycling in the soil was minimally affected by the straw removal rates in the short term [12], and therefore, plant nutrition was little affected. Among the evaluated elements, K and Ca presented the largest variations in both experimental areas. The sugarcane straw can recycle about 80–100 kg ha^{-1} of K [35,36]. According to Rosolem et al. [37], more than 80% of the K is present in a soluble form in plant tissues, which makes this element easily leached from plant tissues [9] for later absorption by the plant root system. Unbalanced absorption of K may also change Ca content in plant tissues, since both elements interfere in the physiological process of transporting sugars into the plant [38].

Although K and Ca were the most variable in the plant tissues as a function of straw removal, N and P were the most sensitive nutrients to the straw removal rates (Figure 9c,d). When compared to other nutrients—such as P, K, Ca, Mg, and S—N presented the largest reduction in the soil with straw removal [39]. On the other hand, straw removal led to small reductions in the level of soil P, since the content of this element in straw is low (Table 3). However, the majority of P is highly adsorbed in the soil particles (especially Fe and Al oxides) [40], thus, small inputs of P in organic forms via straw decomposition can be important to increase plant-availability P in the soils [41]. In addition, the organic acids released during straw decomposition may compete for adsorption sites and increase the availability of phosphate ions in the soil solution [42].

The sugarcane yield varied three times between the two areas evaluated (Figure 8). This difference may be related to the greater soil fertility of Area 1, in which soil P, K, Ca, and Mg contents and base saturation percentage were on the average 42%, 55%, 69%, 67%, and 29%, respectively higher than Area 2. Thus, higher soil fertility in Area 1 likely allowed that plants accumulate higher levels of these nutrients in the leaf tissue, which is directly related to the increase of the sugarcane yield [43]. Sugarcane presents high nutritional needs once it a crop that remains in the field for several years (~5 to 8 years), with high nutrient removal in each annual harvesting cycle. Therefore, improper replenishment of nutrients is one of the reasons for yield loss during the annual harvesting cycles [44,45].

The sugarcane straw removal did not affect biomass production (Figure 9a). These results are in agreement with other studies reporting that the removal of up to 50% of the straw produced did not impact crop yield [18,19,46]. In addition, through an extensive literature review, Carvalho et al. [9] suggested that at least 7 Mg ha^{-1} of straw should remain on the soil surface to sustain suitable soil functioning and plant yield over time. However, as the crop tends to lose the productive potential (stalk and straw) over successive harvesting cycles [19,47], the amount of straw that should be removed after each harvesting cycle may be different.

Over the past few years, studies have brought very relevant understanding about the impacts of straw removal management on soil quality [12,14,36,48], SOC accumulation [49,50], greenhouse gas emissions [51–53], plant growth and stalk yield [18,19,46,54–56]. Despite this fact, future studies are essential to understand the long-term implications of straw removal management on soil and stalk yield. So far, we only have prediction performed by DayCent model of sugarcane straw removal impacts on SOC accumulation across the different soil types within the main core occupied with the crop in Brazil (i.e., central-south region) [49,50]. However, these estimates performed by Carvalho et al. [49] and Oliveira et al. [50] still have to be validated by long-term experiment, still inexistent in Brazil. The same statement is true for the yield prediction models developed in this present study, which gave us relevant insights related to the potentialities of NVDI and leaf-tissue nutrient concentration to predict sugarcane yield in fields impacted by straw removal management.

4.2. NDVI Acquisition by Satellite Image vs Hyperspectral Sensor

The quality of the data used to calculate the *NDVI* is affected by interference from diverse factors, depending on the form of data acquisition. When satellite images are used, although the data is collected with minimum cloud cover, the interference of these few clouds at the moment of data acquisition could alter the quality of the incident and reflected radiation [57]. Besides this, the capture of satellite images tends to suffer a greater influence from the soil than that experienced using the hyperspectral sensor, where readings are taken above the canopy and rows with minimal soil influence. Thus, the *NDVI* values using satellite images are normally lower [58]. These interferences explain the difference of 34% observed between the *NDVI* using the hyperspectral sensor those obtained using satellite images for the sugarcane in the two experimental areas (Figure 5). Although several studies have shown a high correlation between the *NDVI* obtained using hyperspectral sensor and satellite image data [59,60], we suggest caution when choosing the spectral data acquisition platform to calculate the *NDVI*.

The *NDVI* from the satellite images data was included in all the models generated (Table 4) and was efficient for detecting the effects of straw removal rates on sugarcane yield in Areas 1 and 2 (Figure 9b). In spite of limitations attributed to *NDVI* data derived from satellite images, the results showed that acquired data from satellite images better represented the variation in sugarcane yield [61]. This type of data acquisition has the advantage that it can be applied for larger areas in relation to the data acquired using hyperespectral sensors, which are generally limited to smaller areas of the crop canopy.

4.3. Sugarcane Yield Prediction Using the NVDI and Leaf-Tissue Nutrient Concentration

In all the stepwise models established to estimate sugarcane yield as a function of each straw removal rate, the *NDVI* was included (Table 4). This inclusion can be explained by the fact that most of models used to estimate sugarcane yield are based on the estimated biomass production [62]. This would explain, at least in part, the efficiency of the *NDVI* to detect stalk yield changes as a function of the straw removal rates, as this index has a high correlation with crop biomass [63].

Among the nutrients included in the site-specific models (i.e., N, P, and K for Area 1; and N, P, and Ca for Area 2), only N and P were included in the overall model for the two areas (Table 4). Nutrients such as N and P are present in larger quantities in younger plants, being more sensitive to changes caused by different management systems [64,65]. In addition, the relationship of these nutrients with the *NDVI* in the models was expected due to the influence of these elements on the photosynthetic process, which in turn, change the spectral responses by the plant, influencing the *NDVI* [66,67]. In this approach, other studies can be performed to identify the spectral signatures of the nutrients in the leaves so that the model can only be applied using spectral data.

The performance of the models in the validation process were also satisfactory when observed RMSE values between the rates of 0% to 100%, ranged from 0.24 to 10.34 Mg ha^{-1} for the area 1 and 0.99 to 4.15 Mg ha^{-1} for Area 2 (Figure 10). These values agree with those reported by Fernandes et al. [68], which found RMSE values of predicted sugarcane yields ranged from 7.20 to 11.00 Mg ha^{-1}. In addition, the sensitivity analysis showed that overall model for Areas 1 and 2 provides a satisfactory margin for the prediction of sugarcane yield with similar soil and climate conditions of this study. Therefore, the overall model performance showed *NDVI* from satellite images and leaf-tissue nutrient concentrations can be useful and alternative tool to improve the sugarcane yield predictions.

5. Conclusions

The normalized difference vegetation index (*NDVI*) can be a useful tool for predicting sugarcane yield in fields managed with straw removal management in Brazil. Between the methods of data acquisition to calculate *NDVI*, satellite image was more efficient than hyperspectral sensor for detecting straw removal effects on sugarcane yield. In addition, concentrations of leaf-tissue N and P were also important parameters to develop the prediction models of sugarcane yield.

Prediction model approach based on data of *NDVI* and concentration of leaf-tissue nutrients collected in early stages of crop growth can help the Brazilian sugarcane sector to predict crop yield in fields intensively managed for bioenergy production. In addition, these models can be used for monitoring spatio-temporal crop yield changes induced by straw removal and supporting decision making towards a more sustainable crop residue management in Brazilian sugarcane fields. The sugarcane yield prediction models developed in this study have to be further tested and validated using data from long-term straw removal experiments, so far, still inexistent in Brazil.

Author Contributions: I.P.L.: field sampling, writing, review; J.M.D.: data analyses, writing, review; M.R.C.: writing and review; P.P.S.B.: field sampling and writing; P.R.F.: financial support and writing; C.C.C.: this paper is in his memory; C.E.P.C.: financial support, writing and review.

Acknowledgments: We are grateful to Lilian Duarte, Admilson Margato, Ralf Araújo, Sandra Nicolete, Dagmar Vasca and Eleusa Basse for the strong assistance in the experiment conduction and samples analyses in the laboratory.

References

1. FAO—Food and Agriculture Organization. Available online: http://faostat.fao.org/ (accessed on 25 July 2018).
2. Companhia Nacional de Abastecimento-Conab. Available online: https://www.conab.gov.br/ (accessed on 25 July 2018).
3. REN21-Highlights of the REN21 Renewables 2017 Global Status Report in Perspective. Available online: http://www.ren21.net/wp-content/uploads/2017/06/GSR2017_Highlights_FINAL.pdf (accessed on 15 June 2018).
4. Goldemberg, J.; Mello, F.F.; Cerri, C.E.; Davies, C.A.; Cerri, C.C. Meeting the global demand for biofuels in 2021 through sustainable land use change policy. *Energy Policy* **2014**, *69*, 14–18. [CrossRef]
5. de Expansão de Energia 2024. Available online: http://www.epe.gov.br (accessed on 1 May 2018).
6. Cunha, F.M.; Badino, A.C.; Farinas, C.S. Effect of a novel method for in-house cellulase production on 2G ethanol yields. *Biocatal. Agric. Biotechnol.* **2017**, *9*, 224–229. [CrossRef]
7. Lisboa, I.P.; Cherubin, M.R.; Cerri, C.C.; Cerri, D.G.P.; Cerri, C.E.P. Guidelines for the recovery of sugarcane straw from the field during harvesting. *Biomass Bioenergy* **2017**, *96*, 69–74. [CrossRef]
8. Leal, M.R.L.V.; Galdos, M.V.; Scarpare, F.V.; Seabra, J.E.A.; Walter, A.; Oliveira, C.O.F. Sugarcane straw availability, quality, recovery and energy use: A literature review. *Biomass Bioenergy* **2013**, *53*, 11–19. [CrossRef]
9. Carvalho, J.L.N.; Nogueirol, R.C.; Menandro, L.M.S.; Bordonal, R.D.O.; Borges, C.D.; Cantarella, H.; Franco, H.C.J. Agronomic and environmental implications of sugarcane straw removal: A major review. *Glob. Chang. Biol. Bioenergy* **2017**, *9*, 1181–1195. [CrossRef]
10. Cherubin, M.R.; Oliveira, D.M.S.; Feigl, B.J.; Pimentel, L.G.; Lisboa, I.P.; Gmach, M.R.; Varanda, L.L.; Moraes, M.C.; Satiro, L.S.; Popin, G.V.; et al. Crop residue harvest for bioenergy production and its implications on soil functioning and plant growth: A review. *Sci. Agric.* **2018**, *75*, 255–272. [CrossRef]
11. Corrêa, S.T.R.; Carvalho, J.L.N.; Hernandes, T.A.D.; Barbosa, L.C.; Menandro, L.M.S.; Leal, M.R.L.V. Assessing the effects of different amountsof sugarcane straw on temporal variability of soil moisture content and temperature. In Proceedings of the 25th European Biomass Conference and Exhibition, Stockholm, Sweden, 12–15 June 2017; pp. 12–15.
12. Satiro, L.S.; Cherubin, M.R.; Safanelli, J.L.; Lisboa, I.P.; Junior, P.R.R.; Cerri, C.E.E.; Cerri, C.C. Sugarcane straw removal effects on Ultisols and Oxisols in south-central Brazil. *Geod. Reg.* **2017**, *11*, 86–95. [CrossRef]
13. Dos Anjos, J.C.R.; Júnior, A.S.A.; Bastos, E.A.; Noleto, D.H.; Melo, F.B.; De Brito, R.R. Water storage in a Plinthaqualf cultivated with sugarcane under straw levels. *Pesqui. Agropecu. Brasil.* **2017**, *52*, 464–473. [CrossRef]

14. Bordonal, R.O.; Menandro, L.M.S.; Barbosa, L.C.; Lal, R.; Milori, D.M.B.P.; Kolln, O.T.; Franco, H.C.J.; Carvalho, J.L.N. Sugarcane yield and soil carbon response to straw removal in south-central Brazil. *Geoderma* **2018**, *328*, 79–90. [CrossRef]
15. Fortes, C.; Vitti, A.C.; Otto, R.; Ferreira, D.A.; Franco, H.C.J.; Trivelin, P.C.O. Contribution of nitrogen from sugarcane harvest residues and urea for crop nutrition. *Sci. Agric.* **2013**, *70*, 313–320. [CrossRef]
16. Paredes Junior, F.P.; Portilho, I.I.R.; Mercante, F.M. Atributos microbiológicos de um latossolo sob cultivo de cana-de-açúcar com e sem queima da palhada. *Semin. Ciênc. Agrár.* **2015**, *36*, 151–164. [CrossRef]
17. Valim, W.C.; Panachuki, E.; Pavei, D.S.; Sobrinho, T.A.; Almeida, W.S. Effect of sugarcane waste in the control of interrill erosion. *Semin. Ciênc. Agrár.* **2016**, *37*, 1155–1164. [CrossRef]
18. Aquino, G.S.; Medina, C.C.; Costa, D.C.; Shahaba, M.; Santiago, A.D.; Cunha, A.C.B.; Kussaba, D.A.O.; Carvalho, J.B.; Moreira, A. Does straw mulch partial-removal from soil interfere in yield and industrial quality sugarcane? A long term study. *Ind. Crops Prod.* **2018**, *111*, 573–578. [CrossRef]
19. Lisboa, I.P.; Cherubin, M.R.; Lima, R.P.; Cerri, C.C.; Satiro, L.S.; Wienhol, B.J.; Schmer, M.R.; Jin, V.L.; Cerri, C.E.P. Sugarcane straw removal effects on plant growth and stalk yield. *Ind. Crops Prod.* **2018**, *111*, 794–806. [CrossRef]
20. Tucker, C.J.; Holben, B.N.; Elgin, J.H.; Mcmurtrey, J.E. Relationship of spectral data to grain yield variation. *Photogramm. Eng. Remote Sens.* **1980**, *46*, 657–666.
21. Tarnavsky, E.; Garrigues, S.; Brown, M.E. Multiscale geostatistical analysis of AVHRR, SPOT-VGT, and MODIS global *NDVI* products. *Remote Sens. Environ.* **2008**, *112*, 535–549. [CrossRef]
22. Sanches, G.M.; Duft, D.G.; Kölln, O.T.; Luciano, A.C.; De Castro, S.G.; Okuno, F.M.; Franco, H.C. The potential for RGB images obtained using unmanned aerial vehicle to assess and predict yield in sugarcane fields. *Int. J. Remote Sens.* **2018**, *15*, 1–3. [CrossRef]
23. Peralta, N.R.; Assefa, Y.; Du, J.; Barden, C.J.; Ciampitti, I.A. Mid-season high-resolution satellite imagery for forecasting site-specific corn yield. *Remote Sens.* **2016**, *8*, 1–16. [CrossRef]
24. Damian, J.M.; Santi, A.L.; Fornari, M.; Da Ros, C.O.; Eschner, V.L. Monitoring variability in cash-crop yield caused by previous cultivation of a cover crop under a no-tillage system. *Comput. Electron. Agric.* **2017**, *142*, 607–621. [CrossRef]
25. Lofton, J.; Tubana, B.S.; Kanke, Y.; Teboh, J.; Viator, H.; Dalen, M. Estimating sugarcane yield potential using an in-season determination of normalized difference vegetative index. *Sensors* **2012**, *12*, 7529–7547. [CrossRef] [PubMed]
26. Yang, H.; Yang, X.; Heskel, M.; Sun, S.; Tang, J. Seasonal variations of leaf and canopy properties tracked by ground-based *NDVI* imagery in a temperate forest. *Sci. Rep.* **2017**, *1267*, 1–10. [CrossRef] [PubMed]
27. Sardans, J.; Grau, O.; Chen, H.Y.H.; Janssens, I.A.; Ciais, P.; Piao, S.; Peñuelas, J. Changes in nutrient concentrations of leaves and roots in response to global change factors. *Glob. Chang. Biol.* **2017**, *23*, 3849–3856. [CrossRef]
28. USDA-United States Department of Agriculture. Available online: https://www.usda.gov/ (accessed on 18 June 2018).
29. CEPAGRI—Tempo e Clima Unicamp. Available online: https://www.cpa.unicamp.br (accessed on 2 February 2017).
30. Posto Meteorológico "Professor Jesus Marden dos Santos" ESALQ—USP. Available online: http://www.leb.esalq.usp.br/posto/ (accessed on 2 February 2017).
31. Raij, B.; Cantarella, H.; Guaggio, J.A.; Furlani, A.M.C. *Recomendações De Adubação E Calagem Para O Estado De São Paulo*, 1st ed.; Instituto Agronômico and Fundacão IAC: Campinas, Brazil, 1997; pp. 6–13.
32. Vermote, E.F.; Tanre, D.; Deulze, J.L.; Herman, M.; Morcrette, J.J. Second Simulation of the Satellite Signal in the Solar Spectrum, 6S: An overview. *IEEE Trans. Geosci. Remote* **1997**, *35*, 675–686. [CrossRef]
33. Carlotto, M.J. Reducing the effects of space-varying, wavelength-dependent scattering in multispectral imagery. *Int. J. Remote Sens.* **1999**, *20*, 3333–3344. [CrossRef]
34. Neter, J.; Wasserman, W.; Kutner, M.H. *Applied Linear Statistical Methods*, 5th ed.; Irwin Professional Publishing: Chicago, IL, USA, 1985; pp. 50–78.
35. Franco, H.C.J.; Pimenta, M.T.B.; Carvalho, J.L.N.; Magalhães, P.S.G.; Rossell, C.E.V.; Braunbeck, O.A.; Vitti, A.C.; Kolln, O.T.; Neto, J.R. Assessment of sugarcane trash for agronomic and energy purposes in Brazil. *Sci. Agric.* **2013**, *70*, 305–312. [CrossRef]

36. Trivelin, P.C.O.; Franco, H.C.J.; Otto, R.; Ferreira, D.A.; Vitti, A.C.; Fortes, C.; Faroni, C.E.; Oliveira, E.C.A.; Cantarella, H. Impact of sugarcane trash on fertilizer requirements for São Paulo, Brazil. *Sci. Agric.* **2013**, *70*, 345–352. [CrossRef]
37. Rosolem, C.A.; Calonego, J.C.; Foloni, J.S.S. Lixiviação de potássio da palha de espécies de cobertura de solo de acordo com a quantidade de chuva aplicada. *Rev. Bras. Cienc. Sol.* **2003**, *27*, 355–362. [CrossRef]
38. Endres, L.; Cruz, S.J.S.; Vilela, R.D.; Santos, J.M.; Barbosa, G.V.S.; Silva, J.A.C. Foliar applications of calcium reduce and delay sugarcane flowering. *Bioenerg. Res.* **2016**, *9*, 98–108. [CrossRef]
39. Vitti, G.; Luz, P.; Otto, R. *Agrícola Ouro Verde*, 10th ed.; Lençóis Paulista: Lencóis Paulista, Brazil, 2008; pp. 1–19.
40. Campos, M.; Antonangelo, J.A.; Alleoni, L.R.F. Phosphorus sorption index in humid tropical soils. *Soil Tillage Res.* **2016**, *156*, 110–118. [CrossRef]
41. Damon, P.M.; Bowden, B.; Rose, T.; Rengel, Z. Crop residue contributions to phosphorus pools in agricultural soils: A review. *Soil Biol. Biochem.* **2014**, *74*, 127–137. [CrossRef]
42. Pavinato, P.S.; Rosolem, C.A. Disponibilidade de nutrientes no solo—Decomposição e liberação de compostos orgânicos de resíduos vegetais. *Rev. Bras. Cienc. Sol.* **2008**, *32*, 911–920. [CrossRef]
43. Leite, J.M.; Ciampitti, I.A.; Mariano, E.; Vieira-Megda, M.X.; Trivelin, P.C. Nutrient, partitioning and stoichiometry in unburnt sugarcane ratoon at varying yield levels. *Front. Plant Sci.* **2016**, *7*, 1–14. [CrossRef] [PubMed]
44. Paul, G.C.; Bokhtiar, S.M.; Rehman, H.; Kabiraj, R.C.; Rahman, A.B.M.M. Efficacies of some organic fertilizers on sustainable sugarcane production in old Himalayan piedmont plain soil of Bangladesh. *Pak. Sugar J.* **2005**, *20*, 2–5.
45. Dotaniya, M.L.; Datta, S.C.; Biswas, D.R.; Dotaniya, C.K.; Meena, B.L.; Rajendiran, S.; Regar, K.L.; Lata, M. Use of sugarcane industrial by-products for improving sugarcane productivity and soil health. *Int. J. Recycl. Org. Waste Agric.* **2016**, *5*, 185–194. [CrossRef]
46. Aquino, G.S.; Medina, C.C.; Costa, D.C.; Shahaba, M.; Santiago, A.D. Sugarcane straw management and its impact on production and development of ratoons. *Ind. Crops Prod.* **2017**, *102*, 58–64. [CrossRef]
47. Singh, S.N.; Singh, A.K.; Malik, J.P.S.; Kumar, R.; Sharma, M.L. Cultural-practice packages and trash management effects on sugarcane ratoons under sub-tropical climatic conditions of India. *J. Agric. Sci.* **2012**, *150*, 237–247. [CrossRef]
48. Castioni, G.A.; Cherubin, M.R.; Menandro, L.M.S.; Sanches, G.M.; de Oliveira Bordonal, R.; Barbosa, L.C.; Carvalho, J.L.N. Soil physical quality response to sugarcane straw removal in Brazil: A multi-approach assessment. *Soil Tillage Res.* **2018**, *184*, 301–309. [CrossRef]
49. Carvalho, J.L.; Hudiburg, T.W.; Franco, H.C.; DeLucia, E.H. Contribution of above-and belowground bioenergy crop residues to soil carbon. *Glob. Chang. Biol. Bioenerg.* **2017**, *9*, 1333–1343. [CrossRef]
50. Oliveira, D.M.; Williams, S.; Cerri, C.E.; Paustian, K. Predicting soil C changes over sugarcane expansion in Brazil using the DayCent model. *Glob. Chang. Biol. Bioenerg.* **2017**, *9*, 1436–1446. [CrossRef]
51. Pitombo, L.M.; Cantarella, H.; Packer, A.P.C.; Ramos, N.P.; do Carmo, J.B. Straw preservation reduced total N2O emissions from a sugarcane field. *Soil Use Manag.* **2017**, *33*, 583–594. [CrossRef]
52. Vasconcelos, A.L.S.; Cherubin, M.R.; Feigl, B.J.; Cerri, C.E.; Gmach, M.R.; Siqueira-Neto, M. Greenhouse gas emission responses to sugarcane straw removal. *Biomass Bioenergy* **2018**, *113*, 15–21. [CrossRef]
53. Tavares, R.L.M.; Spokas, K.; Hall, K.; Colosky, E.; Souza, Z.M.D.; Scala, N.L. Sugarcane residue management impact soil greenhouse gas. *Ciênc. Agrotec.* **2018**, *42*, 195–203. [CrossRef]
54. da Silva, J. Sistemas de manejo da palhada influenciam acúmulo de biomassa e produtividade da cana-de-açúcar (var. RB855453). *Act. Sci. Agron.* **2010**, *32*, 345–350. [CrossRef]
55. Aquino, G.S.; de Conti Medina, C.; Junior, A.D.O.M.; Pasini, A.; Brito, O.R.; Cunha, A.C.B.; Almeida, L.F. Impact of harvesting with burning and management of straw on the industrial quality and productivity of sugarcane. *Afr. J. Agric. Res.* **2016**, *11*, 2462–2468. [CrossRef]
56. Aquino, G.S.; de Conti Medina, C.; Silvestre, D.A.; Gomes, E.C.; Benitez, A.C.; Cunha, D.A.O.K.; Santiago, A.D. Straw removal of sugarcane from soil and its impacts on yield and industrial quality ratoons. *Sci. Agric.* **2018**, *75*, 526–529. [CrossRef]
57. Wang, Q.; Tenhunen, J.; Dinh, N.Q.; Reichstein, M.; Vesala, T.; Keronen, P. Similarities in ground- and satellite-based *NDVI* time series and their relationship to physiological activity of a scots pine forest in finland. *Remote Sens. Environ.* **2004**, *93*, 225–237. [CrossRef]

58. Balzarolo, M.; Anderson, K.; Nichol, C.; Rossini, M.; Vescovo, L.; Arriga, N.; Wohlfahrt, G.; Calvet, J.C.; Carrara, A.; Cerasoli, S.; et al. Ground-based optical measurements at european flux sites: A review of methods, instruments and current controversies. *Sensors* **2011**, *11*, 7954–7981. [CrossRef] [PubMed]
59. Genc, L.; Turhan, H.; Asar, B.; Smith, S.E. Comparison of spectral indices from QUICKBIRD and ground based hyper-spectral data for winter wheat. *World Appl. Sci. J.* **2009**, *7*, 756–762. [CrossRef]
60. Caturegli, L.; Casucci, M.; Lulli, F.; Grossi, N.; Gaetani, M.; Magni, S.; Bonari, E.; Volterrani, M. GeoEye-1 satellite versus ground-based multispectral data for estimating nitrogen status of turfgrasses. *Int. J. Remote Sens.* **2015**, *36*, 2238–2251. [CrossRef]
61. Bu, H.; Sharma, L.K.; Denton, A.; Franzen, D.W. Comparison of Satellite Imagery and Ground-Based Active Optical Sensors as Yield Predictors in Sugar Beet, Spring Wheat, Corn, and Sunflower. *Agron. J.* **2017**, *109*, 299–308. [CrossRef]
62. Morel, J.; Bégué, A.; Todoroff, P.; Martiné, J.F.; Lebourgeois, V.; Petit, M. Coupling a sugarcane crop model with the remotely sensed time series of fIPAR to optimise the yield estimation. *Eur. J. Agron.* **2014**, *61*, 60–68. [CrossRef]
63. Bégué, A.; Lebourgeois, V.; Bappel, E.; Todoroff, P.; Pellegrino, A.; Baillarin, F.; Siegmund, B. Spatio-temporal variability of sugarcane fields and recommendations for yield forecast using *NDVI. J. Remote Sens.* **2010**, *31*, 5391–5407. [CrossRef]
64. Jarrell, W.M.; Beverly, R.B. The dilution effect in plant nutrition studies. *Adv. Agron.* **1981**, *34*, 197–224. [CrossRef]
65. Muchovej, R.M.; Newman, P.R.; Luo, Y. Sugarcane leaf nutrient concentrations: With or without midrib tissue. *J. Plant. Nutr.* **2005**, *28*, 1271–1286. [CrossRef]
66. Al-Abbas, A.H.; Barr, R.; Hall, J.D.; Crane, F.L.; Baumgardner, M.F. Spectra of normal and nutrient-deficient maize leaves. *Agron. J.* **1974**, *66*, 16–20. [CrossRef]
67. Raper, T.B.; Varco, J.J.; Hubbard, K.J. Canopy-Based normalized difference vegetation index sensors for monitoring cotton nitrogen status. *Agron. J.* **2013**, *105*, 1345–1354. [CrossRef]
68. Fernandes, J.L.; Ebecken, N.F.F.; Esquerdo, J.C.D.N. Sugarcane yield prediction in Brazil using *NDVI* time series and neural networks ensemble. *ISPRS J. Photogramm. Remote Sens.* **2017**, *38*, 4631–4644. [CrossRef]

Permissions

All chapters in this book were first published in MDPI; hereby published with permission under the Creative Commons Attribution License or equivalent. Every chapter published in this book has been scrutinized by our experts. Their significance has been extensively debated. The topics covered herein carry significant findings which will fuel the growth of the discipline. They may even be implemented as practical applications or may be referred to as a beginning point for another development.

The contributors of this book come from diverse backgrounds, making this book a truly international effort. This book will bring forth new frontiers with its revolutionizing research information and detailed analysis of the nascent developments around the world.

We would like to thank all the contributing authors for lending their expertise to make the book truly unique. They have played a crucial role in the development of this book. Without their invaluable contributions this book wouldn't have been possible. They have made vital efforts to compile up to date information on the varied aspects of this subject to make this book a valuable addition to the collection of many professionals and students.

This book was conceptualized with the vision of imparting up-to-date information and advanced data in this field. To ensure the same, a matchless editorial board was set up. Every individual on the board went through rigorous rounds of assessment to prove their worth. After which they invested a large part of their time researching and compiling the most relevant data for our readers.

The editorial board has been involved in producing this book since its inception. They have spent rigorous hours researching and exploring the diverse topics which have resulted in the successful publishing of this book. They have passed on their knowledge of decades through this book. To expedite this challenging task, the publisher supported the team at every step. A small team of assistant editors was also appointed to further simplify the editing procedure and attain best results for the readers.

Apart from the editorial board, the designing team has also invested a significant amount of their time in understanding the subject and creating the most relevant covers. They scrutinized every image to scout for the most suitable representation of the subject and create an appropriate cover for the book.

The publishing team has been an ardent support to the editorial, designing and production team. Their endless efforts to recruit the best for this project, has resulted in the accomplishment of this book. They are a veteran in the field of academics and their pool of knowledge is as vast as their experience in printing. Their expertise and guidance has proved useful at every step. Their uncompromising quality standards have made this book an exceptional effort. Their encouragement from time to time has been an inspiration for everyone.

The publisher and the editorial board hope that this book will prove to be a valuable piece of knowledge for researchers, students, practitioners and scholars across the globe.

List of Contributors

Marco Vizzari, Francesco Santaga and Paolo Benincasa
Department of Agricultural, Food, and Environmental Sciences, University of Perugia, 06121 Perugia, Italy

Georg Röll and Simone Graeff-Hönninger
Department of Agronomy, Institute of Crop Science, University of Hohenheim, 70599 Stuttgart, Germany

William D. Batchelor
Biosystems Engineering Department, Auburn University, Auburn, AL 36849, USA

Ana Carolina Castro
Cerealicultura, Facultad de Ciencias Agrarias y Forestales, Universidad Nacional de La Plata, Av. 60 y 119, La Plata 1900, Argentina
Consejo Nacional de Investigaciones Científicas y Técnicas (CONICET), Centro Científico Tecnológico (CCT), La Plata 1900, Argentina

María Rosa Simón
Cerealicultura, Facultad de Ciencias Agrarias y Forestales, Universidad Nacional de La Plata, Av. 60 y 119, La Plata 1900, Argentina
Comisión de Investigaciones Científicas, Pcia. de Buenos Aires, 526, 10 y 11, La Plata 1900, Argentina

Elia Scudiero
Department of Environmental Sciences, University of California Riverside, 900 University Ave., Riverside, CA 92521, USA
USDA-ARS, United States Salinity Laboratory, 450 West Big Springs Rd., Riverside, CA 92507, USA

Todd H. Skaggs
USDA-ARS, United States Salinity Laboratory, 450 West Big Springs Rd., Riverside, CA 92507, USA

Pietro Teatini
Department of Civil, Environmental, and Architectural Engineering (ICEA), University of Padua, Via Trieste 63, 35121 Padua, Italy

Gabriele Manoli
ETH Zurich, Department of Civil, Environmental and Geomatic Engineering, Stefano-Franscini-Platz 5, 8093 Zürich, Switzerland

Federica Braga
Institute of Marine Sciences (ISMAR), National Research Council, Arsenale Tesa 104, Castello 2737/F, 30122 Venezia, Italy

Francesco Morari
Department of Agronomy, Food, Natural resources, Animals, and Environment (DAFNAE), University of Padua, Viale dell'Università 16, 35020 Legnaro, Italy

Antonio Santos-Rufo
Department of Agronomy, University of Cordoba, Campus de Rabanales, 14071 Córdoba, Spain

Francisco-Javier Mesas-Carrascosa, Alfonso García-Ferrer and Jose Emilio Meroño-Larriva
Department of Graphic Engineering and Geomatics, University of Cordoba, Campus de Rabanales, 14071 Córdoba, Spain

Piero Toscano, Salvatore Filippo Di Gennaro and Alessandro Matese
Institute of BioEconomy (IBE), National Research Council (CNR), Via Caproni 8, 50145 Florence, Italy

Annamaria Castrignanò, Alessandro Vittorio Vonella and Domenico Ventrella
Council for Agricultural Research and Economics, Research Centre for Agriculture and Environment (CREA-AA), Via Celso Ulpiani 5, 70125 Bari, Italy

Stefano Marino and Arturo Alvino
Department of Agricultural, Environmental and Food Sciences (DAEFS), University of Molise, Via De Sanctis, 86100 Campobasso, Italy

Giulia Ronchetti and Giovanna Sona
Department of Civil and Environmental Engineering, Politecnico di Milano, 20133 Milan, Italy

Alice Mayer, Arianna Facchi and Bianca Ortuani
Department of Agricultural and Environmental Sciences - Production, Landscape, Agroenergy, Università degli Studi di Milano, 20133 Milan, Italy

Ephrem Habyarimana, Paolo De Franceschi and Michela Dall'Agata
CREA Research Center for Cereal and Industrial Crops, Bologna 40128, Italy

Isabelle Piccard
Vlaamse instelling voor technologisch onderzoek N.V., MOL 2400, Belgium

Marcello Catellani
CREA Research Center for Cereal and Industrial Crops, Bologna 40128, Italy
Department for Sustainability, Italian National Agency for New Technologies, Energy and Sustainable Economic Development (ENEA), Rotondella (MT) 75026, Italy

Hang Zhu, Hongze Li and Cui Zhang
School of Mechanical and Aerospace Engineering, Jilin University, Changchun 130022, China

Junxing Li
Jilin Academy of Agricultural Machinery, Changchun 130062, China

Huihui Zhang
Water Management and Systems Research Unit, USDA-ARS, Fort Collins, CO 80526, USA

Juan M. Jurado and Francisco R. Feito
Computer Graphics and Geomatics Group of Jaén, University of Jaén, 23071 Jaén, Spain

Luís Pádua and Joaquim J. Sousa
Engineering Department, School of Science and Technology, University of Trás-os-Montes e Alto Douro, 5000-801 Vila Real, Portugal
Centre for Robotics in Industry and Intelligent Systems (CRIIS), INESC Technology and Science (INESC-TEC), 4200-465 Porto, Portugal

Erqi Xu and Hongqi Zhang
Key Laboratory of Land Surface Pattern and Simulation, Institute of Geographic Sciences and Natural Resources Research, Chinese Academy of Sciences, Beijing 100101, China

Yongmei Xu
Institute of Soil and Fertilizer, Xinjiang Academy of Agricultural Sciences, Urumqi 830091, China

Mbulisi Sibanda, Onisimo Mutanga and Paramu L. Mafongoya
Discipline of Geography, School of Agricultural, Earth and Environmental Sciences, University of KwaZulu-Natal, Scottsville, Pietermaritzburg 3209, South Africa

Shirly T. Magwaza
Discipline of Horticulture, School of Agricultural, Earth and Environmental Sciences, University of KwaZulu-Natal, Scottsville, Pietermaritzburg 3209, South Africa

Alfred O. Odindo and Asanda Mditshwa
Discipline of Crop Science, School of Agricultural, Earth and Environmental Sciences, University of KwaZulu-Natal, Scottsville, Pietermaritzburg 3209, South Africa

Lembe S. Magwaza
Discipline of Horticulture, School of Agricultural, Earth and Environmental Sciences, University of KwaZulu-Natal, Scottsville, Pietermaritzburg 3209, South Africa
Discipline of Crop Science, School of Agricultural, Earth and Environmental Sciences, University of KwaZulu-Natal, Scottsville, Pietermaritzburg 3209, South Africa

Timothy Dube
Institute forWater Studies, Department of Earth Sciences, University of the Western Cape, Bellville 7535, South Africa

Izaias Pinheiro Lisboa, Júnior Melo Damian, Maurício Roberto Cherubin, Pedro Paulo Silva Barros, Peterson Ricardo Fiorio, Carlos Clemente Cerri and Carlos Eduardo Pellegrino Cerri
"Luiz de Queiroz" College of Agriculture, University of São Paulo, 11 Pádua Dias Avenue, Piracicaba, SP 13418-900, Brazil

Index

www.ingramcontent.com/pod-product-compliance
Lightning Source LLC
LaVergne TN
LVHW060550210926
848485LV00005B/1289